Lecture Notes in Computer Science 6165

Commenced Publication in 1973
Founding and Former Series Editors:
Gerhard Goos, Juris Hartmanis, and Jan

Editorial Board

David Hutchison
Lancaster University, UK

Takeo Kanade
Carnegie Mellon University, Pittsburgh, PA, USA

Josef Kittler
University of Surrey, Guildford, UK

Jon M. Kleinberg
Cornell University, Ithaca, NY, USA

Alfred Kobsa
University of California, Irvine, CA, USA

Friedemann Mattern
ETH Zurich, Switzerland

John C. Mitchell
Stanford University, CA, USA

Moni Naor
Weizmann Institute of Science, Rehovot, Israel

Oscar Nierstrasz
University of Bern, Switzerland

C. Pandu Rangan
Indian Institute of Technology, Madras, India

Bernhard Steffen
TU Dortmund University, Germany

Madhu Sudan
Microsoft Research, Cambridge, MA, USA

Demetri Terzopoulos
University of California, Los Angeles, CA, USA

Doug Tygar
University of California, Berkeley, CA, USA

Gerhard Weikum
Max-Planck Institute of Computer Science, Saarbruecken, Germany

David Zhang   Milan Sonka (Eds.)

Medical Biometrics

Second International Conference, ICMB 2010
Hong Kong, China, June 28-30, 2010
Proceedings

 Springer

Volume Editors

David Zhang
Department of Computing
The Hong Kong Polytechnic University
Kowloon, Hong Kong, China
E-mail: csdzhang@comp.polyu.edu.hk

Milan Sonka
Department of Electrical and Computer Engineering
The University of Iowa
Iowa City, IA, USA
E-mail: milan-sonka@uiowa.edu

Library of Congress Control Number: 2010928544

CR Subject Classification (1998): I.4, I.5, I.2, I.2.10, H.3, J.3

LNCS Sublibrary: SL 6 – Image Processing, Computer Vision, Pattern Recognition, and Graphics

ISSN   0302-9743
ISBN-10   3-642-13922-1 Springer Berlin Heidelberg New York
ISBN-13   978-3-642-13922-2 Springer Berlin Heidelberg New York

springer.com

© Springer-Verlag Berlin Heidelberg 2010
Printed in Germany

Typesetting: Camera-ready by author, data conversion by Scientific Publishing Services, Chennai, India
Printed on acid-free paper   06/3180

Preface

In the medical field, personal medical feature data, especially digital images, can be referred to as medical biometrics. Such data are produced in ever-increasing quantities and used for diagnostics and therapy purposes. Medical biometric research aims to use personal medical features in different formats such as images, signals and other sources to solve medical problems and to provide high-performance services in the medical field. Medical biometric systems integrate multidisciplinary technologies in biology, medicine, electronics, computing, and statistics. The importance of computer-aided diagnosis and therapy has drawn more and more attention worldwide and laid the foundation for modern medicine with excellent potential for promising applications such as telemedicine and Web-based healthcare.

The 2010 International Conference on Medical Biometrics (ICMB 2010) placed emphasis on efficient and effective medical biometric technologies and systems. It provided a central forum for researchers, engineers and vendors from different disciplines to exchange the most recent results, identify future directions and challenges, and initiate possible collaborative research and system development. We are pleased that this conference attracted a large number of high-quality research papers that reflect the increasing interests and popularity in this fast-growing field. The conference proceedings contain 45 papers which were selected through a strict review process, with an acceptance rate at 38%. Each paper was assessed by three independent reviewers. All of the accepted papers were presented in either oral (20) or poster (25) sessions at the conference in conjunction with three special sessions on State-of-the-Art of Computer-Aided Detection/Diagnosis (CAD), Modernization of Traditional Chinese Medicine (TCM) and Effective Healthcare.

We would like to take this opportunity to thank the five keynote speakers for their inspiring talks at ICMB 2010 and sharing their valuable experience: Ching Suen, Yueting Zhang, Hiroshi Fujita, Bernd Fischer and Lianda Li. In addition, we would like to express our gratitude to all the contributors, reviewers, Program Committee and Organizing Committee members who made their contribution to the success of ICMB 2010 in different ways. Once again, we greatly appreciate the continuing support from the International Association of Pattern Recognition (IAPR), IEEE Computational Intelligence Society (IEEE-CIS), National Natural Science Foundation in China (NSFC), and Springer. Our special thanks also go to our working team Jane You, Xiaoyi Jiang, Prabir Bhattacharya, Lei Zhang, Zhenhua Guo, Xingzheng Wang, Dongmin Guo, Feng Liu, Qin Li and Yan Wu for their dedication and hard work on various aspects of this event. Last but not least, we sincerely wish that the fruitful technical interactions during this conference will benefit everyone concerned.

April 2010

David Zhang
Milan Sonka

Organization

General Chairs

David Zhang   The Hong Kong Polytechnic University, Hong Kong

Milan Sonka   The University of Iowa, USA

Program Chairs

Jane You   The Hong Kong Polytechnic University, Hong Kong

Xiaoyi Jiang   University of Münster, Germany

Prabir Bhattacharya   University of Cincinnati, USA

Program Committee

Michael D. Abramoff	The University of Iowa, USA
Bir Bhanu	University of California, Riverside, USA
Zhaoxiang Bian	Hong Kong Baptist University, Hong Kong
Egon L. van den Broek	University of Twente, The Netherlands
Yung-Fu Chen	China Medical University, Taiwan
Da-Chuan Cheng	China Medical University, Taiwan
Ruwei Dai	Chinese Academy of Science, China
Mohammad Dawood	University of Münster, Germany
David Feng	The University of Sydney, Australia
Bernd Fischer	University of Lübeck, Germany
Hiroshi Fujita	Gifu University, Japan
Joachim Hornegger	University of Erlangen, Germany
Yung-Fa Huang	Chaoyang University of Technology, Taiwan
Xudong Jiang	Nayang Technological University, Singapore
Mohamed Kamel	University of Waterloo, Canada
Rajiv Khosla	La Trobe University, Australia
Jai-Hie Kim	Yonsei University, Korea
Naimin Li	Harbin Institute of Technology, China
Yanda Li	Tsinghua University, China
Meindert Niemeijer	Utrecht University, The Netherlands
Witold Pedrycz	University of Alberta, Canada
Edwige Pissaloux	Université de Rouen, France
Gerald Schaefer	Loughborough University, UK
Dinggang Shen	UNC-CH School of Medicine, USA

Table of Contents

Feature Extraction and Classification

Health Care

Medical Diagnosis

Medical Image Processing and Registration

Fiber Segmentation Using Constrained Clustering

Daniel Duarte Abdala* and Xiaoyi Jiang

Department of Mathematics and Computer Science
University of Münster, Münster, Germany
{abdalad,xjiang}@uni-muenster.de

Abstract. In this work we introduce the novel concept of applying constraints into the fiber segmentation problem within a clustering based framework. The segmentation process is guided in an interactive manner. It allows the definition of relationships between individual and sets of fibers. These relationships are realized as pairwise linkage constraints to perform a constrained clustering. Furthermore, they can be refined iteratively, making the process of segmenting tracts quicker and more intuitive. The current implementation is based on a constrained threshold based clustering algorithm using the mean closest point distance as measure to estimate the similarity between fibers. The feasibility and the advantage of constrained clustering are demonstrated via segmentation of a set of specific tracts such as the cortico-spinal tracts and corpus collosum.

1 Introduction

With the introduction of DWI (Diffusion Weighted Images) around two decades ago, a new helm of possibilities concerning the non-invasive investigation of the human brain was started. However, the advantages provided by this new imaging protocol cannot be directly appreciated simply by inspecting DWI images [1]. DWI by itself has little clinical value. Figure 1(A) shows two examples of DWI images. As one can see, it provides less anatomical details if compared to other MRI (Magnetic Resonance Imaging) protocols such as T1, T2, FLAIR or MPRage. The DWI protocol is an encoding protocol, although it is possible to directly visualize it as anatomical image. The real value of DWI series resides in the information that can be extrapolated from it, allowing the inspection of the micro-structural anatomy of living fibrous tissues. This is possible since water molecules tend to diffuse strongly alongside fibrous tissues. This protocol encodes the apparent diffusivity of water molecule samples in N principal directions. The measures are usually made in $N = 6$ or more independent diffusion directions in each voxel. A tensor volume is generated by combining the N-dimensional voxels and each tensor can be interpreted as the probability movement distribution of the water

---

* Daniel D. Abdala thanks the CNPq, Brazil-Brasilia for granting him a Ph.D. scholarship under the process number 290101-2006-9.

D. Zhang and M. Sonka (Eds.): ICMB 2010, LNCS 6165, pp. 1–10, 2010.

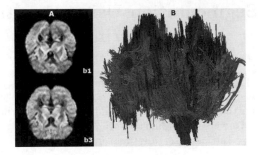

Fig. 1. (A) DWI sample images. Weighted b1 image and b3 from the 6 weighted directions. (B) Fiber tracking produced with the MedINRIA [8] software. Colors are assigned based on the principal diffusion direction.

molecules to behave from its center to the voxel periphery. This tensor representation allows the combination of the N diffusion directions into a single object.

The simplest way of post-processing a tensor volume is to compute indexes such as FA (Fractional Anisotropy), MD (Mean Diffusivity), and so on. Computer generated images based on those indexes can be overlaid to high-resolution anatomical images (e.g. T1 and T2) in order to show additional information. For instance, FA color maps were used to aid identification of hippocampus atrophy [2] and in studies of early detection of multiple sclerosis [3]. However, the real value of DWI resides elsewhere. The tensor volumes can be further processed in order to generate detailed mappings of the white matter fiber connectivity of the brain [4], a process called fiber tracking [5]. Fiber trackings show a very detailed representation of the white matter, allowing differentiation of its inner structures, which are difficult to be perceived by other protocols. The second improvement refers to the fact that individual fibers or bundles, also called fiber bundles, represent the connection between regions of the brain and/or the nerve system. Although presenting a beautiful aspect, fiber trackings can be tricky to visualize and interpret as one can see in Figure 1(B). Given its three-dimensional nature and the huge amount of data (potentially thousands of fibers), it becomes hard to identify anatomical structures by visual inspection only. A subsequent task of fiber segmentation (grouping of fibers into meaningful subsets) comes into play here in order to assign semantics to the extracted subsets of fibers.

In this paper we propose to apply constrained clustering to the fiber segmentation problem. First, we offer the user a suggestion of segmentation produced by a clustering algorithm without considering any constraints. Subsequently, this suggestion is refined interactively by defining pairwise linkage constraints between fibers. A new clustering is produced, taking into account the newly defined constraints. The process repeats itself until the user is satisfied with the segmentation result.

The remainder of this paper is organized as follows. Section 2 reviews the existing fiber segmentation methods. Section 3 gives a short introduction of constrained clustering. Section 4 describes in detail the proposed method. Finally, the empirical evaluation of our approach is presented in Section 5, followed by our final remarks.

2   Existing Fiber Segmentation Methods

There are three main research lines dealing with the fiber segmentation problem: a) interactive; b) clustering; and c) atlas-based. In this section we briefly review the most relevant related works.

2.1   Interactive Segmentation

Interactive fiber segmentation methods provide visual tools to allow the manual selection of subsets of fibers as well as to assign semantic labels. Manual segmentation methods were successfully used to generate comprehensive 3D atlases of the white matter's structure [6]. The most basic tool regarding manual segmentation refers to the definition of ROIs (Regions of Interest) [7]. ROIs are geometrical shapes defined by the user. Fibers are only selected within such regions. A good example of interactive tools is the DTI track module of the Med-INRIA toolset [8]. Although being the most commonly used method, interactive fiber segmentation requires extensive knowledge about the white matter's three-dimensional structure and considerable human efforts. This fact alone motivates the development of quicker and less demanding fiber segmentation solutions.

2.2   Clustering Based Segmentation

Clustering based segmentation uses clustering algorithms to group fibers into clusters based on their similarity. This approach relies on the idea that fibers sharing similar geometry and similar spatial positioning should belong to the same structure. In order to accomplish it, a suitable similarity measure between fibers needs to be defined. In fact, a number of similarity measures were proposed in the literature. Ding et al. [9] introduced the idea of using a descriptor based on the mean Euclidean distance between pair of points of two fibers. Corouge [10] proposed a fiber descriptor based on the re-parametrization of each fiber as a Frenet space curve in order to compute the normal, curvature and torsion in different points, allowing accurate representation of the fiber's shape. By uniformly sampling points, a direct comparison between shape descriptors gives the similarity measure. This method succeeds in describing well the shape of fibers, but fails to address their location. Finally, Geric [11] introduced the mean closest point distance (MCPD). It became popular since it accumulates reliable information about the shape and location of fibers. Today, this similarity measure is by far the mostly used in the literature. Using the similarity measures described above, a number of clustering methods were applied in order to solve the fiber segmentation problem. In [10,11,12] hierarchical clustering is used to produce the segmentation. Spectral clustering methods are also popular choices as shown in [13,14].

2.3 Atlas Based Segmentation

Atlas based segmentation relies on the idea of mapping a fiber tracking to a standard coordinate space where they can be properly correlated with structures or regions. This approach is divided in two parts: a) atlas creation; and b) fiber segmentation. Maddah et al. [15] created an atlas via manual/interactive methods. It is also possible to automate to some degree the atlas creation task. O'Donnell [16] proposed a clustering based method to generate an initial segmentation of various subjects. Subsequently, a human interaction step is taken in order to assign labels to each of the clusters. Afterwards, corresponding clusters are merged in order to create the atlas. Once the atlas is available, the fiber segmentation is done by realigning the fiber tracking to the atlas space.

3 Constrained Clustering

In this section we briefly introduce constrained clustering which is fundamental to our fiber segmentation method. Constrained clustering uses side information in terms of constraints to aid the clustering process. There are a number of ways to constrain a clustering procedure. The most popular method refers to pairwise linkage constraints. Constraints are formulated to control the way patterns will be assigned to clusters. There are basically two types of constraints: (i) must-link (ML), indicating that the constrained patterns must be clustered together, and (ii) cannot-link (CL), meaning patterns must be assigned to different clusters. Wagstaff [17] proposed a constrained version of k-means considering pairwise constraints. An extension of hierarchical clustering was investigated in [18]. It was proved that a complete dendrogram cannot be generated in all cases when constraints are considered. It is important to note that the consideration of constraints does not substantially increase the computational time.

4 Fiber Segmentation Method

Automatic fiber segmentation, e.g. by means of clustering, is highly prone to errors. We propose to integrate human knowledge into the clustering problem within the framework of constrained clustering. Initially, a fiber tracking is produced, generating a set F of M fibers $f_i, i = 1, \ldots, M$, each represented by an ordered list of 3D points. This raw fiber tracking is clustered without considering any constraints and the clustering result is presented using a 3D visualization tool. Subsequently, this initial segmentation suggestion is refined interactively by defining pairwise ML and CL linkage constraints. A new clustering is produced, taking the newly defined constraints into account. The process repeats itself until the user is satisfied with the segmentation result. To implement the general approach described above we need to specify the details of: 1) similarity measure between fibers; 2) how to define constraints; and 3) which constrained clustering algorithm to use. In the following we present such details for our current implementation.

4.1   Computing Similarity between Fibers

As reviewed in Section 2, there are a number of ways to define the similarity between two fibers. In our experiments we have adopted the mean closest point distance [11]. This choice was made based on the fact that this measure encompasses information both from shape and location into a single measurement. It is defined as follows:

$$d_{MCPD}(f_1, f_2) = \frac{1}{n_1} \sum_{i=1}^{n_1} \min_{q_j \in f_2} \|p_i - q_j\| \tag{1}$$

where $\| \cdot \|$ is the Euclidean norm and n_1 is the number of points in fiber f_1. It is computed by averaging the minimum distance between each point of a reference fiber f_1 to the closest point in a second fiber f_2; see Figure 2(B) for an illustration. It is important to point out that MCPD is not symmetric as one can see in the example given in Figure 2(B) as well. A way to circumvent such difficulty is to take the minimum of the two possible distance values:

$$d_{simMCPD}(f_1, f_2) = \min(d_{MCPD}(f_1, f_2), d_{MCPD}(f_2, f_1)) \tag{2}$$

There are certainly other options to achieve symmetry. For our experiments we have selected this version.

4.2   Constraint Specification

The definition of constraints is proposed as a way to enforce the merge or separation of fibers that otherwise would be clustered erroneously. Figure 2(A) presents two examples where such constraints are helpful. It is possible to specify whole subsets of fibers to be constrained by manually defining ROIs comprehending specific regions. However, this is not necessary, since the definition of a single ML constraint between two fibers belonging to different clusters usually suffices to merge

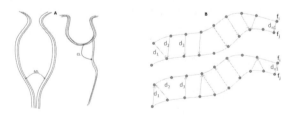

Fig. 2. (A) Example of fiber constraint specification. On the left a must-link constraint is specified between fibers of the two clusters in order to ensure they will be merged into a single one; on the right side, a cannot-link constraint is specified in order to ensure separation of the two clusters. (B) Example showing the computation of MCPD between fibers. On top, the computation of $MCPD(f_i, f_j)$ is illustrated. The arrows indicate the direct closest distance from the individual points of fiber i to the closest point in fiber j. On bottom, the computation of $MCPD(f_j, f_i)$. Generally, $MCPD(f_i, f_j) \neq MCPD(f_j, f_i)$ holds.

them. Similarly, in order to keep two clusters separated, we usually only need to specify CL constraints between pairs of fibers, each from one of the clusters.

It is also important to check the transitive closure of ML and CL constraints to ensure the triangle inequality property is not violated. Consider the following two ML constraints: $ml(f_1, f_2)$ and $ml(f_2, f_3)$. This means f_1 is ML constrained to f_2 and f_2 is ML constrained to f_3. These two constraints imply that fibers f_1 and f_3 must be collocated into the same cluster as well. Consequently, a new constraint $ml(f_1, f_3)$ must be inferred. Similarly, consider $ml(f_1, f_2)$ and $cl(f_2, f_3)$. It is easy to see that f_1 and f_3 cannot be put into the same cluster, otherwise the $cl(f_2, f_3)$ constraint will be violated. In this case a new constraint $cl(f_1, f_3)$ must be inferred. The computation of such transitive closures is an important component of any constrained clustering algorithms and the readers are referred to [17] for more details.

4.3   Constrained Clustering Algorithm

The pseudo code presented Table 1 describes in detail the envisioned constrained clustering method. In fact, it is a constrained extension of the threshold based clustering proposed in [10]. It receives as input the set of fibers FT, two sets of constraints, ML and CL, and a minimum similarity threshold value T. It starts by computing the transitive closure of CL and ML, inferring new constraints if needed. The next step creates a list of cluster representatives, which are fibers that can potentially evolve into clusters. Initially, they are set to be all elements from CL, since fibers subject to CL constraints are expected to be in different clusters. Step 3 selects the next cluster representative among the ones not yet assigned to any cluster. In the next step, each not yet clustered fiber g is visited, and if no constraint is violated (see [17] for a detailed discussion of constraint violation check) it proceeds by checking if the actual fiber can be assigned to the current cluster. This is done by testing if the nearest fiber in the current cluster is within a distance up to T from the candidate fiber g.

Table 1. Constrained Clustering Algorithm

```
FIBER_CLUSTERING (FT, ML, CL, T)
1.    Compute the transitive closure of ML and CL
2.    Create list L of cluster representatives using CL
3.    Select next cluster representative f and build a singleton cluster
      CC = f
4.    FOR each not yet clustered fiber g
4.1     IF g does not violate any constraint when adding g to CC
            and its distance to the nearest fiber in CC < T, THEN
            add g to CC, i.e. CC = CC ∪ {g}
4.2     IF any fiber in CC occurs in L, THEN
            remove it from L
        GOTO(3)
```

4.4   Threshold Definition

The proposed algorithm requires a single parameter in order to define the minimum similarity between fibers. This threshold T ultimately controls the number of final clusters. However, it does not guarantee that a minimum number of fibers will be assigned to each cluster. A small T value would produce many clusters, since only very similar fibers would be clustered together. On the other hand, a large value would potentially lead to clusters comprising of different anatomical structures. Experimentally, $T = 20$ seems to be a reliable choice for most fiber trackings. This value provides a good trend between the total number of clusters and the amount of outlier clusters. Outlier clusters are those formed by a very small amount of fibers, being most likely wrongly clustered ones. In cases where two structures are erroneously merged by the algorithm, it is easy to decrease T's value in order to separate them. This adjustment, however, may provoke some over-clustering in other areas of fibers. In such cases, we can simply introduce ML constraints to enforce merging the over-clustered fibers.

4.5   Outlier Removal

Outlier clusters are obviously undesired and thus we need a way of outlier removal. It is possible to define a minimum size acceptance level, requiring the size to exceed a minimum percentage of the total number of fibers. This value controls which clusters will be accepted. The relationship between T values and the number of identified clusters (blue curve) can be seen in Figure 3. These diagrams also show the relationship between the clusters found and the number of accepted clusters for two different acceptance levels (2% and 0.5%). All fibers not belonging to the accepted clusters will be presented to the user as un-clustered fibers, allowing the definition of additional constraints in order to deal with them in subsequent interactions.

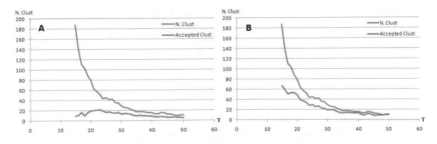

Fig. 3. Threshold values ranging from 5 to 50 vs. number of clusters. In both diagrams, N.Clust represents the number of produced clusters. Accepted clusters refer to the number of clusters accepted after outlier removal. In (A), the outlier removal is set to 2% of the initial size of the fiber set and in (B) to 0.5%.

4.6   Experimental Results

In order to assess the feasibility of the proposed approach, we demonstrate the effective segmentation of a set of tracts. The DWI volumes used in our experiments were produced using a Siemens Magneton Sonata $1.5T$, $TR = 9.75$, $TE = 4s$, with field of view set to $220mm$. The series comprises of 7 volumes (one un-weighted anatomical reference series, and six weighted series encoding 6 principal diffusion directions) of 19 images each. The size of the acquired images is 128×128 pixels with voxel size $1.79 \times 1.79 \times 6.5mm$. Fiber tracking is generated using the MedINRIA software [8]. An initial clustering such as the ones presented in Figure 4 is produced without any constraints and presented to the user. CL and ML constraints are defined manually allowing further refinement. Figure 5(A-B) shows the refinement process from a real example of segmenting the spinal cord fibers, where it is erroneously divided in two clusters (coded by red and yellow, respectively). For the refinement the user needs to visually identify these two clusters. Then, CL constraints are defined between the main bundles and fibers that should not be considered as part of this structure in order to remove them in a subsequent iteration. (B) shows the two clusters with the undesired fibers removed. Finally, a single ML constraint enforces to merge the two main bundles to properly segment the spinal cord in (C). Note that not necessarily all undesired fibers need to be CL constrained since once some of them are removed, there is a chance that the algorithm will potentially create another cluster with the undesired fibers. For this example, only one ML and nine CL constraints were used to perform the segmentation. The final tract is composed of 713 spinal fibers.

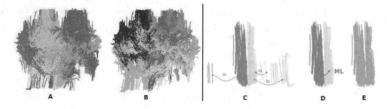

Fig. 4. Left - Complete plot of initial unconstrained clusterings. (A) shows the results of threshold $T = 30$ and cluster min. size 0.5%; (B) was generated with threshold $T = 10$ and cluster min. size 2.0%. Right - Constraint definition. (C) and (D) show two clusters produced by our algorithm referring the spinal cord. In (E) the final segmented bundle.

Another interesting example refers to the segmentation of the corpus collosum 5(C-K). This particular region is relatively difficult to segment since it represents the inner core of the brain connecting most of its functional regions. The clustering algorithm produced a series of clusters containing colossal fibers. They are visually identified by its position in relation to the anatomical reference images. For each cluster CL constraints are defined in order to remove subsets of fibers that do not belong to this structure. Finally, ML constraints are defined to segment the whole structure in a subsequent iteration.

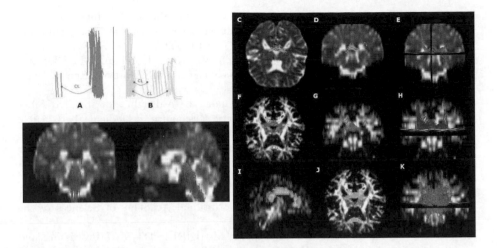

Fig. 5. (A,B) Iterative segmentation of spinal cord fibers. (C-K) Iterative segmentation of colossal fibers. (C), (D), (F) and (G) show axial and coronal views of an identified cluster. (E) and (H) show a 3D plot of such fibers. (I) and (J) show the sagital and axial projections of 7 colossal clusters. (K) is the 3D plot of the whole tract.

5 Conclusion

This paper proposed to a novel approach to fiber segmentation by applying constrained clustering. It consists of a baseline constrained clustering algorithm that allows a reliable segmentation in an interactive manner. The user only needs to formulate high-level knowledge in form of pairwise relationships. In the current implementation a constrained threshold based clustering algorithm was developed. This method allows automatically deciding the number of clusters, depending on the similarity threshold. The MCPD distance was used as similarity measure between fibers. The feasibility of our approach is shown through the segmentation of the spinal and colossal fibers. The last example is especially interesting since it poses a major difficulty given the complex spatial distribution of the fibers. The popular segmentation methods based on clustering alone will hardly produce high-quality segmentations, especially in abnormal cases. Our approach provides a means to alleviate this problem. In addition, the usage of pairwise constraints seems to be more intuitive than the traditional ROI-based methods.

References

1. Bihan, D.L., Mangin, J.-F., Poupon, C., Clark, C.A., Pappata, S., Molko, N., Chabriat, H.: Diffusion Tensor Imaging - Concepts and Applications. Journal of Magnetic Resonance Imaging 13, 534–546 (2001)
2. Müller, M., Greverus, D., Weibrich, C., Dellani, P., Scheurich, A., Stoeter, P., Fellgiebel, A.: Diagnostic Utility of Hippocampal Size and Mean Diffusivity in Amnestic MCI. Neurobiology of Aging 28, 398–403 (2006)

3. Cercignani, M., Bozzali, M., Iannucci, G., Comi, G., Filippi, M.: Intra-Voxel and Inter-Voxel Coherence in Patients with Multiple Sclerosis Assessed Using Diffusion Tensor MRI. Journal of Neurology 249, 875–883 (2002)
4. Poupon, C., Mangin, J.-F., Clark, C., Frouin, V., Régis, J., Bihan, D.L., Bloch, I.: Towards Inference of Human Brain Connectivity from MR Diffusion Tensor Data. Medical Image Analysis 5, 1–15 (2001)
5. Basser, P.J., Pajevic, S., Pierpaoli, C., Duda, J., Aldroubi, A.: In Vivo Fiber Tractography Using DT-MRI Data. Magnetic Resonance in Medicine 44, 625–632 (2000)
6. Wakana, S., Jiang, H., Nagae-Poetscher, L.M., van Zijl, P.C.M., Mori, S.: Fiber Tract-based Atlas of Human White Matter Anatomy. Radiology 230, 77–87 (2004)
7. Conturo, T.E., Lori, N.F., Cull, T.S., Akbudak, E., Snyder, A.Z., Shimony, J.S., McKinstry, R.C., Burton, H., Raichle, M.E.: Tracking Neuronal Fiber Pathways in the Living Human Brain. Proc. of the National Academy of Sciences of the U.S.A. 96, 10422–10427 (1999)
8. Toussaint, N., Souplet, J.-C., Fillard, P.: MedINRIA: DT-MRI Processing and Visualization Software. In: Workshop on Interaction in medical image analysis and visualization (2007)
9. Ding, Z., Gore, J.C., Anderson, A.W.: Classification and Quantification of Neuronal Fiber Pathways Using Diffusion Tensor MRI. Magnetic Resonance in Medicine 49, 716–721 (2003)
10. Corouge, I., Gouttard, S., Gerig, G.: Towards a Shape Model of White Matter Fiber Bundles Using Diffusion Tensor MRI. In: IEEE Int. Symposium on Biomedical Imaging: Nano to Macro., vol. 1, pp. 344–347 (2004)
11. Gerig, G., Gouttard, S., Corouge, I.: Analysis of Brain White Matter via Fiber Tract Modeling. In: Proc. of the 26th An. Int. Conf. of the IEEE, EMBS (2004)
12. Zhang, S., Laidlaw, D.H.: DTI Fiber Clustering and Cross-subject Cluster Analysis. Int. Soc. of Mag. Res. in Med. (2005)
13. O'Donnell, L., Westin, C.-F.: White Matter Tract Clustering and Correspondence in Populations. In: 8th Int. Conf. on Med. Im. Comp. and Computer-Assisted Intervention, pp. 140–147 (2005)
14. Jonasson, L., Hagmann, P., Thiran, J., Wedeen, V.: Fiber Tracts of High Angular Resolution Diffusion MRI are Easily Segmented with Spectral Clustering. In: Proc. of 13th An. Meeting ISMRM, Miami, p. 1310, SPIE (2005)
15. Maddah, M., Mewes, A., Haker, S., Grimson, W., Warfield, S.: Automated Atlas-Based Clustering of White Matter Fiber Tracts from DTMRI. Med. Img. Comp. Assist. Interv. (2005)
16. O'Donnell, L., Westin, C.-F.: Automatic Tractography Segmentation Using a High-Dimensional White Matter Atlas. IEEE Transac. on Med. Img. 26, 1562–1575 (2007)
17. Wagstaff, K.: Intelligent Clustering with Instance-Level constraints. Ph.D. dissertation, Cornell University (2002)
18. Davidson, I., Ravi, S.S.: Agglomerative Hierarchical Clustering with Constraints: Theoretical and Empirical Results. In: Jorge, A.M., Torgo, L., Brazdil, P.B., Camacho, R., Gama, J. (eds.) PKDD 2005. LNCS (LNAI), vol. 3721, pp. 59–70. Springer, Heidelberg (2005)

A New Multi-Task Learning Technique to Predict Classification of Leukemia and Prostate Cancer

Austin H. Chen[1,2] and Zone-Wei Huang[2]

[1] Department of Medical Informatics, Tzu-Chi University, No. 701, Sec. 3, Jhongyang Rd.
Hualien City, Hualien County 97004, Taiwan
[2] Graduate Institute of Medical Informatics, Tzu-chi University, No. 701, Sec. 3, Jhongyang Rd.
Hualien City, Hualien County 97004, Taiwan
achen@mail.tcu.edu.tw

Abstract. Microarray-based gene expression profiling has been a promising approach in predicting cancer classification and prognosis outcomes over the past few years. In this paper, we have implemented a systematic method that can improve cancer classification. By extracting significant samples (which we refer to as support vector samples because they are located only on support vectors), we allow the back propagation neural networking (BPNN) to learn more tasks. The proposed method named as the multi-task support vector sample learning (MTSVSL) technique. We demonstrate experimentally that the genes selected by our MTSVSL method yield superior classification performance with application to leukemia and prostate cancer gene expression datasets. Our proposed MTSVSL method is a novel approach which is expedient and perform exceptionally well in cancer diagnosis and clinical outcome prediction. Our method has been successfully applied to cancer type-based classifications on microarray gene expression datasets, and furthermore, MTSVSL improves the accuracy of traditional BPNN technique.

1 Background

Microarray-based gene expression profiling has been a promising approach in predicting cancer classification and prognosis outcomes over the past few years [1]. In most cases, cancer diagnosis depends on using a complex combination of clinical and histopathological data. However, it is often difficult or impossible to recognize the tumour type in some atypical instances [2]. In the past decade, DNA microarray technologies have had a great impact on cancer genome research; this technology measures the expression level for thousands of gene expressions simultaneously. It is a powerful tool in the study of functional genomes [3, 4]. Large scale profiling of genetic expression and genomic alternations using DNA microarrays can reveal the differences between normal and malignant cells, genetic and cellular changes at each stage of tumour progression and metastasis, as well as the difference among cancers of different origins. Cancer classification is becoming the critical basis in patient therapy. Researchers are continuously developing and applying the most accurate classification algorithms based on gene expression profiles of patients.

Prostate cancer and leukemia are very common cancers in the United States. In 2007 alone, approximately 24,800 new cases and 12,320 deaths among males were attributed to leukemia. Among males age 40 and below, leukemia is the most common fatal cancer. Meanwhile, 19,440 new cases and 9,470 deaths among females were attributed to leukemia, and it is the leading cause of cancer death among females below age 20. Acute lymphocytic leukemia (ALL) is the most common cancer in children ages 0 to 14. As for prostate cancer, it alone accounts for almost 29% (218,890) of incident cases in men in 2007. For men age 80 and older, prostate cancer is the second most common cause of cancer death. Based on cases diagnosed between 1996 and 2002, an estimated 91% of these new cases of prostate cancer are expected to be diagnosed at local or regional stages, for which 5-year relative survival approaches 100% [5]. Therefore, the identification of significant genes is of fundamental and practical interest. The examination of the top ranking genes may be helpful in confirming recent discoveries in cancer research or in suggesting new methods to be explored.

There are several machine learning techniques such as k-nearest neighbours, clustering methods, self organizing maps (SOM) [6], back-propagation neural network [7, 8] , probabilistic neural network, decision tree [9], random forest [10], support vector machines (SVM) [11, 12], multicategory support vector machines (MC-SVM) [1], Bayesian network [13], and the module–network [14] that have been applied in cancer classification. Among these techniques, SVM is touted as having the best accuracy in the prediction of cancer classification [1, 15]. Of the classification algorithms mentioned above, most suffer from a high dimensional input space problem due to a large amount of DNA microarrary gene expression data. Reducing the number of genes used for analysis while maintaining a consistently high accuracy is one method to avoid this problem. Likewise, how to reduce the number of genes needed for analysis while then increasing the prediction accuracy becomes a key challenge.

In this paper, we propose a new method to select genes that can produce a very high prediction accuracy of classification outcomes and use the second task to improve traditional BPNN. We hypothesize that samples (i.e. patient's profiles) which are incorrectly classified may significantly affect the selection of significant genes. By removing these samples, one should have a better chance of targeting relevant genes and, therefore, increasing classification performance. We call this approach the multi-task support vector sample learning (MTSVSL) technique. To validate the benefits of this new method, we applied our approach to both leukemia and prostate cancers. Our method produces a super classification performance for small gene subsets by selecting genes that have plausible relevance to cancer diagnosis.

2 Materials and Methods

2.1 Theoretical Basis of the SVS Technique

The SVS technique is briefly described as follows. A binary SVM attempts to find a hyperplane which maximizes the "margin" between two classes (+1/-1). Let

$$\{X^i, Y^i\}, i = 1...j, Y^i \in \{-1,1\}, X \in R$$

be the gene expression data with positive and negative class labels, and the SVM learning algorithm should find a maximized separating hyperplane

$$W \bullet X + b = 0$$

where W is the n-dimensional vector (called the normal vector) that is perpendicular to the hyperplane, and b is the bias. The SVM decision function is showed in formula (1), where α_i are positive real numbers and ϕ is mapping function

$$W^T \phi(X) + b = \sum_{i=1}^{j} \alpha_i Y_i \phi(X_i)^T \phi(X) + b \tag{1}$$

Only $\phi(X_i)$ of $\alpha_i > 0$ would be used, and these points are support vectors. The support vectors lay close to the separating hyperplane (shown in figure 2). α_i represents non-negative Lagrange multipliers, and it is used to discriminate every piece of training data which has different influences to the hyperplane in high dimension feature spaces. To explain the meanings of α_i, we first maximize the Lagrange problem:

$$L_D \equiv \sum_{i=1}^{j} \alpha_i - \frac{1}{2} \sum_{i,m=1}^{j} \alpha_i \alpha_m Y_i Y_m X_i \cdot X_m \tag{2}$$

When $\alpha_i = 0$ then $L_D = 0$ in formula (2), in this case, α_i means that the i th data that has no influence to the hyperplane; therefore, this sample is correctly classified by the hyperplane (such as point A in Fig. 1). When $0 < \alpha_i < C$, where $C > 0$ is the penalty parameter of the error term, the Lagrange problem L_D is subject to

$$\sum_{i=1}^{j} \alpha_i Y_i = 0$$

therefore, $L_D = \alpha_i$, under this circumstance, α_i means that the ith data has the degree of influence to the hyperplane (such as point B in Fig. 1).

When $\alpha_i = C$, the Lagrange problem L_D is subject to

$$\frac{1}{2} \sum_{i,m=1}^{j} \alpha_i \alpha_m Y_i Y_m X_i \cdot X_m > \alpha_i$$

L_D is negative, and therefore, α_i means the ith data is incorrectly classified by the hyperplane (such as point C in Fig. 1).

Each α_i determines how much each training example influences the SVM function. Because the majority of the training examples do not affect the SVM function, most of the α_i are 0. We can then infer that these support vectors should contain the desired strong classification information. By extracting only the samples (such as point B) located on the hyper plane, we can run a gene selection algorithm that better identifies biologically relevant genes.

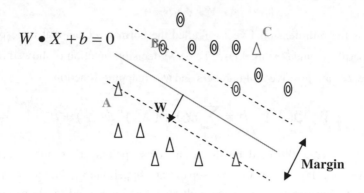

Fig. 1. The hyper plane and the support vectors. The red line is the hyperplane which separates two classes of samples. The blue points are the support vectors. A is the right classified sample but has less influence to the hyperplane. B is the right classified sample and has influence to the hyperplane. C is the incorrectly classified sample.

We then applied our method to two microarray datasets of leukemia and prostate cancers. We found 32 support vector samples in 72 leukemia samples and 44 support vector samples in 102 prostate cancer samples. After reducing the original samples by almost 55%, we used these samples to find the most informative genes through gene selection algorithms.

2.2 Multi-Task Learning Concept

The concept of multi-task learning (MTL) can be referred to [16]. The principle goal of multi-task learning is to improve the performance of classifier. The multi-task learning can be considered as an inductive transfer mechanism. The inductive transfer is leverage additional sources of information to improve the performance of learning on current task. Considering some variables which cannot be used as useful inputs because they will be missing for making predictions; however, they may be useful as extra outputs. Instead of discarding these variables, MTL get the inductive transfer benefit from discarded variables by using them as extra output.

To take back-propagation neural network as example, the figure 2 shows the three back-propagation network structures. In Figure 2, (A) is a standard back-propagation network and only one main task as output. (B) adds the extra variable as the extra input to learn in the main task. (C) presents that the network uses the extra variable as extra output to learn in parallel with main task and second task and it's a common structure for MTL.

MTL uses the inductive bias contained in training signals of related tasks. It is implemented by having extra outputs and learn task in parallel while using shared perception (see figure 2 C), what is learned for each task can help other task to learn better.

We propose a new approach, multi-task support vector sample learning (MTSVSL), which combines the SVS method with MTL concept together to classify the gene expression data. The pseudo code is displayed in Table 1. By this approach, classifier need to simultaneously learn two tasks, the main task is "which kind of

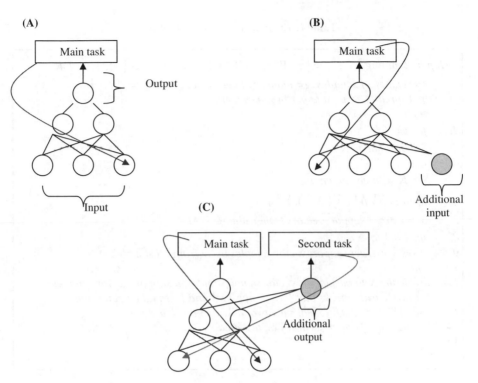

Fig. 2. Three Back-propagation network structures for learning

sample is this?" and the second task is "is this sample a support vector sample?". We propose to categorize the samples into four classes, namely:

I: the sample which is both class 1 and support vector sample.
II: the sample which is both class 2 and support vector sample.
III: the sample which is class 1 but not support vector sample.
IV: the sample which is class 2 but not support vector sample.

The support vector sample's feature is mentioned in section 2.1. Most classifier just learned the main task; however in our experiment, we find that the bias which is generated by learning the second task can improve the classifier performance. The ratio of these two biases is set as $0.9 * main\_bias + 0.1 * second\_bias$. The ratio of second bias cannot be set too high since it may harm the performance of main task.

2.3   Gene Selection Method

Gene selection is widely used to select target genes in the diagnosis of cancer. One of the prime goals of gene selection is to avoid the over-fitting problems caused by the high dimensions and relatively small number of samples of microarray data. Theoretically, in cancer classification, only informative genes which are highly related to particular classes (or subtypes) should be selected. In this study, we use signal-to-noise ratio (SNR) as our gene selection method. For each gene, we

Table 1. The Pseudo Code of MTSVSL

Input: Training data $X = \{x_G^S, Y^S\}, S = 1...s, G = 1...g, Y^S \in \{-1,1\}, X \in R$

    s=number of samples, g= number of genes, base learner= BP network

Output: predict result and performance metrics

1.   *begin*
2.   *for i = 1 to* S
3.   *do normalize* X
4.   *end*
5.   *Set* K *= linear function*
6.   *do train* $SVM(K(X^S), Y^S)$
7.   *sv = extract support vectors from training SVM*
8.   *for i = 1 to* S
9.   *svs = extract support vector samples by sv from all samples*
10.   *end*
11.   *divide the X to be support vector samples and non support vector samples*
12.   *set two learning tasks that {svs or non-svs} and {normal or abnormal}*
13.   *use MTL approach to learn the two tasks by base learner*
14.   *estimate the predict accuracy on test data*
15.   *end*

---

normalized the gene expression data by subtracting the mean and then dividing by the standard deviation of the expression value. Every sample is labeled with {+1,-1} as either a normal or a cancer sample. We can use the formula (3) to calculate each gene's F score.

$$F(g_i) = \frac{|\mu^{+1}(g_i) - \mu^{-1}(g_i)|}{\sigma^{+1}(g_i) + \sigma^{-1}(g_i)} \tag{3}$$

The μ and σ are mean and standard deviation of samples in each class (either +1or -1) individually. We rank these genes with F score and then select the top 25, 50, 100, and 150 gene set as the features.

2.4 Back Propagation Neural Network (BPNN) Classifier

The typical back-propagation neural network consists of one input layer, one output layer, and at least one hidden layer. The number of neurons presented in the input and output layer depends on the number of variables. Furthermore, the number of neurons used for the hidden layer is optimized by trial-and-error training assays. The behaviour of a back-propagation neural network is mainly determined by the transfer functions of its neurons. Here, we develop the BPNN codes using the Matlab NN tool box. The tan-sigmoid transfer function (formula 4) and log-sigmoid transfer function (formula 5) were applied for hidden and output neurons individually.

$$\tan{-sig}(n) = \frac{2}{(1+e^{-2n})-1} \qquad (4)$$

$$\log{-sig}(n) = \frac{1}{(1+e^{-n})} \qquad (5)$$

We set the output threshold to be 0.5. If the output value is over 0.5, the sample will be set as a normal sample; otherwise, the sample will be as an abnormal sample.

2.5 Cross Validation Method

The main purpose of cross validation methods is used to estimate the model performance based on the re-sampling technique. This method separates data into a training dataset and a validation dataset. A model is trained by the training dataset and then verified by the validation dataset. The following two common cross validation methods are usually used.

(1) K-fold cross validation method: this method divides the data into k subsets of equal or approximately size. It chose one subset from k subsets as the validation dataset for verifying the model, the remainder (k-1) dataset as the model's training dataset. The cross validation process is then repeated K times with each of the K subsets used only once as the validation data. The K results from the folds then can be averaged to produce a single estimation of the model performance.

(2) Leave-one-out cross validation method: As the name of this method, if we have n dataset, this method extracts one dataset as the validation dataset; the remainder (n-1) is used as the training dataset. Repeat this approach until all dataset was used. The results can be averaged to produce a single estimation of the model performance.

As Ambroise and McLachlan pointed out that the performance of classification maybe overestimated by using Leave-out-out method, therefore, we verify our experiment using random average 3-fold method. This random average 3-fold method randomly separates dataset into 3-fold and repeat validation 100 times in order to get the impartial performance results for our model.

3   Results and Discussions

In this section, we compare the results from MTSVSL with the results from the basic BPNN. To demonstrate the benefits of the MTSVSL method, we experiment it for leukemia and prostate cancer gene expression data. We applied random average 3-fold cross validation method to get the impartial performance results. In this experimentation the gene selection method is SNR.

3.1   Application of MTSVSL to the Leukemia Dataset

In the first set of experiment, we carried out a comparison between our MTSVSL method and BPNN method on the leukemia data. BPNN method has a high accuracy (94%) in the smaller gene number and a lower accuracy (89%) in 100 and 150 genes. Our MTSVSL generates a better accuracy with a 96% in the smaller gene number and a 91% in 100 and 150 genes. Our model improves the accuracy approximately 1% to 3% (see table 2). In general, the average sensitivity and specificity of BPNN is improved by MTSVSL which then improve the accuracy. One reason of this improvement is that the BPNN is very sensitive to selected genes and the higher number of genes (100 and 150 genes) may contain lower score genes. Therefore, the performance of traditional BPNN approach is worse in higher number of genes. Our MTSVSL method, however, is less affected by the higher number of genes.

Table 2. Comparison of leukaemia classification performance between BPNN and MTSVSL with random average 3-fold and repeat validation 100 times

Number of top genes	BPNN			MTSVSL		
	Sensitivity	specificity	accuracy	sensitivity	specificity	accuracy
25	99.00	94.21	95.83	97.64	95.98	96.67
50	93.71	93.36	93.33	95.06	96.53	95.83
100	88.27	93.24	89.58	86.25	92.38	90.00
150	76.94	91.96	88.33	90.67	95.42	92.92

3.2 Application of MTSVSL to the Prostate Cancer Dataset

In the second set of experiment, we carried out a comparison between our MTSVSL method and BPNN method on the prostate cancer data. Due to the prostate cancer is considered as a more difficult data to be classified. Traditional BPNN method has an average accuracy of 83% in the range of gene number. Our MTSVSL generates a better accuracy with an 86% average that is above 3% better. Apparently, MTSVSL method can improve all classification performance (sensitivity, specificity, and accuracy) in all the range of gene numbers selected.

Table 3. Comparison of prostate cancer classification performance between BPNN and MTSVSL with random average 3-fold and repeat validation 100 times

Number of top genes	BPNN			MTSVSL		
	sensitivity	specificity	accuracy	sensitivity	specificity	accuracy
25	86.15	86.17	85.88	89.00	91.24	88.53
50	83.14	80.17	81.47	88.21	84.25	82.65
100	82.87	86.31	82.35	88.92	87.85	87.65
150	76.58	79.71	80.59	80.77	85.22	82.06

4 Conclusions

Microarray-based gene expression profiling has been a promising approach in predicting cancer classification and prognosis outcomes over the past few years. In this paper, we have implemented a systematic method that can improve cancer classification. By extracting significant samples (which we refer to as support vector samples because they are located only on support vectors), we allow the back propagation neural networking (BPNN) to learn more tasks. The proposed method named as the multi-task support vector sample learning (MTSVSL) technique. We let the basic BPNN to learn more than one task and regulate these learning errors to improve the performance of basic BPNN. This novel approach can improve BPNN's classification performance metrics (accuracy, sensitivity, and specificity) with a small set of genes selected.

We demonstrate experimentally that the genes selected by our MTSVSL method yield superior classification performance with application to leukemia and prostate cancer gene expression datasets. Our proposed MTSVSL method is a novel approach which is expedient and perform exceptionally well in cancer diagnosis and clinical outcome prediction. Our method has been successfully applied to cancer type-based classifications on microarray gene expression datasets, and furthermore, MTSVSL improves the accuracy of traditional BPNN technique.

Acknowledgements

The author thanks the National Science Council for their financial support for projects NSC 98-2221-E-320-005.

References

1. Statnikov, A., Aliferis, C.F., Tsamardinos, I., Hardin, D., Levy, S.: A comprehensive evaluation of multicategory classification methods for microarray gene expression cancer diagnosis. Bioinformatics 21, 631–643 (2005)
2. Ramaswamy, S., et al.: Multiclass cancer diagnosis using tumour gene expression signatures. Proc. Natl Acad. Sci. USA 98, 15149–15154 (2001)
3. Chen, J.J., et al.: Global analysis of gene expression in invasion by a lung cancer model. Cancer Research 61, 5223–5230 (2001)
4. Morley, M., et al.: Genetic analysis of genome-wide variation in human gene expression. Nature 430, 743–747 (2004)
5. Jemal, A., Siegel, R., Ward, E., Murray, T., Xu, J., Thun, M.J.: Cancer statistics 2007. CA Cancer J. Clin. 57, 43–66 (2007)
6. Tamayo, P., Slonim, D., Mesirov, J., Zhu, Q., Kitareewan, S., Dmitrovsky, E., Lander, E., Golub, T.: Interpreting patterns of gene expression with self-organizing maps. Proc. Natl. Acad. Sci. USA 96, 2907–2912 (2000)
7. Greer, B.T., Khan, J.: Diagnostic classification of cancer using DNA microarrays and artificial intelligence. Ann. N. Y. Acad. Sci. 1020, 49–66 (2004)
8. Berthold, F., Schwab, M., Antonescu, C.R., Peterson, C., Meltzer, P.S.: Classification and diagnostic prediction of cancers using gene expression profiling and artificial neural networks. Nature Medicine 7, 673–679 (2001)
9. Tan, A.C., Gilbert, D.: Ensemble machine learning on gene expression data for cancer classification. Applied Bioinformatics 2, S75–S83 (2003)
10. Prinzie, A., Van den Poel, D.: Random forests for multiclass classification: Random multinomial logit. Expert Systems with Applications 34, 1721–1732 (2008)
11. Cortes, Vapnik: Support vector networks. Mach. Learning 20, 273–297 (1995)
12. Ramirez, L., Durdle, N.G., Raso, V.J., Hill, D.L.: A support vector machines classifier to assess the severity of idiopathic scoliosis from surface topology. IEEE Trans. Inf. Technol. Biomed. 10(1), 84–91 (2006)

13. Helman, P., et al.: A Bayesian network classification methodology for gene expression data. J. Comput. Biol. 11, 581–615 (2004)
14. Segal, E., et al.: A module map showing conditional activity of expression modules in cancer. Nature Genet. 36, 1090–1098 (2004)
15. Statnikov, A., Wang, L., Aliferis, C.: A comprehensive comparison of random forests and support vector machines for microarray-based cancer classification. BMC Bioinformatics 9, 319 (2008)
16. Caruana, R.: Multitask learning. Machine Learning 28(1), 41–75 (1997)

A Benchmark for Geometric Facial Beauty Study

Fangmei Chen[1,2] and David Zhang[2]

[1] Graduate School at Shenzhen, Tsinghua University, China
[2] Biometrics Research Center, Hong Kong Polytechnic University, Hong Kong

Abstract. This paper presents statistical analyses for facial beauty study. A large-scale database was built, containing 23412 frontal face images, 875 of them are marked as beautiful. We focus on the geometric feature defined by a set of landmarks on faces. A normalization approach is proposed to filter out the non-shape variations – translation, rotation, and scale. The normalized features are then mapped to its tangent space, in which we conduct statistical analyses: Hotelling's T^2 test is applied for testing whether female and male mean faces have significant difference; Principal Component Analysis (PCA) is applied to summarize the main modes of shape variation and do dimension reduction; A criterion based on the Kullback-Leibler (KL) divergence is proposed to evaluate different hypotheses and models. The KL divergence measures the distribution difference between the beautiful group and the whole population. The results show that male and female faces come from different Gaussian distributions, but the two distributions overlap each other severely. By measuring the KL divergence, it shows that multivariate Gaussian model embodies much more beauty related information than the averageness hypothesis and the symmetry hypothesis. We hope the large-scale database and the proposed evaluation methods can serve as a benchmark for further studies.

1 Introduction

The human face plays an important role in the daily life. As the demand for aesthetic surgery has increased tremendously over the past few years, an understanding of beauty is becoming utmost important for the medical settings.

Facial beauty study has fascinated human beings since ancient times. Philosophers, artists, psychologists have been trying to capture the nature of beauty through their observations or deliberately designed experiments. As far as we know, there are several hypotheses and rules of what makes a face attractive [1][2][3]. For example, faces with the following features are thought to be more attractive: general characteristics like symmetry and averageness, or specific characteristics like high forehead and small nose. However, the published studies have not yet reached consensus on those hypotheses or rules.

In the last few years, computational approaches have been developed, which bring new methods and new applications to the facial beauty study. Morphing and

D. Zhang and M. Sonka (Eds.): ICMB 2010, LNCS 6165, pp. 21–32, 2010.

wrapping programs are used to make composite faces or change the appearance of a face, which are further used as stimuli in the experiments [4][6][8][9]. Machine learning techniques such as Support Vector Regression (SVR), kernel regression, and K Nearest Neighbors (KNN) are used for beauty score prediction, virtual plastic surgery and automatic digital face beautification [10][11][12][13][14].

Despite the increasing interest and the extensive research in this field, the nature of human's aesthetic perception is far from discovered. There's no public database or benchmark for the purpose of facial beauty analysis. Although researchers tried to find some numeric expressions of beauty by extracting geometric related features such as distances or proportions [5][6][7], it was not usually possible to generate graphical representations of shape from these features because the geometric relationships among the variables were not preserved [16].

In this study, we have built a large-scale database which contains 15393 female and 8019 male photographs of Chinese people. All the photographs are frontal with neural expressions. In order to learn the attributes of what make a face attractive, we select beautiful faces manually from this database, including 587 female and 288 male images. We automatically extract 68 landmarks from each face image as our geometric feature. The geometric features are normalized and projected to tangent space, which is a linear space and Euclidean distance can be used to measure differences between shapes [21]. After preprocessing, the statistics of the geometric features are calculated. Hotelling's T^2 test shows that the mean faces of female and male have a significant difference in shape. However, compared to pair-wise face shape distance, the mean shapes are much closer to each other. Principle Component Analysis (PCA) is used for summarizing the main modes of variation and dimension reduction. Then we model the shapes as a multivariate Gaussian distribution. Kullback-Leibler (KL) divergence is used for measuring the difference between distributions of attractive faces and the whole population. This difference contains the information of what make a face attractive. By measuring this information, we quantitatively evaluate the popular hypotheses. The results show multivariate Gaussian model embodies much more beauty related information than the averageness hypothesis and the symmetry hypothesis.

The remaining paper is organized as follows: Sect.2 shows the preliminary work, including data collection, geometric feature extraction and normalization. In Sect.3, we give the statistical analysis and introduce the model evaluation method based on KL divergence. The experiment results and discussion are provided in Sect.4.

2   Preliminary

2.1   Data Collection

According to earlier studies, beautiful individuals make up a small percentage of the population [15]. The features representing a face are high dimensional due to the complexity of human face structures. To study facial beauty, we need a large-scale database which includes enough beautiful faces. Because of the highly non-rigid characteristic and intrinsic 3-D structure of face, automatic geometric feature extraction requires face images of the same pose and expression. The data used in our experiments are frontal photographs with neutral expression collected from Northeast China. The volunteers include 15393 females and 8019 males aged between 20 and 45. Each subject corresponds to one color image with 441×358 pixels in size. A sub-database is constructed by selecting more attractive faces manually. The resulting beauty data set contains 587 female images and 288 male images, which account for 3.7% of the total images.

2.2 Feature Extraction

The attractiveness of a face is affected by a number of factors, e.g., facial skin texture, symmetry, color, shape and so on. To quantify and analyze all these factors together is very difficult. Geometric feature plays an important role in facial beauty perception. As it is shown in Fig. 1, people will have different attractiveness responds when changing the geometric features while keeping other factors constant. As a result, we will focus on the geometric feature to learn how it affects the perception of facial beauty.

Geometric Feature Definition. The geometric feature is defined by a set of landmarks on the face. The 68 landmarks used in this study are shown in Fig.2. They locate on the major facial organs, including facial outline, eye-brows, eyes, nose, and mouth. The Cartesian coordinates of the landmarks form a configuration matrix, representing the geometric feature on the face, i.e.,

$$\mathbf{G} = \begin{bmatrix} x_1 \ x_2 \ldots x_{68} \\ y_1 \ y_2 \ldots y_{68} \end{bmatrix}^T, \tag{1}$$

where 'T' denotes the transpose operator. In this way, the face images reduce to configuration matrices.

Fig. 1. Images with same texture but different geometric feature

Fig. 2. Landmarks on the face are used to define its geometric feature

Automatic Geometric Feature Extraction. The large-scale database requires automatic feature extraction. First, we use Viola and Jones detector for face detection [17]. This method is known for its speed and precision on detecting frontal faces. On the detected face region, we then use the method proposed by Rowley [18] to locate the eyes on the face. Based on the positions of the face region and eyes, the landmarks are located by the Active Shape Model (ASM) [19][20] search. The geometric feature of the face is finally generated from these extracted landmarks.

2.3   Geometric Feature Normalization

Before statistical analysis, we normalize the geometric features by removing the non-shape variations, including translation, in-plane rotation and scale. The proposed normalization method is data driven and carried out once for all, i.e., female and male face shapes are treated together as a whole population. After normalization, we rewrite the configuration matrices into 136-D feature vectors and project them into tangent space.

Filtering translation. Translation can be filtered out by moving the center of landmarks to the origin of the coordinate system. It can be achieved by premultiplying $\mathbf{G}$ with a matrix

$$\mathbf{C} = \mathbf{I}_{68} - \frac{1}{68}\mathbf{1}_{68}\mathbf{1}_{68}^T. \tag{2}$$

The centered configuration matrix

$$\mathbf{G}_c = \mathbf{C}\mathbf{G}. \tag{3}$$

Filtering scale. Scale invariance is also desired in shape analysis. The size of a centered configuration matrix is defined as its Euclidean norm

$$\|\mathbf{G}_c\| = \sqrt{trace\left(\mathbf{G}_c^T\mathbf{G}_c\right)}. \tag{4}$$

Scale can be filtered out by make the configuration matrix unit size, i.e.,

$$\widetilde{\mathbf{G}} = \mathbf{G}_c/\|\mathbf{G}_c\|. \tag{5}$$

Filtering rotation. In-plane rotation doesn't change the shape of an object. The pre-shapes need alignment before statistical analysis. The alignment process can be achieved with an iterative scheme. First the mean shape is calculated, then we align all the shapes onto the mean shape by finding the fittest rotation matrix. After that, the mean shape is updated and all the shapes are aligned to the updated mean shape until convergence.

Projection to Tangent space. Tangent space is a linear space and Euclidean distance can be used to measure the difference between shapes. The configuration matrix can be rewritten in vector form,

$$\mathbf{g} = [G_{1,1}, G_{1,2}, G_{2,1}, G_{2,2}, \dots, G_{68,1}, G_{68,2}]^T \ . \tag{6}$$

The constraint of tangent space is $(\widetilde{\mathbf{g}}_i - \overline{\mathbf{g}}) \cdot \overline{\mathbf{g}} = 0$, where $\overline{\mathbf{g}} = \frac{1}{n} \sum_{i=1}^{n} (\mathbf{g}_i)$. When $\|\overline{\mathbf{g}}\| = 1$,

$$\widetilde{\mathbf{g}}_i = \frac{\mathbf{g}_i}{\mathbf{g}_i \cdot \overline{\mathbf{g}}} \ . \tag{7}$$

In this way, we project all the geometric feature vectors to its tangent space. Fig.3 shows the scatter plots of landmarks before (3a) and after (3b) normalization.

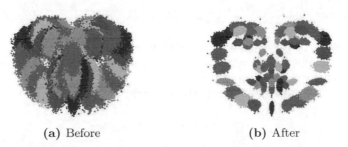

(a) Before   (b) After

Fig. 3. Landmarks plots before and after normalization

3 Methodologies

3.1 Basic Statistical Analysis

One basic statistic we get from the database is mean shape. We are interested in how much difference between female and male mean shapes and whether this difference is statistically significant.

We consider female and male face shapes as two independent random samples $X_1, X_2, \dots, X_{n_1}$ and $Y_1, Y_2, \dots, Y_{n_2}$ from mean shapes μ_1 and μ_2. To test between

$$H_0 : \mu_1 = \mu_2 \quad versus \quad H_1 : \mu_1 \neq \mu_2 \ ,$$

we carry out Hotelling's T^2 two sample test [21].

3.2   Principal Component Analysis (PCA)

PCA can be used in tangent space to summarize the main modes of variation in a database, from which we can have an intuitive understanding of the database. Besides, PCA is often applied for dimension reduction, under the criterion of minimum mean square error (MSE) [22]. We do PCA analysis only on female data and denote them as a sample of random vector X. The covariance matrix of X is defined as

$$\Sigma_X = E\left\{[X - E(X)][X - E(X)]^T\right\} = \boldsymbol{\Phi}\boldsymbol{\Lambda}\boldsymbol{\Phi}^T, \tag{8}$$

with

$$\boldsymbol{\Phi} = [\phi_1, \phi_2, \dots, \phi_{136}],$$

$$\boldsymbol{\Lambda} = diag\{\lambda_1, \lambda_2, \dots, \lambda_{136}\},$$

where $\boldsymbol{\Phi}$ is the eigenvector matrix and $\boldsymbol{\Lambda}$ is the diagonal eigenvalue matrix with diagonal elements in decreasing order.

For visually showing the modes of variation, we reconstruct the shape vector along the direction of each PC (principal component). A series of reconstructed shape vectors for PCi are calculated by

$$X_c^* = E(X) + c\phi_i \tag{9}$$

with different c. In this way, we can visually tell what effect each PC has on the variation of shapes. Further more, we can study the relationship between this variation and the perception of beauty.

For dimension reduction, we use the first 25 PCs, i.e.,

$$P = [\phi_1, \phi_2, \dots, \phi_{25}],$$

whose correspondent eigenvalues take 99.9949% of the total energy. Hence, there's almost no information lost in the dimension reduction process

$$Y = P^T X = [Y_1, Y_2, \dots, Y_{25}]^T. \tag{10}$$

3.3   Multivariate Gaussian Model

We are interested in distributions of two groups of people, one is the attractive group, and the other is the whole population. For the data after dimension reduction,

$$\begin{matrix} Y_b : \text{attractive sample} \\ Y : \text{the whole population} \end{matrix}, \quad \{Y_b\} \subset \{Y\},$$

we make an assumption that each element in vector Y follows Gaussian distribution. Take the first element Y_1 as an example, the Gaussian assumption is an approximate to the real distribution shown in Fig. 4, where 4a is the histogram of Y_1 and 4b is its quantile-quantile plot [23]. If the distribution of Y_1 is Gaussian, the plot will be close to linear. According to Fig.4, the Gaussian assumption is

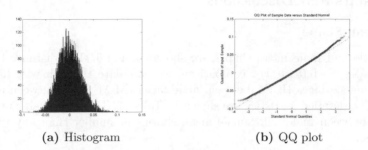

(a) Histogram   (b) QQ plot

Fig. 4. Histogram and quantile-quantile plot of Y_1

reasonable. Since the covariance matrix of Y is diagonal, the elements of Y are independent with each other. So, we have

$$Y_i \sim N(\mu_i, \sigma_i^2), Y_{bi} \sim N(\mu_{bi}, \sigma_{bi}^2)$$

$$p(Y) = p(Y_1)p(Y_2)\cdots p(Y_{25}) \tag{11}$$

$$p(Y_b) = p(Y_{b1})p(Y_{b2})\cdots p(Y_{b25}) \tag{12}$$

3.4   Model Evaluation

Given $p(Y)$ and $p(Y_b)$, we will focus on the difference of them, which embodies the beauty related information brought by people who selected the more attractive images. Kullback-Leibler divergence (KL divergence) is a measure of difference between two probability distributions. For two univariate Gaussian distributions, the KL divergence is defined as

$$\begin{aligned} D(Y_{bi}\|Y_i) &= \int_{-\infty}^{\infty} p(Y_{bi}) \log \frac{p(Y_{bi})}{p(Y_i)} dy_i \\ &= \frac{(\mu_{bi}-\mu_i)^2}{2\sigma_i^2} + \frac{1}{2}\left(\frac{\sigma_{bi}^2}{\sigma_i^2} - 1 - \ln \frac{\sigma_{bi}^2}{\sigma_i^2}\right) \end{aligned} \tag{13}$$

We replace $p(Y_{bi})$ and $p(Y_i)$ with $p(Y_b)$ and $p(Y)$ given by (11) and (12) and get the KL divergence

$$D(Y_b\|Y) = \sum_{i=1}^{25} D(Y_{bi}\|Y_i). \tag{14}$$

For comparison, we randomly choose n_b samples from the whole population 1000 times and calculate the KL divergence between sample distribution $p(Y_{random})$ and $p(Y)$. Then we define the information gain as

$$IG = D(Y_b\|Y) - E\left(D(Y_{random}\|Y)\right), \tag{15}$$

which is the information gain of the multivariate Gaussian model on our database. Under other Facial beauty hypotheses and rules [7], we can also calculate their correspondent IG. By comparing the information gain, different hypotheses, rules or features have quantitative evaluations.

4   Results and Discussions

4.1   Mean Shape

The female and male mean shapes are shown in Fig.5. The distance between them is $d_{\mu_1\mu_2} = 0.0161$. For comparison, we calculate the pair-wise distances between female faces (F-F), between male faces (M-M), and between all faces (ALL) .The detailed statistics are shown in Table 1. The results show that the distance between male and female mean shapes is smaller than any pair-wise face shape.

Now we are interested in whether this distance is caused by chance or statistically significant. We carry out the Hotelling's T^2 two sample test.We have the statistic $F = 140.85$ and the null distribution $F_{132,23279}$. Since $P(F > 1.55) = 0.0001$, and our F is much larger than 1.55, there is strong evidence for a significant difference in female and male shapes.

4.2   Main Modes of Variation

The main modes of variation are summarized by applying PCA. Fig.6 shows the reconstructed shape vector series along directions of the first 5 PCs, each row for one PC. The columns represent different deviations,$[-4\sigma, \ldots, -\sigma, 0, \sigma, \ldots, 4\sigma]$, from mean.

The first PC includes the variation of face width, the second PC includes the variations of eyebrow length and face shape, the third PC includes the variation of configuration of facial organs, and the fifth PC includes slight out-of-plane rotation. As the shape variability is complicated, we can only make some coarse descriptions.

Table 1. Statistics of pair-wise shape distance

Statistics	F-F	M-M	ALL
MAX	0.2284	0.2223	0.2254
MIN	0.0201	0.0208	0.0193
MEAN	0.0717	0.0749	0.074
STD	0.0165	0.0175	0.0168

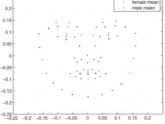

Fig. 5. Overlaid female and male mean shapes

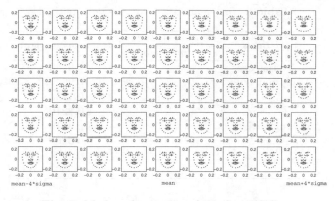

Fig. 6. Modes of variation of the first 5 PCs

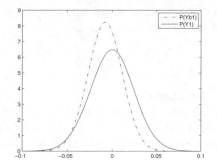

Fig. 7. Probability density functions of Y_1 and Y_{b1}

However, Fig.6 provides visual explanation for beauty studies based on PCA coefficients. For example, the distributions of the first element of Y and Y_b are shown in Fig.7. With reference to the first row in Fig.6, we observe that attractive faces have a tendency to be thinner than the mean face.

Results in this part are descriptive. They explain the differences between two groups of faces Y and Y_b visually, which gives some cues for facial beauty perception study.

4.3 Information Gain of Multivariate Gaussian Model, Averageness Hypothesis, and Symmetry Hypothesis

For quantitative study of facial beauty, we calculate the information gain defined by (15) under multivariate Gaussian model, averageness hypothesis, and symmetry hypothesis.

Multivariate Gaussian Model. The information gain of multivariate Gaussian model is calculated by (11)(12) and (15). During the process, we have two findings. First, the KL divergence between $p(Y_{random})$ and $p(Y)$ is much smaller than that between $p(Y_b)$ and $p(Y)$. Second, different elements contribute different amount of information. Some detailed results are listed in Table 2.

Table 2. Information gain of different methods

Methods	Dependent variable dimension	Information gain
Averageness	1	0.0585
Symmetry	1	0.0046
	1	0.1288
	2	0.2210
Multivariate	3	0.3110
Gaussian Model	5	0.3867
	10	0.5191
	25	0.5907

Averageness hypothesis. The averageness hypothesis assumes that a beautiful face is simply one that is close to an average of all faces. We define the averageness score s_a of subject face shape $\mathbf{g}$ as

$$s_a = -\|\mathbf{g} - \overline{\mathbf{g}}\| \ , \tag{16}$$

where $\overline{\mathbf{g}}$ is the average face shape.

The beautiful group X_b and the whole population X have different distributions of s_a. We model them as Gaussian distributions and denote them as $p(s_a^{X_b})$ and $p(s_a^X)$. The information gain can be calculated by (15). The result is shown in Table 2.

Symmetry Hypothesis. The symmetry hypothesis assumes the more symmetric a face, the more beautiful it will be. In the present study, the symmetry is narrowed down to the symmetry of landmarks. We define the symmetry score s_s in the following way. For each face, we first estimate the symmetry axis by fitting the least square regression line through the eight landmarks, which are along the middle of a face, defined in Fig.2. Then we calculate the symmetry points of the left 60 landmarks according to the symmetry axis. We call these new generated points pseudo-landmarks and denote them as L^{pseudo}. Each pseudo-landmark has a corresponding landmark on the same side of the face. We denote these landmarks as L^{cor}, and

$$s_s = -\|L^{pseudo} - L^{cor}\| \ . \tag{17}$$

Similar to the averageness hypothesis, we have two distributions $p(s_s^{X_b})$ and $p(s_s^X)$. The information gain calculated by (15) is shown in Table 2.

The results show averageness hypothesis and symmetry hypothesis reveal much less beauty related information than multivariate Gaussian model. In Gaussian model, the information is concentrated in a few elements. Although the multivariate model is not proved to be optimal, it can serve as a benchmark under the information gain criterion.

5 Conclusion

In this paper, we study facial beauty with geometric features, which are defined by 68 landmarks on faces. The geometric features are then normalized and mapped to tangent space, which is a linear space and Euclidean distance can be used to measure the shape difference. The statistical analyses show that male and female faces are generated from different distributions, but the two distributions overlap each other and the two mean face shapes are very similar. PCA is applied to summarize the main modes of face shape variation and do dimension reduction. Kullback-Leibler divergence between beautiful group and the whole population is used to measure the beauty related information embodied in the selected features. In this sense, multivariate Gaussian model performs better than the averageness hypothesis and symmetry hypothesis. We hope the findings in this paper and the proposed evaluation method provide useful cues for further research on facial beauty.

There are still some limitations to the present study. The landmarks we use in the experiments can not describe the full geometry of a face. Human face are 3-D in nature, so landmarks from 2-D frontal face images can not capture the height information, for example, the heights of nose and cheekbones, which are often considered useful in assessing facial beauty. Other features like texture and color are not considered in the previous study. However, the database and the proposed evaluation method are not constrained to geometric features. The above-mentioned issues will be explored in our future study.

References

1. Etcoff, N.L.: Beauty and the beholder. Nature 368, 186–187 (1994)
2. Langlois, J.H., Roggman, L.A.: Attractive Faces are only Average. Psychological Science 1, 115–121 (1990)
3. Perrett, D.I., Burt, D.M., Penton-Voak, I.S., Lee, K.J., Rowland, D.A., Edwards, R.: Symmetry and Human Facial Attractiveness. Evolution and Human Behavior 20, 295–307 (1999)
4. Perrett, D.I., May, K.A., Yoshikawa, S.: Facial Shape and Judgements of Female Attractiveness. Nature 368, 239–242 (1994)
5. Wang, D., Qian, G., Zhang, M., Leslie, G.F.: Differences in Horizontal, Neoclassical Facial Cannons in Chinese (Han) and North American Caucasian Populations. Aesthetic Plastic Surgery 21, 265–269 (1997)
6. Pallett, P.M., Link, S., Lee, K.: New 'golden' ratios for facial beauty. Vision Research 50(2), 149–154 (2010)
7. Schmid, K., Marx, D., Samal, A.: Computation of a face attractiveness index based on neoclassical canons, symmetry and golden ratios. Pattern Recognition (2007)
8. Rhodes, G., Yoshikawa, S., Clark, A., Lee, K., McKay, R., Akamatsu, S.: Attractiveness of Facial Averageness and Symmetry in non-Western Cultures: In Search of Biologically based Standards of Beauty. Perception 30, 611–625 (2001)
9. Valenzano, D.R., Mennucci, A., Tartarelli, G., Cellerino, A.: Shape Analysis of female facial attractiveness. Vision Research 46, 1282–1291 (2006)
10. Leyvand, T., Cohen-Or, D., Dror, G., Lischinski, D.: Data-Driven Enhancement of Facial Attractiveness. In: ACM SIGGRAPH (2008)

11. Davis, B.C., Lazebnik, S.: Analysis of Human Attractiveness Using Manifold Kernel Regression. In: ICIP (2008)
12. Liu, J., Yang, X.-b., Xi, T.-t., Gu, L.-x., Yu, Z.-y.: A Novel Method for Computer Aided Plastic Surgery Prediction (2009)
13. Whitehill, J., Movellan, J.R.: Personalized Facial Attractive Prediction (2008)
14. Eisenthal, Y., Dror, G., Ruppin, E.: Facial Attractiveness: Beauty and the Machine. Neural Computation 18, 119–142 (2006)
15. Atiyeh, B.S., Hayek, S.N.: Numeric Expression of Aesthetics and Beauty. Aesthetic Plastic Surgery 32, 209–216 (2008)
16. Adams, D.C., James Rohlf, F., Slice, D.E.: Geometric morphometrics: ten years of progress following the 'revolution'. Ital. J. Zool. 71, 5–16 (2004)
17. Viola, P., Jones, M.J.: Robust Real-Time Face Detection. International Journal of Computer Vision 57(2), 137–154 (2004)
18. Rowley, H.A., Baluja, S., Kanade, T.: Neural Network-Based Face Detection. IEEE Trans. Pattern Analysis and Machine Intelligence 20, 23–38 (1998)
19. Cootes, T.F., Taylor, C.J., Cooper, D.H., Graham, J.: Active Shape Models - Their Training and Application. Computer Vision and Image Understanding 61(1), 38–59 (1995)
20. Sukno, F.M., Ordas, S., Butakoff, C., Cruz, S., Frangi, A.F.: Active Shape Models with Invariant Optimal Features: Application to Facial Analysis. IEEE Trans. Pattern Analysis and Machine Intelligence 29(7), 1105–1117 (2007)
21. Dryden, I.L., Mardia, K.V.: Statistical Shape Analysis. John Wiley, Chichester (1998)
22. Liu, C., Wechsler, H.: Robust Coding Schemes for Indexing and Retrival from Large Face Databases. IEEE Trans. Image Processing 9, 132–137 (2000)
23. Thode, H.C.: Tesing for normality. Marcel Dekker, New York (2002)

An Effective Feature Extraction Method Used in Breath Analysis

Haifen Chen[1], Guangming Lu[1], Dongmin Guo[2], and David Zhang[2]

[1]Shenzhen Graduate School,
Harbin Institute of Technology, Shenzhen, China
luguangm@hit.edu.cn
[2]Biometrics Research Centre, Department of Computing
The Hong Kong Polytechnic University, Kowloon, Hong Kong
csdguo@comp.polyu.edu.hk

Abstract. It has been reported that human breath could represent some kinds of diseases. By analyzing the components of breath odor, it is easy to detect the diseases the subjects infected. The accuracy of breath analysis depends greatly on what feature are extracted from the response curve of breath analysis system. In this paper, we proposed an effective feature extraction method based on curve fitting for breath analysis, where breath odor were captured and processed by a self-designed breath analysis system. Two parametric analytic models were used to fit the ascending and descending part of the sensor signals respectively, and the set of best-fitting parameters were taken as features. This process is fast, robust, and with less fitting error than other fitting models. Experimental results showed that the features extracted by our method can significantly enhance the performance of subsequent classification algorithms.

Keywords: Breath odor, Feature extraction, Curve fitting.

1 Introduction

It has been found that breath compositions contain molecules that originated from normal or abnormal physiology[1]. In recent years, breath odors have been used to diagnose diabetes, renal disease, bacterial infection, dental disease, etc. Breath analysis is a non-invasive and real-time approach by contrast with blood, urine, and other bodily fluids and tissues test and a well-accepted technique because the breath sample is easy to be collected. Hence, it draws more and more research attention recently.

Gas chromatography (GC)[2] and electronic nose (e-nose)[3] are two main tools for capturing signals in the breathe. GC-MS can recognize the volatile organic compounds (VOCs) in the breath accurately, but it is too much expensive and not convenient to use. On the other hand, e-nose is playing a growing role in detecting VOCs for breath analysis. An e-nose is mainly composed of an array of sensors and some relevant signal processing circuits. Sensors interact with gas and create a unique response signal. The signal is processed by denoising, baseline manipulation, and normalization, and then characteristic features are extracted from the processed signal for pattern recognition and classification.

D. Zhang and M. Sonka (Eds.): ICMB 2010, LNCS 6165, pp. 33–41, 2010.

In all of these processes, feature extraction is the most important stage. Original signal from the breath analysis system is too large to be processed and it maybe includes notoriously redundant data but no much useful information. Hence, it is necessary to transform the original signal into a reduced representation, i.e. feature extraction. If the features are extracted properly, the features set will represent the most relevant information from original data. The accuracy of breath analysis depends greatly on what features are extracted from the response curve of breath analysis system.

Currently, there are three main kinds of features are extracted in the signal of odor response, geometric features[4-7], frequency features[8-9], and curve features extracted by fitting the signal curve[10]. Geometric features are features like the peak value of the signal, the time from the beginning to the peak, the area under the curve, gradient, and curvatures. These kinds of features are easy to obtain and they are satisfactory for some applications. However, they can not represent the genuine information of signal since they are just the segments of signal. Frequency features contain more signal information than geometric features, and are better for classification. But the feature extraction method is complicated and time-consuming, and hence not suitable for on-line classification. One more flexible feature extraction method is curve fitting. It formed parametric model for each signals and used the parameter of model as characteristic feature. Features extracted by this method are easy to get and contain more signal information. The most important is, the feature set is very small.

In this paper, we present a new feature extraction method based on curve fitting. The idea of this method based on Planck model and Lorentzian model. The aim is to find an analytic expression to model the time-dependent response. For the response of each channel, we use two different models to fit the ascending and descending part of the signal curve respectively. The combination of these two models can better fit the signal curve and hence yield less fitting error. Experimental results show that the proposed method can get better classification results than other methods.

The remainder of this paper is organized as follows. Section 2 introduces our breath analysis system and the breath samples. Section 3 describes the models we proposed to fit the curves and the feature extraction procedure. Section 4 presents the classification experiment results. Section 5 gives the conclusion.

2   Breath Analysis System and Breath Samples

The samples used in this paper were collected by a self-designed breath analysis system. In this system, we employed 12 chemical sensors to capture the signals of breath odors. These chemical sensors are sensitive to a large number of gaseous molecules, but their dominantly sensitive subjects are different. Hence, they can create a unique pattern for each sample, which includes 12 response curves.

We collected diabetes samples and healthy samples from hospital by using a self-designed breath analysis system. The responses are shown in Fig. 1. In each figure, the horizontal axis stands for the sampling time and the vertical axis shows the amplitude of the sensor output in volts. Two figures in Fig. 1(a) are the response of samples from diabetics and figures in Fig. 1(b) are the responses of samples from healthy subjects. From these figures, we can see the distinguishable response patterns of diabetes samples and healthy samples. The differences indicate that it is possible to recognize sick samples from healthy samples by using our breath analysis system.

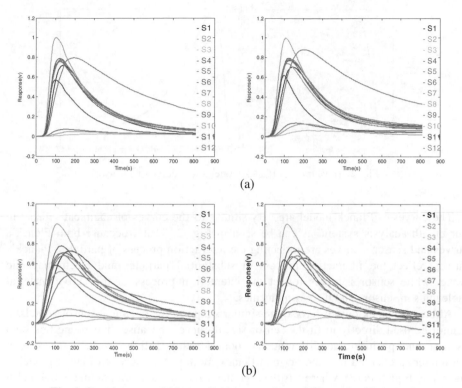

Fig. 1. Signal curves of (a) two diabetes samples and (b) two healthy samples

3 Feature Extraction Based on Curve-Fitting Models

In this section we present two curve-fitting models for feature extraction. First, a brand new model named Planck model is proposed. Then we introduce another model named Lorentzian model and give some improvement to it. By combining these two models, we obtain much less fitting error and at the same time keep the number of features small. Finally a comparison between our models and several other models is made and proves that the combination of Planck and the improved Lorentzian model outperforms the other models.

Planck model is from Planck's law in physics. Planck's law describes the spectral radiance of electromagnetic radiation at all wavelengths emitted in the normal direction from a black body at temperature T. As a function of wavelength λ, Planck's law is written as

$$I(\lambda, T) = \frac{2hc^2}{\lambda^5} \frac{1}{e^{\frac{hc}{\lambda kT}} - 1},$$

(1)

where h, c and k are physical constants, and e is the base of Napierian logarithm.

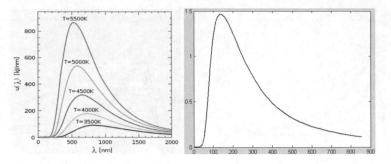

Fig. 2. The curves of Planck model and a curve of a sensor

The curves of Planck model are very similar to the curves of chemical sensors in our breath analysis system, as can be seen in Fig. 2. That's because both Planck's curves and sensor's curves are reflecting the interaction process of particles. Planck's curves reflect the interaction process of substantial particles and electromagnetic wave, while sensor's curves reflect the interaction process of the active material molecules in sensors and the gas molecules.

Although Planck curves have quite similar shape to the chemical sensor signals, it cannot be used directly to fit the sensor signal curves, because their properties such as size and position are different. To obtain better fitting result, some special parameters are added into the original Planck model. An improved Planck model is shown by Equation (2), where x is the argument, t, a, b and c are parameters, A and B are constants whose values are 1.191×10-16 and 0.0144 respectively.

$$f(x;t,a,b,c) = c\frac{A}{(ax+b)}\frac{1}{e^{\frac{B}{t(ax+b)}}-1},$$ (2)

The new parameters a, b, and c are used to control the shape and position of the fitting curve. If we use the Planck model to fit the whole curve, the result is shown in Fig. 3(c). From the figure we can see that the whole fitting is not satisfying, and improvement can be made by adding more parameters. So we use the technique of segment fitting, where the breakpoint is the peak point. As is shown in Fig. 3(d), the effect of segment fitting is much better than the whole fitting.

Although the segment fitting of Planck model has been improved greatly, the fitting effect is still not very good in the descending part of the sensor response. To solve this problem, we introduce a fitting model named Lorentzian model. Proposed by L. Carmel[10], this model was derived according to the response process of the gases and sensors. The expression of Lorentzian model is showed in Equation. (3), where $R_i(t)$ is the response of the i-th sensor at time t, t_i is the time when the sensors are exposed to the gases, T is the time when the gas stop going in the breath analysis system, β_i and τ_i are parameters associate with sensors.

$$R_i(t) = \begin{cases} 0, t < t_i \\ \beta_i \tau_i \tan^{-1}(\frac{t-t_i}{\tau_i}), t_i \leq t \leq t_i + T \\ \beta_i \tau_i \left[\tan^{-1}(\frac{t-t_i}{\tau_i}) - \tan^{-1}(\frac{t-t_i-T}{\tau_i}) \right], t > t_i + T \end{cases} ,$$

(3)

The expression of Lorentzian model has some deficiencies. Firstly, it neglects the response of the sensors after the gases stop going in the breath analysis system. Another problem is that there are many subjective estimates during the building of this model.

On the base of experiments, we give some improvement to Lorentzian model, which is shown as follows:

$$f(x) = \begin{cases} 0, x < 0 \\ a_1 \tan^{-1}(\frac{x-a_2}{a_3}) + a_4, 0 \leq x \leq x_{max} \\ d_1 \tan^{-1}(\frac{x-d_2}{d_3}) - d_4 \tan^{-1}(\frac{x-d_2-x_{max}}{d_3}), x > x_{max} \end{cases} ,$$

(4)

where x_{max} is the abscissa value of the curve's peak. The fitting effect of the improved Lorentzian (marked as iLorentzian) model is shown in Fig. 3(e). It can be seen that the iLorentzian model fits the sensor's curve very well, though there are still some shortcomings in the beginning part of the curve and the part around the peak. However, the rest of the line segments are fitted perfectly, especially the descending part of the curve.

Combining the advantages of Planck model and the iLorentzian model, we obtain a new curve fitting model (named PL model), which is shown by Equation (5). By using the Planck model to fit the ascending part of the curve and the iLorentzian model to fit the descending part, we can achieve a very good fitting effect. As can be seen in Fig. 3(f), PL model fits the sensor's response really well, except for a little insufficiency around the curve's peak.

$$f(x) = \begin{cases} 0, x < 0 \\ c \dfrac{A}{(ax+b)} \dfrac{1}{e^{\frac{B}{t(ax+b)}} - 1}, 0 \leq x \leq x_{max} \\ d_1 \tan^{-1}(\frac{x-d_2}{d_3}) - d_4 \tan^{-1}(\frac{x-d_2-x_{max}}{d_3}), x > x_{max} \end{cases} ,$$

(5)

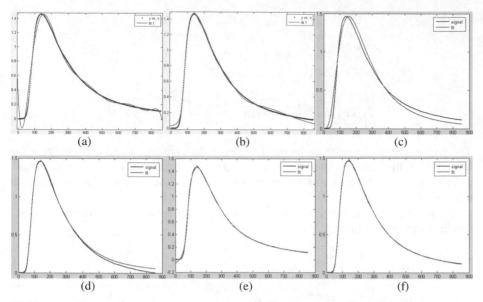

Fig. 3. The fitting effect of the following models: (a) 9-rank polynomial model; (b) 5-rank Gaussian model; (c)Planck model; (d) Planck model, piecewise; (e)improved Lorentzian model, piecewise; (f) PL model

Table 1. The comparison result of curve-fitting models

Models	Number of parameters	SSE
Poly9	10	1.7041
Poly5+Poly5	12	0.1102
Guass5	15	0.4166
Guass2+Guass2	12	0.0364
Planck	4	3.7562
Planck+Planck	8	0.7054
iLorentzian+iLorentzian	8	0.1051
Planck+iLorentzian	8	0.0206

We compare the fitting models based on two aspects: the number of parameters and SSE (Sum of Square Error). The number of parameters represents the complexity of the model and also the number of features associate with this model. The definition of SSE is given by Equation (6), where R and Y refer to the original curve and the fitting curve respectively, n is the number of points in a curve and m is the number of curves in a sample. SSE is the measurement criteria of how well the model fits a curve. Table 1 shows the comparison results of those models, where Ploy9 means the 9-rank polynomial model, Guass5 means the 5-rank Gaussian model, Poly5+Poly5 means piecewise fitting using two models to fit the ascending and descending part of the curve respectively. It can be seen from Table 1 that Planck+iLorentzian(PL model), can achieve the least SSE while has small number of parameters, which means PL model can fits the sensor's curve well while keeps the complexity down.

$$SSE = \sum_{i=1}^{n} \sum_{j=1}^{m} (R_{ij} - Y_{ij})^2 , \qquad (6)$$

4   Experiments and Analysis

We have collect breath samples of about 150 people, including 100 healthy samples and 50 diabetes samples, see Table 2. Samples in Group I are from students and teachers, and Group II samples are from confirmed diabetics who have high blood sugar level.

Table 2. The basic information of the samples collected

Group No.	Type	Quantity	Age	Male-to-female ratio
I	healthy	100	23~60	58:42
II	diabetes	50	32~70	27:23

In this section, we use PL model to extract features from the response of the sensors and perform clustering and classification experiments, then, compare the experimental results of PL features with geometric features, respectively.

For the sake of visualization, only 20 healthy samples and 20 diabetic samples from Table 2 are used in clustering. Those samples are chosen randomly. Since PL model has 8 parameters, there are 8 PL features for each sensor curve, which make 96 features for each breath sample. At the same time, 8 geometric features are extracted from each sensor curve, which are maximum response, position of maximum, integral value and maximum slopes of the ascending part, integral value, maximum and minimum slopes of the descending part, and the final response. The clustering results are shown in Fig. 4, where the numbers which are not bigger than 20 denote healthy samples while the rest are diabetic samples. We can see that samples are all clustered correctly into two classes in Fig. 4(a) by using PL features, while in Fig. 4(b), the clustering based on geometric features, there are about 17% misclassification.

For comparison, here we extract other 4 groups of geometric features defined by references [4], [5], [6] and [7], respectively. Each group has 5 or 4 geometric features extracted from each signal curve. They are: (1) baseline, final response, maximum response, maximum derivative and minimum derivative [4]; (2) maximum response, minimum response, response at time 20 and maximum derivative [5]; (3) maximum response, maximum slope, integral value and average value [6]; (4) maximum response, position of maximum, maximum derivative and minimum derivative [7].

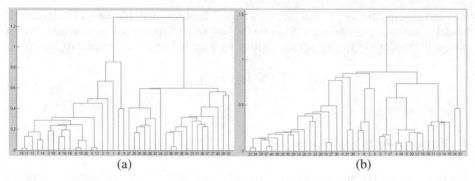

Fig. 4. The clustering results of samples: (a) with 96 features extracted by curve fitting using PL model (PL features); (b) with 96 geometric features

After feature extraction, SVM is adopted to classify breath samples. All samples in Table 2 are used in the following experiments, which includes 50 diabetic samples and 100 healthy samples. For classification experiments, we randomly choose 30 diabetic samples and 60 healthy samples for training, so there are totally 90 training and 60 testing samples. Table 3 gives the results of the experiment, it can be seen that the performance of classification has been greatly improved by the features extracted through curve fitting using PL model.

Table 3. Results of SVM experiment

Feature	Geometric features				Curve fitting
Method	Reference 4	Reference 5	Reference 6	Reference 7	Proposed PL model
Feature length	5*12	4*12	4*12	4*12	8*12
Accuracy	89.17%	88.89%	91.945%	93.89%	**94.79%**

5 Discussions and Conclusions

In this paper we propose an effective feature extraction method for breath analysis. This method uses two different models (Planck model and the improved Lorentzian model) to fit the ascending and descending part of the sensor's response, respectively. The combination of the two models can greatly reduce the error of fitting while keep the number of parameters small. From the testing results, we can see that the proposed feature extraction method is very efficient, and can get higher classification accuracy comparing with other methods by using SVM.

Acknowledgements

The work is supported by the NSFC/SZHK-innovation funds under Contract No. 60803090 and SG200810100003A in China, and the Natural Scientific Research Innovation Foundation in Harbin Institute of Technology, Key Laboratory of Network Oriented Intelligent Computation, Shenzhen.

References

1. Risby, T.H., Solga, S.F.: Current status of clinical breath analysis. Applied Physics B: Lasers and Optics 85, 421–426 (2006)
2. Phillips, M., Herrera, J., et al.: Variation in volatile organic compounds in the breath of normal humans. Journal of Chromatography B 729, 75–88 (1999)
3. Pearce, T.C., Schiffman, S.S., Nagle, H.T., Gardner, J.W.: Handbook of Machine Olfaction - Electronic Nose Technology. John Wiley & Sons, Chichester (2006)
4. Eklov, T., Martensson, P., Lundstrom, I.: Enhanced selectivity of MOSFET gas sensors by systematical analysis of transient parameters. Analytica Chimica Acta 353, 291–300
5. Sundic, T., Marco, S., Perera, A., et al.: Potato creams recognition from electronic nose and tongue signals: feature extraction/selection and RBF Neural Networks classifiers. In: 5th seminar on Neural Network Application in Electrical Engineering, IEEE NEUREL-2000, Yugoslavia, pp. 69–74 (2000)
6. Yu, H., Wang, J., Zhang, H., et al.: Identification of green tee grade using different feature of response signal from E-nose sensors. Sensors and Actuators B 128, 455–461 (2008)
7. Zhang, S., Xie, C., Hu, M., et al.: An entire feature extraction method of metal oxide gas sensors. Sensors and Actuators B 132, 81–89 (2008)
8. Leone, A., Distante, C., Ancona, N., et al.: A powerful method for feature extraction and compression of electronic nose responses. Sensors and Actuators B 105, 378–392 (2005)
9. Hristozov, I., Iliev, B., Eskiizmirliler, S.: A combined feature extraction method for an electronic nose. In: Modern Information Processing: From Theory to Applications, pp. 453–465. Elsevier, Amsterdam (2006)
10. Carmel, L., Levy, S., Lencet, D., Harel, D.: A feature extraction method for chemical sensors in electronic noses. Sensors and Actuators B 93, 67–76 (2003)

Classification of Diabetics with Various Degrees of Autonomic Neuropathy Based on Linear and Nonlinear Features Using Support Vector Machine

Chuang-Chien Chiu[1], Shoou-Jeng Yeh[2], and Tai-Yue Li[1]

[1] Department of Automatic Control Engineering, Feng Chia University
Taichung, Taiwan, R.O.C.
[2] Section of Neurology and Neurophysiology, Cheng-Ching General Hospital
Taichung, Taiwan, R.O.C.
chiuc@fcu.edu.tw, {seanyeh1011,jason110052}@hotmail.com

Abstract. In this study, we investigate the feasibility of using the linear cross-correlation function (CCF) and nonlinear correlation dimension (CD) features of mean arterial blood pressure (MABP) and mean cerebral blood flow velocity (MCBFV) as a signature to classify diabetics with various degrees of autonomic neuropathy. 72 subjects were recruited. For each subject, continuous CBFV was measured using a Transcranial Doppler ultrasound, and continuous ABP recorded using a Finapres device in supine position. The CCFs and CDs of pre-filtered spontaneous MABP and MCBFV were computed. Twelve CCF features and one CD feature were extracted from three frequency ranges: very low frequency (VLF, 0.015-0.07Hz), low frequency (LF, 0.07-0.15 Hz), and high frequency (HF, 0.15-0.40Hz). The feature vectors are classified using a support vector machine (SVM) classifier; and a classification rate of 91.67% is obtained under the leave-one-out cross validation evaluation scheme. This very encouraging result indicates that the proposed linear and nonlinear features between MABP and MCBFV can be effective features to discriminate diabetic patients with autonomic neuropathy.

Keywords: Diabetic, autonomic neuropathy, cross-correlation function, correlation dimension, support vector machine.

1 Introduction

The cerebral autoregulation (CA) mechanism refers to the cerebral blood flow (CBF) tendency to maintain relatively constant in the brain despite changes in mean arterial blood pressure (MABP) in the interval from 50-170 mmHg [1]. A technique using a transcranial Doppler (TCD) was introduced to evaluate the dynamic response of CA in humans [2]. Rapid drops in arterial blood pressure (ABP) caused by the release of thigh blood pressure cuffs were used as an autoregulatory stimulus. The ABP and CBF velocity (CBFV) were compared during the autoregulatory process. Some investigators using the same paradigm [3] validated this approach. They demonstrated that the relative changes in CBFV had an extremely close correlation reflecting relative changes

D. Zhang and M. Sonka (Eds.): ICMB 2010, LNCS 6165, pp. 42–51, 2010.

in CBF during autoregulation testing. ABP can also be acquired non-invasively using a finger cuff device (Finapres BP monitor). A high-speed servo system in the Finapres inflates and deflates the cuff rapidly to maintain the photoplethysmographic output constant at the unloaded state. CA is more a concept rather than a physically measurable entity [4]. Noninvasive CA assessment has been developed and studied using either static or dynamic methods [5]. It is a challenge to find appropriate methods to assess CA non-invasively and reliably using simple and acceptable procedures. Recent investigations have shown that the autoregulatory dynamic response can be identified from spontaneous fluctuations in MABP and CBFV [6]. Some investigators assessed the dynamic relationship between spontaneous MABP and CBFV using transfer function analysis in either normal subjects [7][8] or in autonomic failure patients [9]. Some investigators used spontaneous blood pressure changes as input signals to test CA [10][11]. Spectral and transfer function analyses of CBFV and ABP were performed using fast Fourier transform (FFT) in their experiments. However, the stationary property and time resolution are two critical problems for spectral analysis. Another study was made to explore spontaneous beat-to-beat fluctuations in MABP and breath-by-breath variability in end-tidal CO_2 (EtCO$_2$) in continuous recordings obtained from normal subjects at rest to estimate the dynamic influences of arterial blood pressure and CO_2 on CBFV [12].

Time domain cross-correlation function (CCF) has been recently used by previous investigators as a method to characterize the dynamics of CA [13-15]. Previously, in studying cerebral autoregulation time domain behaviors [16], we have observed that diabetic patients with autonomic neuropathy often exhibit distinct time domain CA patterns compared to normal patients. Also, linear features extracted from CCFs and using support vector machine (SVM) as the classifier achieved 87.5% of classification rate in classifying normal subjects and three groups of diabetic patients with varying degrees of autonomic neuropathy [17].

The chaotic analysis will be implemented to assess dynamic CA in diabetics by a nonlinear measure, correlation dimension (CD) in this study. CD focuses on the system's complexity. By using the parameters of chaotic analysis to quantify CA, we expect normal autoregulation system would tend to lower chaos and higher regular, predictable behavior of the system to initial conditions, but impaired autoregulation would be more chaotic and less predictable in terms of chaotic analysis. Motivated by preliminary observations, we investigate the feasibility of using the both linear CCF features and nonlinear CD feature of MABP and MCBFV as a signature to distinguish diabetic patients with varying degrees of autonomic neuropathy.

2   Materials and Methods

2.1   Subjects and Measurements

Four groups of subjects were recruited in this study including 14 healthy adults for normal subjects (4 men, 10 women) with a mean age of 30.3±8 years, 15 diabetics without autonomic neuropathy (10 men, 5 women) with a mean age of 52.5±16.27 years, 25 diabetics with mild autonomic neuropathy (15 men, 10 women) with a mean age of 67.5±8.8 years, and 18 diabetics with severe autonomic neuropathy (12 men, 6 women) with a mean age of 61.6±10.9 years. The subjects in the healthy group were included only if they had no history of vascular disease, heart problems, hypertension, migraine, epilepsy, cerebral aneurysm, intra-cerebral bleeding or other pre-existing neurological conditions. None of the subjects were receiving any medication during the time of the study. Here we describe abnormalities in autonomic functions determined using Low's non-invasive techniques for the assessment of autonomic functions [18, 19] with minor modification by replacing the sudomotor examination with the sympathetic skin response (SSR) test. Low's criteria incorporated a composite score (maximum score 10) that gives an overall indication of the severity of autonomic abnormalities [19]. Three components that make up the composite score are, the adrenergic index (AI, an indicator of sympathetic function) maximum score 4, the cardiovascular heart rate index (CVHRI, indicator of parasympathetic function) maximum score 3, and the sudomotor index, maximum score 3. In our study, the patient with composite score 0-3 is determined as without autonomic failure, 3-6 as mild autonomic failure, and 6-10 as severe autonomic failure. CBFV was measured in the right middle cerebral artery using TCD (transcranial Doppler ultrasound, EME TC2020) in conjunction with a 5-MHz transducer fixed over the temporal bones using an elastic headband. Continuous ABP recordings were obtained through the Finapres (Ohmeda 2300) device with the cuff attached to the middle finger of the right hand. Data acquisition was started after a 10-min relaxation in supine position. Spontaneous ABP and CBFV were recorded simultaneously to a PC for off-line analysis. The acquisition periods were approximately 5 minutes in supine position using a custom-developed data acquisition system. The personal computer combined with a general purpose data acquisition board and LabVIEW environment for acquiring signals correctly was developed in our previous study [20]. The sampling rate needed to acquire the analog data from TCD and Finapres is adjustable in this system. In our experiments, the sampling rate was set to 60 Hz.

2.2 Preprocessing

The Finapres device was fully automated. The blood volume under an inflatable finger cuff was measured with an infrared plethysmograph, and kept constant to a set point value by controlling the cuff pressure in response to volume changes in the finger artery. Using a built-in servo adjustment mechanism, a proper volume-clamped set point was established and adjusted at regular intervals. However, this procedure interrupted the blood pressure recording (usually for 2-3 beats every 70 beats). We called the artifacts caused by the regular servo adjustment in the continuously acquired pulse signal "*servo components*" because both ABP and CBFV signals were simultaneously acquired, displayed, and stored into a PC. The direct removal of servo artifacts from the ABP was not appropriate because it would result in different time duration in comparison with the CBFV signal. Therefore, a signal relaxation algorithm was presented [16] to compensate for the artifacts caused by the servo components.

The mean ABP value was calculated for each heart beat as follows.

$$MABP_i = \frac{1}{V_i - V_{i-1}} \sum_{k=V_{i-1}}^{V_i - 1} x(k) \qquad (1)$$

Where $x(\cdot)$ is the ABP pulse signal acquired continuously from the Finapres analog output port. V_{i-1} is the wave-trough time index in the $(i\text{-}1)$th pulse beat. V_i is the time index of the wave-trough in the ith pulse beat. Therefore, $MABP_i$ is the calculated mean APB value for the ith pulse beat. Similarly, the mean CBFV value was calculated for each heart beat as follows.

$$MCBFV_i = \frac{1}{D_i - D_{i-1}} \sum_{k=D_{i-1}}^{D_i - 1} y(k) \tag{2}$$

Where $y(\cdot)$ is the CBFV signal continuously acquired from the analog output port of the TCD. D_{i-1} is the time index of the wave-trough in the CBFV signal corresponding to the $(i\text{-}1)$th pulse beat and D_i is the time index of the wave-trough in the CBFV signal corresponding to the ith pulse beat. $MCBFV_i$ is the mean CBFV value for the ith pulse beat. The wave-peak and wave-trough of each cardiac cycle can be marked from the ABP and CBFV signals using the approach proposed in the previous study [20]. Afterward, the MABP and MCBFV time series calculated using Equations (1) and (2) are placed at regular intervals equal to their mean heart period.

2.3   Cross-Correlation Function

Before calculating the CCF between the MABP and MCBFV time series, the MABP and MCBFV were normalized using their mean values. Assume that the normalized MABP and MCBFV time series are $f(n)$ and $g(n)$ respectively. $f(n)$ and $g(n)$ signals were then bandpass filtered in the very low-frequency range (VLF, 0.015-0.07 Hz), low-frequency range (LF, 0.07-0.15 Hz) and high-frequency range (HF, 0.15-0.40 Hz) before applying the CCF. A third-order digital bandpass Chebyshev filter provided in the LabVIEW signal analysis functions was applied. The sampling period of the filter was set equal to the mean heart period of each subject. The low cutoff frequency and high cutoff frequency were assigned at the beginning and ending passband frequencies at each bandpass frequency range of interest. The passband ripple error of each bandpass Chebyshev filter was limited to 1 dB. Assume that the bandpass filtered $f(n)$ and $g(n)$ time series are $\hat{f}(n)$ and $\hat{g}(n)$ respectively. The CCF between $\hat{f}(n)$ and $\hat{g}(n)$ is calculated as follows.

$$CCF_i(k) = \frac{R_{\hat{f}\hat{g}}^i(k)}{\left[R_{\hat{f}\hat{f}}^i(0)R_{\hat{g}\hat{g}}^i(0)\right]^{\frac{1}{2}}} \quad k = 0, \pm 1, \pm 2, \cdots, i=1 \text{ to } N\text{-}W\text{+}1 \tag{3}$$

Where $R_{\hat{f}\hat{g}}^i(k)$ is an estimate of the cross-covariance in ith time window and defined as

$$R_{\hat{f}\hat{g}}^i(k) = \begin{cases} \dfrac{1}{W}\displaystyle\sum_{j=i}^{i+W}\hat{f}(j)\hat{g}(j+k), k = 0,1,2,\cdots \\ \dfrac{1}{W}\displaystyle\sum_{j=i}^{i+W}\hat{f}(j-k)\hat{g}(j), k = 0,-1,-2,\cdots \end{cases} \tag{4}$$

Also, $R^i_{\hat{f}\hat{f}}(0) = \dfrac{1}{W}\sum\limits_{j=i}^{i+W}[\hat{f}(j)]^2$, and $R^i_{\hat{g}\hat{g}}(0) = \dfrac{1}{W}\sum\limits_{j=i}^{i+W}[\hat{g}(j)]^2$. N is the total number of cardiac cycles, W is the window width and k is the time lag. $CCF_i(\cdot)$ is the result of the CCF between $\hat{f}(n)$ and $\hat{g}(n)$ in the ith time window. Mean CCF patterns were obtained for each subject and for the entire population. In this study, the cross-correlation functions (CCFs) were estimated using a 64 beat wide moving window. The total number of cardiac cycles for both MABP and MCBFV to compute the CCF was 256 beats. Thus, the total number of CCFs included for each subject was 193. As mentioned, the MABP and MCBFV signals were bandpass filtered in the VLF, LF, and HF ranges before extracting features from CCFs for the purpose of studying the effect of dynamic CA. A third-order digital bandpass Chebyshev filter was applied. The sampling period of the filter was set equal to the mean heart period of each subject. The low cutoff frequency and high cutoff frequency were assigned at the beginning and ending passband frequencies at each bandpass frequency range of interest. The passband ripple error of each bandpass Chebyshev filter was limited to 1 dB. Only time lags of CCFs ranged between -10 and +5 with positive extreme values existed were taken into account to extract the features in order to eliminate the outliers of CCFs caused by artifacts in each frequency band. Afterward, the means and standard deviations of the valid CCFs in each frequency band and the means and standard deviations of their corresponding time lags were calculated as the features to represent each subject. There are 12 linear features extracted for each subject.

2.4   Correlation Dimension

The nonlinear parameter, correlation dimension (CD) [21], was calculated to evaluate the complexity of MABP and MCBFV. The signal was reconstructed from time series before measuring the chaos properties to determinate the complexity and regularity. A new coordinate system, phase space, was used to reconstruct an attractor. Phase space is an abstract mathematical space that can expresses the behavior of a dynamic system. In the phase space, the behavior of a dynamic system (trajectory or orbit) finally converges to the stationary state, called attractor. Due to the number of phase space coordinates cannot be calculated immediately from the time data, it is necessary to reconstruct the n-dimensional phase space from a signal record of time series measurement. The n-dimensional attractor can be reconstructed from one-dimensional projected data of time series using embedding method. The n-dimensional vector is constructed as follows.

$$X(t) = \{x(t), x(t+\tau), \cdots, x(t+(dm-1)\tau)\} \tag{5}$$

Fig. 2 is a two-dimensional reconstruction in MABP phase space diagram. Where $C_d(R)$ is correlation function.

$$C_d(R) = \lim_{N\to\infty}[\frac{1}{N^2}\sum_{i,j=1,i\neq j}H_E(R-|X_i-X_j|)] \tag{6}$$

N is the total number of time series; H_E: Heaviside step function; $H_E=1$, if $R - |X_i - X_j| \geq 0$; $H_E=0$,otherwise; R is radius. The CD value is obtained from the slope of the curve that $C_d(R)$ is plotted against R because CD is unknown in the beginning, a series of calculations with gradually increasing an embedding dimension has to be performed until the slope tends not to increase. Note that one nonlinear CD feature extracted from MCBFV in LF range is used in this study.

2.5 Support Vector Machine

Support vector machine (SVM) is a very useful technique for pattern classification. The goal of SVM is to produce a model which predicts the class labels of patterns in the testing set which are given only the features. The training patterns are mapped into a higher dimensional space and SVM finds a linear separating hyperplane with the maximal margin in this higher dimensional space to classify different groups [22]. Assumed that the training patterns are linearly separable, and given the training data S as follows.

$$S = \{(x_1, y_1), \cdots, (x_l, y_l)\}, x_i \in R^n, i = 1,2,\cdots,l, y_i \in \{+1,-1\} \tag{7}$$

Where x_i stands for the ith training pattern, y_i stands for the class label of the ith training pattern, l denotes the total number of the training data, and n denotes the dimensions of the feature spaces. Sometimes, the training data may not be completely separable by a hyperplane. The slack variables, denoted by $\xi, \xi_i \geq 0, i = 1,2,3,\cdots,l$, can be introduced to relax the constraints and $C > 0$ is the penalty parameter of the error. Therefore, the SVM require the solution of the following optimization problem as follows.

$$\min_{w,b,\xi} \frac{1}{2}\|w\|^2 + C\sum_i \xi_i \quad \text{subject to} \quad y_i[w^T\phi(x_i)+b] \geq 1 - \xi_i, i = 1,\cdots,l \tag{8}$$

Note that $K(x_i, x_j) \equiv \phi(x_i)^T \phi(x_j)$ is called the kernel function. Here training vectors x_i are mapped into a higher (maybe infinite) dimensional space by the function $\phi(x_j)$.Though new kernels are being proposed by researchers, we may find in SVM books the following four basic kernels [22]: linear, polynomial, radial basis function (RBF), and sigmoid kernels. The kernel applied in this study was the radial basis function (RBF) in which $K(x_i, x_j) = \exp(-\gamma\|x_i - x_j\|^2), \gamma > 0$.

3 Results and Discussion

Cross-correlation analysis provides correlation and phase relationship between MABP and MCBFV. Results indicated that the correlation values for normal subjects are generally higher than that for diabetics without and with autonomic neuropathy [16]. The peak MABP-MCBFV CCF values indicate significant relationships between these variables. Usually, the mean CCF values are not significant without distinctive peaks in the VLF range. However, significant peaks could be obtained in both the LF and HF ranges [17]. Because CBF remains constant despite cerebral perfusion pressure changes, the region in this constant level is nonlinear. Some useful nonlinear information maybe missed and cannot be characterized by using linear analysis. Therefore, it might be beneficial to analyze CA using nonlinear methods. Chaos theorem, the nonlinear approach, has been applied to many physiological studies, such as heart rate variability (HRV), electroencephalograph (EEG), respiratory rhythm, speech recognition, blood pressure and cardiovascular system. They provided the chaotic nature in physiological systems. Nevertheless, nonlinear dynamic CA in diabetics dealt with chaotic analysis was still rarely discussed. Non-linear control mechanism is also essential to the control mechanism of CA. In this study, the correlation dimension analysis is applied. The correlation dimension analysis of MABP and MCBFV by taking the average value of 20 iterations of the dimension analysis is listed in Table 1.

Table 1. The correlation dimension analysis of MABP and MCBFV by taking the average value of 20 iterations of the dimension analysis

	Normal			
	MCBFV		MABP	
	Supine	Tilt-up	Supine	Tilt-up
VLF	1.553±0.860	0.628±0.322	1.949±1.285	0.880±0.492
LF	0.828±0.235	0.458±0.349	0.875±0.472	0.445±0.269
HF	0.810±0.241	0.468±0.359	0.793±0.392	0.441±0.243
	Without DAN			
VLF	2.163±1.835	1.119±0.781	1.436±1.071	0.742±0.546
LF	1.098±0.611	0.517±0.366	0.816±0.343	0.440±0.197
HF	1.148±0.639	0.548±0.405	0.839±0.353	0.470±0.221
	Mild DAN			
VLF	2.418±2.657	0.931±0.715	1.557±1.300	0.576±0.364
LF	1.087±0.880	0.520±0.427	0.833±0.524	0.378±0.278
HF	1.157±0.987	0.502±0.386	0.831±0.528	0.397±0.292
	Severe DAN			
VLF	2.544±2.898	1.367±0.939	1.763±2.050	0.544±0.249
LF	1.121±0.668	0.733±0.376	0.936±0.438	0.456±0.255
HF	1.087±0.693	0.711±0.335	0.892±0.442	0.499±0.273

In diabetics with severe autonomic neuropathy, the CD values of the MCBFV did not differ significantly between supine and tilt-up positions. On the other hand, higher CD values with statistical significance of both MABP and MCBFV in normal subjects, and diabetics without and mild autonomic neuropathy in supine position compared to that in tilt-up position. Also note that all CD values in tilt-up position are less than that in supine position.

The SVM software-LIBSVM developed by C.C. Chang and C.J. Lin [23] was adopted in our classification task. 13 features (i.e., 12 linear CCF features and one nonlinear CD feature) extracted from VLF, LF, and HF ranges obtained the best classification performance. There were two parameters while using RBF kernels: C and γ. The best (C, γ) is set to be $(2^9, 2^{-3})$ to obtain the best cross-validation rate. It made no significant difference whether we normalized the features or not. After applying these features to SVM and using leave-one-out cross validation to classify these four groups of 72 subjects, 91.67% of classification rate was achieved as shown in Table 2. 66 subjects out of 72 subjects could be classified correctly.

Table 2. SVM classification results using leave-one-out cross validation

Leave-one-out cross validation	Classification results			
	Normal	Without	Mild	Severe
Normal (14)	13	0	1	0
Without (15)	1	14	0	0
Mild (25)	0	1	22	2
Severe (18)	0	0	1	17

The sensitivity, specificity, and accuracy of the classification results for each group of subjects using the SVM classifier are listed in Table 3. The results indicate that our method could discriminate among normal, diabetics without autonomic neuropathy, diabetics with mild autonomic neuropathy, and diabetics with severe autonomic neuropathy with an average sensitivity of 92.2%, an average specificity of 97.1%, and an average accuracy of 95.8%.

Table 3. Classification results using ART neural network

	Normal	Without	Mild	Severe	Average
Sensitivity	92.9%	93.3%	88.0%	94.4%	92.2%
Specificity	98.3%	98.2%	95.7%	96.3%	97.1%
Accuracy	97.2%	97.2%	93.1%	95.8%	95.8%

4 Conclusion

This study focused on using both the linear cross-correlation function features and nonlinear correlation dimension feature to assess the effects of changes on MABP and MCBFV without any special maneuvers needed for the measurement. The MABP and MCBFV signals were bandpass filtered before applying linear and nonlinear analyses in studying the effect of different bandwidths. Means and standard deviations of the valid CCFs in each frequency band, the means and standard deviations of their corresponding time lags, and CD feature of MCBFV in LF range were calculated as features to represent each subject in only supine position. 91.67% of classification rate was achieved in classifying 4 groups of subjects after applying SVM classifier with leave-one-out cross validation. This study suggests that both linear and nonlinear features extracted from CCFs and CD respectively using SVM as the classifier outperforms our previous approach with only linear features used. Our method can be a noninvasive, simple, easy, and effective method to classify the dynamic cerebral autoregulation in diabetics with autonomic neuropathy. Our future work involves reproducing our findings in a larger population including patients that might show the absence or impairment of pressure autoregulation.

Acknowledgments. The authors would like to thank the National Science Council, Taiwan, ROC, for supporting this research under Contracts NSC 96-2628-E-035-083-MY3.

References

1. Lassen, N.A.: Cerebral Blood Flow and Oxygen Consumption in Man. Physiological Reviews 39, 183–238 (1959)
2. Aaslid, R., Lindegaard, K.F., Sorteberg, W., Nornes, H.: Cerebral Autoregulation Dynamics in Humans. Stroke 20, 45–52 (1989)
3. Newell, D.W., Aaslid, R., Lam, A., Mayberg, T.S., Winn, H.R.: Comparison of Flow and Velocity During Dynamic Autoregulation Testing in Humans. Stroke 25, 793–797 (1994)
4. Panerai, R.B.: Assessment of Cerebral Pressure Autoregulation in Humans-A Review of Measurement Methods. Physiological Measurement 19, 305–338 (1998)
5. Tiecks, F.P., Lam, A.M., Aaslid, R., Newell, D.W.: Comparison of Static and Dynamic Cerebral Autoregulation Measurements. Stroke 26, 1014–1019 (1995)
6. van Beek, A.H., Claassen, J.A., Rikkert, M.G., Jansen, R.W.: Cerebral Autoregulation: An Overview of Current Concepts and Methodology with Special Focus on the Elderly. Journal of Cerebral Blood Flow & Metabolism 28, 1071–1085 (2008)
7. Diehl, R.R., Linden, D., Lücke, D., Berlit, P.: Spontaneous Blood Pressure Oscillations and Cerebral Autoregulation. Clinical Autonomic Research 8, 7–12 (1998)
8. Zhang, R., Zuckerman, J.H., Giller, C.A., Levine, B.D.: Transfer Function Analysis of Dynamic Cerebral Autoregulation in Humans. American Journal of Physiology 274, H233–H241 (1998)
9. Blaber, A.P., Bondar, R.L., Stein, F., Dunphy, P.T., Moradshahi, P., Kassam, M.S., Freeman, R.: Transfer Function Analysis of Cerebral Autoregulation Dynamics in Autonomic Failure Patients. Stroke 28, 1686–1692 (1997)

10. Kuo, T.B.J., Chern, C.M., Sheng, W.Y., Wong, W.J., Hu, H.H.: Frequency Domain Analysis of Cerebral Blood Flow Velocity and Its Correlation with Arterial Blood Pressure. Journal of Cerebral Blood Flow and Metabolism 18, 311–318 (1998)
11. Chern, C.M., Kuo, T.B., Sheng, W.Y., Wong, W.J., Luk, Y.O., Hsu, L.C., Hu, H.H.: Spectral Analysis of Arterial Blood Pressure and Cerebral Blood Flow Velocity During Supine Rest and Orthostasis. Journal of Cerebral Blood Flow & Metabolism 19, 1136–1141 (1999)
12. Panerai, R.B., Simpson, D.M., Deverson, S.T., Mathony, P., Hayes, P., Evans, D.H.: Multivariate Dynamic Analysis of Cerebral Blood Flow Regulation in Humans. IEEE Transactions on Biomedical Engineering 47, 419–423 (2000)
13. Panerai, R.B., Kelsall, A.W.R., Rennie, J.M., Evans, D.H.: Analysis of Cerebral Blood Flow Autoregulation in Neonates. IEEE Transactions on Biomedical Engineering 43, 779–788 (1996)
14. Steinmeier, R., Bauhuf, C., Hübner, U., Bauer, R.D., Fahlbusch, R., Laumer, R., Bondar, I.: Slow Rhythmic Oscillations of Blood Pressure, Intracranial Pressure, Microcirculation, and Cerebral Oxygenation. Stroke 27, 2236–2243 (1996)
15. Chiu, C.C., Yeh, S.J.: Assessment of Cerebral Autoregulation Using Time-Domain Cross-Correlation Analysis. Computers in Biology & Medicine 31, 471–480 (2001)
16. Chiu, C.C., Yeh, S.J., Liau, B.Y.: Assessment of Cerebral Autoregulation Dynamics in Diabetics Using Time-Domain Cross-Correlation Analysis. Journal of Medical and Biological Engineering 25, 53–59 (2005)
17. Chiu, C.C., Hu, Y.H., Yeh, S.J., Chou, D.Y.: Classification of Dynamic Cerebral Autoregulation in Diabetics with Autonomic Neuropathy Using Support Vector Machine. In: Proceedings of the International Conference on Bioinformatics & Computational Biology (BIOCOMP 2009), vol. 1, pp. 353–358 (2009)
18. Low, P.: Autonomic Nervous System Function. Journal of Clinical Neurophysiology 10, 14–27 (1993)
19. Low, P.: Composite Autonomic Scoring Scale for Laboratory Quantification of Generalized Autonomic Failure. Mayo Clinic Proceedings 68, 748–752 (1993)
20. Chiu, C.C., Yeh, S.J., Lin, R.C.: Data Acquisition and Validation Analysis for Finapres Signals. Chinese Journal of Medical and Biological Engineering 15, 47–58 (1995)
21. Grassberger, P., Procaccia, I.: Measuring the Strangeness of Strange Attractors. Physica D 9, 189–208 (1983)
22. Cortes, C., Vapnik, V.: Support-Vector Network. Machine Learning 20, 273–297 (1995)
23. Chang, C.C., Lin, C.J.: LIBSVM: a library for support vector machines, Software available at http://www.csie.ntu.edu.tw/~cjlin/libsvm

Diabetes Identification and Classification by Means of a Breath Analysis System

Dongmin Guo[1], David Zhang[1,*], Naimin Li[2], Lei Zhang[1], and Jianhua Yang[3]

[1] Department of Computing, The Hong Kong Polytechnic University, Hong Kong
csdzhang@comp.polyu.edu.hk
[2] Department of Computer Science and Engineering, HIT, Harbin, P.R. China
[3] School of Automation, Northwestern Polytechnical University, Xian, P.R. China

Abstract. This article proposes a breath analysis system that makes use of chemical sensors to detect acetone in human breath, and hence detect the diabetes and measure the blood glucose levels of diabetics. We captured the breath samples from healthy persons and patients known to be afflicted with diabetes and conducted experiments on disease identification and simultaneous blood glucose measurement. SVM classifier was used to identify diabetes from healthy samples and three models were built to fit the curves that can represent the blood glucose levels. The results show that the system is not only able to distinguish between breath samples from patients with diabetes and healthy subjects, but also to represent the fluctuation of blood sugar of diabetics and therefore to be an evaluation tool for monitoring the blood glucose of diabetes.

1 Introduction

Endogenous molecules in human breath such as acetone, nitric oxide, hydrogen, ammonia, and carbon monoxide are produced by metabolic processes and partition from blood via the alveolar pulmonary membrane into the alveolar air [1,2,3]. These molecules are present in breath relative to their types, concentrations, volatilities, lipid solubility, and rates of diffusion as they circulate in the blood and cross the alveolar membrane [4]. Therefore, changes in the concentration of these molecules could suggest various diseases or at least changes in the metabolism [5]. To take a few examples, nitric oxide can be measured as an indicator of asthma or other conditions characterized by airway inflammation [6,7,8]. Breath isoprene is significantly lower in patients with acute respiratory exacerbation of cystic fibrosis [9]. Increased pentane and carbon disulfide have been observed in the breath of patients with schizophrenia [10]. The concentration of volatile organic compounds (VOCs) such as cyclododecatriene, benzoic acid, and benzene are much higher in lung cancer patients than in control groups [11], [12]. Ammonia is significantly elevated in patients with renal diseases [13], [14].

For diabetes, the original cause is that the glucose produced by the body cannot enter the bloodstream to provide energy to the cells. Glucose enters the body cells with the help of insulin. If the body is not producing insulin

---

* Corresponding author.

D. Zhang and M. Sonka (Eds.): ICMB 2010, LNCS 6165, pp. 52–63, 2010.
© Springer-Verlag Berlin Heidelberg 2010

(Type I diabetes), glucose cannot get into the cells. As a result, cells have to use fat as an energy source. In the process of metabolizing fat for energy, one of the by-products are ketones. When ketones is accumulated in the blood, ketoacidosis occurs which is characterized by the smell of acetone on the patient's breath [15]. Therefore, the abnormal concentration of acetone in exhaled air is an indicator of diabetes. Additionally, the mean acetone concentration in exhaled air rises progressively with the blood glucose for the diabetics with high blood glucose levels [16]. For Type I diabetes, the measurement of the breath acetone can detect the mild ketosis of diabetes which indicates the inadequacies of insulin treatment but can be not revealed by simply measuring the blood glucose level [17]. Hence, analyzing acetone concentration in the breath of people with diabetes is a meaningful investigation and will be helpful to detect diabetes, monitor the blood glucose level, and diagnosis the ketosis of diabetes in a early stage.

In this paper, we propose a breath analysis system to detect diabetes and monitor the level of blood glucose. The system makes use of chemical sensors that are particularly sensitive to the biomarkers, triggering responses to the breath sample when the biomarkers are detected. The response signal serves to subsequent sampling and processing, and the signal features are extracted and classified in groups. A measurement result is generated in the end in the computer to show the user's body conditions. We captured healthy samples and diabetes samples by this system and conducted experiments. The experimental results show that our system can fairly accurately identify diabetes and measure the blood glucose level of diabetics with quite a high level of accuracy.

The remainder of this paper is organized as follows. Section 2 describes the breath analysis system. Section 3 explains the experimental details and gives the experimental results and discussion. Section 4 offers our conclusion.

2   Description of the System

The prototype of our system is shown in Fig.1. It operates in three phases (Fig.2), gas collection, sampling, and data analysis, with a subject first breathing into a Tedlar gas sampling bag. This gas is then injected into a chamber containing a sensor array where a measurement circuit measures the interaction between the breath and the array. The responses are then filtered, amplified, and sent to computer for further analysis.

The framework for the proposed breath analysis system is made up of three modules: a signal measurement module, a signal conditioning module, and a signal acquisition module. The signal measurement module contains a sensor array, temperature control circuit, and measurement circuit. The temperature control circuit provides negative feedback voltage to heater of sensors so as to guarantee that the sensors are at stable working temperature.

The sensor array is composed of 12 metal oxide semiconductor gas sensors (from FIGARO Engineering Inc.) set in a 600 ml stainless steel chamber. Breath samples from subjects are collected with a Tedlar gas sampling bag that has the same volume with the chamber and then injected into the chamber through an

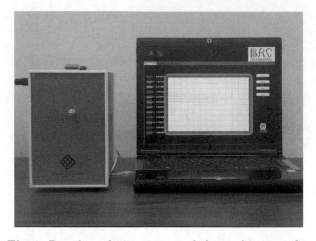

Fig. 1. Breath analysis system and the working interface

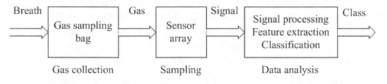

Fig. 2. The working flow defined in our system

auto-sampler. The resistances of the sensors change from R_0 to R_s when they are exposed to sampled gas. The output voltage is

$$V_{Out} = \frac{1}{2}V_{CC}(1 - \frac{R_s}{R_0}),\qquad(1)$$

where V_{CC} is the transient voltage crossing the sensor and V_{Out} is the transient output voltages of the measurement circuit.

The signal measurement module measures the output voltage and converts it into analog electrical signal. The analog signal is subsequently conditioned by signal filtering and amplifying. Finally, the signal is sampled and transmitted through a USB interface to a computer for future analysis.

3   Experiments

In this section, we conduct experiments on diabetes detection and the measurement of blood glucose levels of diabetics. We collected both diabetes samples (Type I diabetes) and healthy samples from Harbin Hospital. Subjects were assumed to be healthy on the basis of a recent health check. Diabetes were all inpatient volunteers. Their conditions were confirmed and correlated by comparing their levels with standard clinical blood markers. Totally, We selected

Table 1. Composition of the subject database

Type of subjects	Number	Male/Female	Age
Healthy subjects	108	58/50	23-60
Subjects with diabetes	90	49/41	25-77

Table 2. The subjects' blood glucose level and the corresponding number

Level stage	Name	Blood glucose level (mg/dL)	Number
Level 1	Low level	81-100	4
Level 2	Borderline	101-150	49
Level 3	High level	151-200	20
Level 4	Very high level	201-421	17
Total		81-421	90

90 diabetes subjects with simultaneous blood glucose level and 108 healthy subjects. Table 1 details the composition of the subject database, and Table 2 list the subjects' blood glucose levels and the corresponding number. The samples were grouped by different blood glucose levels defined in [17]. In Section 3.1, we carry out experiment to identify diabetes and in Section 3.2 we perform experiment about the blood glucose measurement.

3.1 Diabetes Identification

In this section, we apply our system to evaluate the performance of system with respect to disease diagnosis. Since acetone has been found to be more abundant in the breath of diabetics, the responses of our system to the breath of healthy subjects and diabetics are distinguishable. Analyzing and identifying these responses can provide doctor with important additional information and help them make an initial clinical diagnosis.

Fig. 3 shows the responses of the twelve different sensors (S1-S12) to the two different air samples over the 90 s sampling period. Fig. 3(a) is a typical response to a healthy sample. Fig. 3(b) is a typical response to a diabetes sample. The horizontal axis stands for the sampling time (0-90 s) and the vertical axis denotes the amplitude of sensor output in volts. The curves in each figure represent the output of each sensor S1-S12. These responses have been preprocessed using baseline manipulation and normalization introduced in Appendix.

To observe the discrimination between healthy samples and diabetes samples clearly, the PCA two-dimensional plot of the responses from the two classes with the first principal components (PC1) plotted against the second (PC2) is shown in Fig. 4. The *blue+* stands for the samples classified as being from diabetics and *red·* stands for the samples classified as being from healthy subjects. From

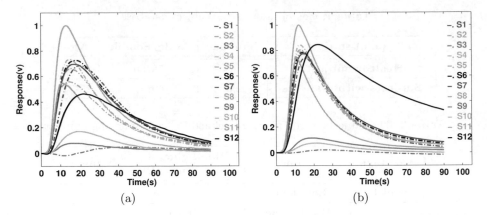

Fig. 3. Typical responses from one healthy sample and one diabetes sample: (a) healthy sample, (b) diabetes sample

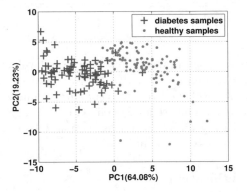

Fig. 4. PCA two-dimensional plot of the sensor signals corresponding to two classes: (a) diabetes samples (*blue+*) and healthy samples (*red·*)

the figure we can see that the two classes are distinguishable even though the two dimensions only explains about 83.31% of the variation in the data, 64.08% for PC1 and 19.32% for PC2.

There are two classes containing 90 diabetes samples and 108 healthy samples separately. We selected randomly half of the samples in each classes as training set. Therefore, the training set includes 99 samples, 45 labeled as diabetes and 54 labeled as healthy. The rest in each class is used as test set, which will be diagnosed later by a classifier trained on the training set. Since our samples are with small total number, but high dimension, SVM classifier is a perfect choice.

Before we report the classification result, we firstly introduce the SVM classifier [18]. We have the training data $\{x_i, y_i\}, i = 1, \cdots, l, y_i \in \{-1, 1\}, x_i \in R^d$, where the y_i is either 1 or -1, indicating the class to which the point x_i belongs. We want to find the maximum-margin hyperplane that divides the points having $c_i = 1$ from those having $c_i = -1$. Any hyperplane can be written as the set of points satisfying

$$y(x) = w^{\mathrm{T}} \cdot x - b, \tag{2}$$

where w is normal to the hyperplane, $\frac{b}{\|w\|}$ is the perpendicular distance from the hyperplane to the origin, and $\|w\|$ is the Euclidean norm of w. We want to choose the w and b to maximize the margin, so we obtain the optimization problem that is to minimize

$$\frac{1}{2}\|w\|^2, \quad s.t. \ \ c_i(w \cdot x_i - b) \geq 1, i = 1, \cdots, l. \tag{3}$$

This problem can be expressed by non-negative Lagrange multipliers α_i as

$$\min_{w,b,\alpha}\{\frac{1}{2}\|w\|^2 - \sum_{i=1}^{l}\alpha_i[c_i(w \cdot x_i - b) - 1]\}. \tag{4}$$

The problem is also transformed to the so-called dual problem of maximization of the function $L(\alpha)$, defined in the way

$$L(\alpha) = \sum_{i=1}^{l}\alpha_i - \frac{1}{2}\sum_{i,j}\alpha_i\alpha_j c_i c_j K(x_i, x_j) \tag{5}$$

with the constraints

$$\sum_{i=1}^{l}\alpha_i c_i = 0, \quad 0 \leq \alpha_i \leq C, \tag{6}$$

where $K(x, x_i) = x_i \cdot x_j$ is the kernel function and C is the user-defined constant, which is the regularizing parameter, characterized by the weight vector w and the classification error of the data.

We train the classifier by using C-SVC (Classification SVM Type 1) with polynomial kernel function $K(x, x_i) = (x'x_i + \gamma)^p$, where $\gamma = 1$ and $p = 3$. We let $C = 100$ and apply the classifier to the test set. This procedure is run for 50 times and the average classification rate over all runs is computed for each class. In the 45-sample diabetes test set, the average of accuracy is 93.52%. In the 54-sample healthy test set, the average of accuracy is 92.66%. One of resulting vote matrix is presented in Fig. 5. Samples that are classified correctly most of the time are plotted as *green∗*, the others as *blue+*. The first 45 numbers in X axis represent the ID of healthy samples and the last 54 numbers stand for the ID of healthy samples.

3.2   Blood Glucose Measurement

In this section, we introduce the experiment in detail showing our system is able to differentiate between samples with different simultaneous blood glucose levels. Diabetics should check their blood glucose levels several times each day by drawing their blood samples. This process is invasive and unsafe and requires

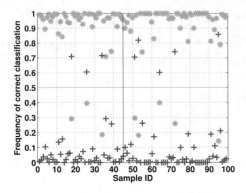

Fig. 5. The resulting vote matrix. Samples that are classified correctly most of the time are plotted as *green∗*, the others as *blue+*. The first 45 numbers in X axis represent healthy samples and the rest stand for the healthy samples.

considerable skill. Therefore, not everyone is suitable for this approach. An alternative is to collect a breath sample with the breath analysis system which generates distinguishable responses to samples with different blood glucose levels [19]. This technology is likely to increase the acceptance of frequent blood glucose monitoring and reduce the danger caused during drawing blood samples.

Fig. 6 shows the responses of the twelve different sensors (S1-S12) to the samples of two diabetics over the 90 s sampling period. The horizontal axis stands for the sampling time (0-90 s) and the vertical axis shows the amplitude of the sensor output in volts. The curves represent the output of each sensor, S1-S12. Fig. 6(a) shows a typical response of diabetes sample which blood glucose level is 86 mg/dL, i.e, in Level 1, Fig. 6(b) shows a typical response of diabetes sample which blood glucose level is 122 mg/dL, i.e, in Level 2, Fig. 6(c) shows a typical response of diabetes sample which blood glucose level is 183 mg/dL, i.e, in Level 3, and Fig. 6(d) shows a typical response of diabetes sample which blood glucose level is 411 mg/dL, i.e, in Level 4, i.e, in Level 4.

As we mentioned previously, the indicator of diabetes is acetone. In our sensor array including 12 sensors, sensor No.1-6 and No.11 are specially sensitive to VOCs. Hence, these sensors have obvious responses to diabetes samples, especially when the blood glucose level is very high. Besides, from these figures, it is clear that when the blood glucose of one patient is in low level(Fig. 6(a)), the response is not distinguishable from the healthy response (Fig. 3(a)). However, when the patient is with high level blood glucose(Fig. 6(c) and Fig. 6(d)), the differences between the response of healthy subject and the patient is discernible.

The response amplitude metal oxide semiconductor gas sensor is the dominant feature to represent the concentration of analyte. To observe the relationship between the responses of our system and the blood glucose levels clearly, we extract the amplitude as feature. The feature defined by

$$SF = \frac{1}{N_s} \sum_{s=1}^{N_s} max(R_{e,s}(t_k)), \quad \forall e, s, k. \tag{7}$$

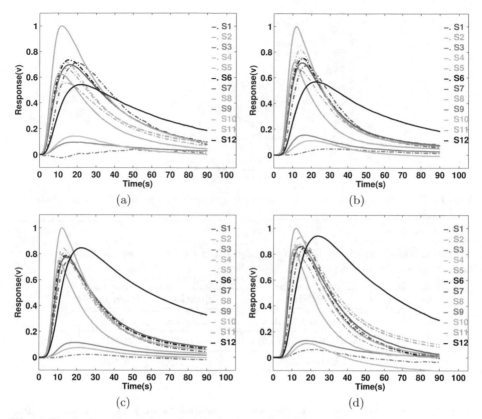

Fig. 6. Typical responses from four diabetics with different simultaneous blood glucose levels: (a) Level 1, (b) Level 2, (c) Level 3, and (d) level 4

Assume one data set has e samples, where $e = 1, ..., N_e$. Each sample consists of s sensor transients, where $s = 1, ..., N_s$. It's worth noting that we just selected the response of sensor No.1-6 and No.11. Other sensors are not specially sensitive to acetone, their responses therefore can not reflect the concentration of acetone in human breath correctly. There are k dimensions per transient, where $k = 1, ..., N_k$.

Fig. 7 shows the graph of the sum of maximum responses of sensors (The feature is defined by Equation 7) versus the simultaneous blood glucose of 90 diabetics. $blue\times$ stands for the subjects whose blood glucose is in Level 1, $red\cdot$ stands for the subjects whose blood glucose is in Level 2, $green*$ stands for the subjects whose blood glucose is in Level 3, and $magenta+$ stands for the subjects whose blood glucose is in Level 4. The details about the levels and the subjects are listed in 2. From the figure we can see that the mean maximum response is increased gradually with the blood glucose level, even though there are still some overlaps between the four levels.

However, it is hard to say there is a linear correlation between the two items. To evaluate if our system have distinguishable responses to samples with different

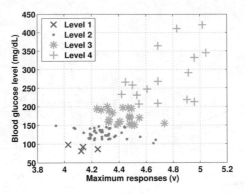

Fig. 7. The sum of maximum responses of sensors versus the simultaneous blood glucose levels

Table 3. Three models to fit the relationship between the mean maximum response and the blood glucose level

No.	Model	Equation	R-square
1	Linear polynomial	$y = ax + b$	0.513
2	Quadratic polynomial	$y = ax^2 + b$	0.611
3	Exponential	$y = ae^{bx} + ce^{dx}$	0.620

simultaneous blood glucose levels, we fit the points in Fig. 7 by using three models: linear polynomial, quadratic polynomial, and exponential equation, as shown in Table 3. In the three models, exponential model gives the largest R-square, however, the linear polynomial model gives the least. The fitted curves by the three models have been shown in Fig.8.

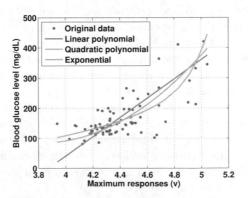

Fig. 8. The original data and fitted curve

Table 4. The classification results calculated by linear polynomial model

Level	Number	Test outcome				Accuracy
		Level 1	Level 2	Level 3	Level 4	
Level 1	4	3	1	0	0	75%
Level 2	49	6	32	8	3	65.31%
Level 3	20	0	5	11	4	55%
Level 4	17	0	1	5	11	64.71%

Table 5. The classification results calculated by quadratic polynomial model

Level	Number	Test outcome				Accuracy
		Level 1	Level 2	Level 3	Level 4	
Level 1	4	1	3	0	0	25%
Level 2	49	7	33	6	3	67.34%
Level 3	20	0	8	11	1	55%
Level 4	17	0	1	6	10	58.82%

Table 6. The classification results calculated by exponential model

Level	Number	Test outcome				Accuracy
		Level 1	Level 2	Level 3	Level 4	
Level 1	4	0	4	0	0	0%
Level 2	49	0	33	10	6	67.34%
Level 3	20	0	7	12	1	60%
Level 4	17	0	0	6	11	64.71%

We use the features defined by Equation 7 as input variables and calculate the output of the three models to judge which levels the output belongs to. The result is listed in Table 4, Table 5, and Table 6. Each table shows the corresponding classification accuracies of the three models.

The classification accuracies of the four levels computed by linear polynomial model are 75%, 65.31%, 65.31%, and 55% separately. The accuracy of Level 1 computed by quadratic polynomial model is 25%, it is quite bad comparing with the first model. For the exponential model, the classification of Level 2,3,4 are good, but Level 1 gives 0% accuracy. Comparing the three models, the linear model is much more suitable for fitting the data.

4   Conclusion

This article proposed a breath analysis system. For the purpose of evaluating the system performance, breath samples were captured and two experiments were conducted. The results show that the system was not only able to distinguish between breath samples from subjects suffering from diabetes and breath samples from healthy subjects, but was also helpful in measuring the blood glucose levels of diabetics.

Acknowledgment

The work is partially supported by the RGF fund from the HKSAR Government, the central fund from Hong Kong Polytechnic University, and the NSFC fund under contract No. 60803090 in China.

References

1. DAmico, A., Di Natale, C., Paolesse, R., Macagnano, A., Martinelli, E., et al.: Olfactory systems for medical applications. Sensors & Actuators: B. Chemical 130(1), 458–465 (2007)
2. Schubert, J., Miekisch, W., Geiger, K., Noldge-Schomburg, G.: Breath analysis in critically ill patients: potential and limitations. Expert Rev. Mol. Diagn. 4(5), 619–629 (2004)
3. Miekisch, W., Schubert, J., Noeldge-Schomburg, G.: Diagnostic potential of breath analysis-focus on volatile organic compounds. Clinica Chimica Acta 347(1-2), 25–39 (2004)
4. Sehnert, S., Jiang, L., Burdick, J., Risby, T.: Breath biomarkers for detection of human liver diseases: preliminary study. Biomarkers 7(2), 174–187 (2002)
5. Amann, A., Schmid, A., Scholl-Burgi, S., Telser, S., Hinterhuber, H.: Breath analysis for medical diagnosis and therapeutic monitoring. Spectroscopy Europe 17(3), 18–20 (2005)
6. Deykin, A., Massaro, A., Drazen, J., Israel, E.: Exhaled nitric oxide as a diagnostic test for asthma: online versus offline techniques and effect of flow rate. American journal of respiratory and critical care medicine 165(12), 1597–1601 (2002)
7. Silkoff, P., Robbins, R., Gaston, B., Lundberg, J., Townley, R.: Endogenous nitric oxide in allergic airway disease. The Journal of Allergy and Clinical Immunology 105(3), 438–448 (2000)
8. Heffler, E., Guida, G., Marsico, P., Bergia, R., Bommarito, L., Ferrero, N., Nebiolo, F., De Stefani, A., Usai, A., Bucca, C., et al.: Exhaled nitric oxide as a diagnostic test for asthma in rhinitic patients with asthmatic symptoms. Respiratory medicine 100(11), 1981–1987 (2006)
9. McGrath, L., Patrick, R., Mallon, P., Dowey, L., Silke, B., Norwood, W., Elborn, S.: Breath isoprene during acute respiratory exacerbation in cystic fibrosis. European Respiratory Journal 16(6), 1065–1069 (2000)
10. Phillips, M., Sabas, M., Greenberg, J.: Increased pentane and carbon disulfide in the breath of patients with schizophrenia. Journal of clinical pathology 46(9), 861–864 (1993)
11. Phillips, M., Altorki, N., Austin, J., Cameron, R., Cataneo, R., Greenberg, J., Kloss, R., Maxfield, R., Munawar, M., Pass, H., et al.: Prediction of lung cancer using volatile biomarkers in breath. Cancer Biomarkers 3(2), 95–109 (2007)
12. Phillips, M., Cataneo, R., Cummin, A., Gagliardi, A., Gleeson, K., Greenberg, J., Maxfield, R., Rom, W.: Detection of Lung Cancer With Volatile Markers in the Breath, Chest 123(6), 2115–2123 (2003)

13. Davies, S., Spanel, P., Smith, D.: Quantitative analysis of ammonia on the breath of patients in end-stage renal failure. Kidney international 52(1), 223–228 (1997)
14. Spanel, P., Davies, S., Smith, D.: Quantification of ammonia in human breath by the selected ion flow tube analytical method using H3O+ and O2+ precursor ions. Rapid communications in mass spectrometry 12(12), 763–766 (1998)
15. http://diabetes.about.com/od/whatisdiabetes/u/symptomsdiagnosis.htm
16. Tassopoulos, C., Barnett, D., Russell Fraser, T.: Breath-acetone and blood-sugar measurements in diabetes. The Lancet. 293(7609), 1282–1286 (1969)
17. Wang, C., Mbi, A., Shepherd, M.: A study on breath acetone in diabetic patients using a cavity ringdown breath analyzer: Exploring correlations of breath acetone with blood glucose and glycohemoglobin a1c. IEEE Sensors Journal 10(1), 54–63 (2010)
18. Burges, C.J.C.: A tutorial on support vector machines for pattern recognition. Data mining and knowledge discovery 2(2), 121–167 (1998)
19. Melker, R., Bjoraker, D., Lampotang, S.: System and method for monitoring health using exhaled breath. uS Patent App. 11/512, 856, August 29 (2006)

Appendix

The responses mentioned in this paper have been preprocessed by baseline manipulation and normalization, as shown in Equation $A.1$ and $A.2$. Assume that one data set has e samples, where $e = 1, ..., N_e$. Each sample consists of s sensor transients, where $s = 1, ..., N_s$. There are k dimensions per transient, where $k = 1, ..., N_k$. The dynamic response of one sample at time t_k is denoted as $R_{e,s}(t_k)$. There are b dimensions in baseline stage, where $b = 1, ..., N_b$. The baseline response of this sample is $B_{e,s}(t_b)$. The relative change for a particular sensor is defined as the preprocessed response

$$R^B_{e,s}(t_k) = R_{e,s}(t_k) - \frac{1}{N_b} \sum_{t_b=1}^{N_b} B_{e,s}(t_b), \quad \forall e, s, k, b. \qquad (A.1)$$

Denote $R^B_{e,s}(t_k)$ is the response of the sensor N_s to the N_e sample, which has been processed by baseline manipulation. The normalized response is defined as

$$R^{BN}_{e,s}(t_k) = \frac{R^B_{e,s}(t_k)}{max(R^B_{e,s}(t_k))}, \quad \forall e, m. \qquad (A.2)$$

Automatic Measurement of Vertical Cup-to-Disc Ratio on Retinal Fundus Images

Yuji Hatanaka[1], Atsushi Noudo[2], Chisako Muramatsu[2], Akira Sawada[3],
Takeshi Hara[2], Tetsuya Yamamoto[3], and Hiroshi Fujita[2]

[1] Department of Electronic Systems Engineering, The University of Shiga Prefecture,
Hikone 522-8533, Japan
[2] Department of Intelligent Image Information, Graduate School of Medicine, Gifu University,
Gifu 501-1194, Japan
[3] Department of Ophthalmology, Graduate School of Medicine, Gifu University,
Gifu 501-1194, Japan
hatanaka.y@usp.ac.jp, {noudo,chisa,hara}@fjt.info.gifu-u.ac.jp,
fujita@fjt.info.gifu-u.ac.jp

Abstract. Glaucoma is a leading cause of permanent blindness. Retinal fundus image examination is useful for early detection of glaucoma. In order to evaluate the presence of glaucoma, the ophthalmologist may determine the cup and disc areas and diagnose glaucoma using a vertical cup-to-disc ratio. However, determination of the cup area is very difficult, thus we propose a method to measure the cup-to-disc ratio using a vertical profile on the optic disc. The edge of optic disc was then detected by use of a canny edge detection filter. The profile was then obtained around the center of the optic disc in the vertical direction. Subsequently, the edge of the cup area on the vertical profile was determined by thresholding technique. Lastly, the vertical cup-to-disc ratio was calculated. Using seventy nine images, including twenty five glaucoma images, the sensitivity of 80% and a specificity of 85% were achieved with this method.

Keywords: Glaucoma, Retinal fundus image, Cup-to-disc ratio, Optic disc extraction, Determination of cup edge, Computer-aided diagnosis, Blindness.

1 Introduction

In a population-based prevalence survey of glaucoma conducted in Tajimi City, Japan, 1 in 20 people over the age of 40 was diagnosed with glaucoma [1, 2]. The number of people affected with glaucoma globally is estimated to be 60.5 million by 2010 and reaching 79.6 million by 2020 [3]. Glaucoma is the leading cause of blindness in Japan. Although it cannot be cured, glaucoma can be treated if diagnosed early. Mass screening for glaucoma using retinal fundus images is simple and effective. In view of this, the fundus is examined selectively in the diagnosis performed by physicians as part of a specific health checkup scheme initiated in Japan in April 2008. Although this has improved the ocular healthcare, the number of ophthalmologists has not increased, thus increasing their workload. Computer-aided diagnosis (CAD) systems, developed for analyzing retinal fundus images, can assist in

D. Zhang and M. Sonka (Eds.): ICMB 2010, LNCS 6165, pp. 64–72, 2010.

reducing the workload of ophthalmologists and improving the screening accuracy. We have been developing a CAD system for analyzing retinal fundus images [4-13]. Our CAD system was targeted for three diseases, hypertensive retinopathy [4-6], diabetic retinopathy [7-9] and glaucoma [10-13].

The purpose of this study is to analyze the optic disc on a retinal fundus image, which is important for diagnosis of glaucoma. First, as a basic study for analysis of the optic disc, we developed a method to extract the optic disc from a retinal fundus image [11-13]. A number of studies has reported on automated localization of optic discs; several studies have also reported on segmentation of optic discs [11-22]. Also Nakagawa et al. had previously proposed a disc detection scheme using the P-tile thresholding method [11]. Since this method was proposed for detecting rather than extracting the optic disc, a more precise method is required for analysis of the optic disc features. The proposed method was intended to improve the conventional method of extracting the optic disc. Three dimensional images captured on stereo fundus camera [13, 17-20] and HRT [21] were used in several studies. However, it is difficult to use such 3D fundus camera in the screening. Thus, we attempted to measure C/D ratio automatically using two dimensional images. But, to our knowledge, none achieved high performance because it was difficult to extract the cup region [22]. The ophthalmologists determine the outline of cup by detection of kinks in the blood vessels (as shown in Fig.1). Wong, et al. proposed an automated C/D ratio measuring method by detecting kinks in the blood vessels on the optic disc [22], but it was very difficult. Therefore, by analyzing a color profile of the optic disc, we developed a computerized scheme that recognizes glaucoma cases. That method can be effective in recognizing glaucoma cases, without detecting kinks, by focusing on the profile of the optic disc on a retinal fundus image.

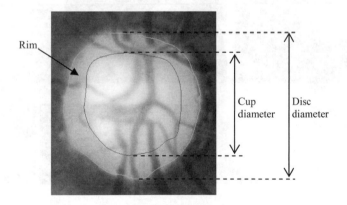

Fig. 1. An example of an ophthalmologist's sketch. A green line shows an outline of a disc, a blue line shows an outline of a cup.

2   Method

Retinal images were captured using a fundus camera (Kowa VX-10i). The photographic angle of the fundus camera was set to 27 degrees. The retinal fundus images were obtained with an array size of 1600 × 1200 pixels and 24-bit color. Our method consists of two steps. First, the optic disc is extracted based on features such as color and shape. Further, we attempt to distinguish between the glaucoma and non-glaucoma cases by calculating the C/D ratio. More details are given below.

2.1 Extraction of Optic Disc

The presence of blood vessels that run on the outline of the optic disc makes accurate extraction of its outline difficult. Thus, to reduce the effect of blood vessels, the method proposed by Nakagawa et al. was applied to create a "blood-vessel-erased" image [11]. The optic disc region tends to be comparatively brighter than the retina. Thus, the approximate region of the optic disc was extracted by using the P-tile thresholding method [23] on each red, green, and blue component image, and the optic disc region was determined by using an image combining these three binary images. The result of this approach is shown in Fig. 2 (b). Then, the area of the 600×600 pixel image, centered on the optic disc thus determined was extracted as the region for analysis. The resulting extracted region is shown in Fig. 2 (c).

An RGB color image was converted to an intensity image, and the pixel values were normalized. The change in intensity (brightness) is usually high at the outline of the optic disc; thus, we applied a canny edge detector [24] to enhance the edge. We determined the outline using the spline interpolation method based on the locations of the outline suggested by edge detection. The results of these approaches are shown in Fig. 2.

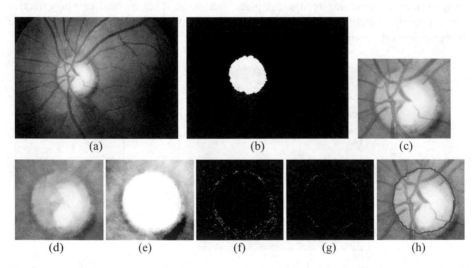

Fig. 2. Example of optic disc extraction. (a) Original retinal image, (b) Extracted optic disc, (c) Rectangular region surrounding the optic disc, (d) Blood-vessel-erased image, (e) a normalized intensity image, (f) Edge image, (g) Extracted outline of the optic disc, (h) Resulting image of the extracted optic disc.

2.2 Determination of Cup-to-Disc Ratio

The vertical C/D ratio is most important factor for diagnosis of glaucoma, because the contrast of the cup region and the rim one was high. Thus, we attempted to measure

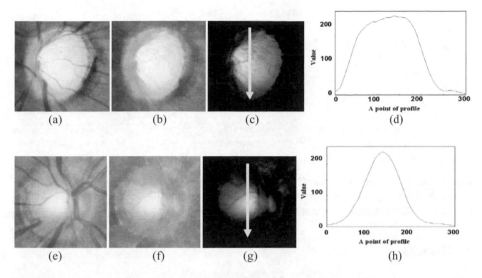

Fig. 3. The examples of profile lines. (a) Optic disc of a glaucoma case, (b) an image erased the blood vessels from (a), (c) a blue bit image of (b) and an arrow shows the location of the obtained profile. (d) profile of (c). (e) to (h) show the images of normal case.

C/D ratio automatically. Glaucoma cases tend to have enlarged cup regions as against the normal cases. (In other words, the bright region is extended in many cases.) Thus, while comparing the optic disc profiles, the profile for normal cases tends to appear as a narrow mountain with long skirts, while that for glaucoma cases appears as a broad mountain with short skirts, as shown in Fig. 3. Our present study focuses on this difference.

The images were preprocessed before obtaining the profile. First, a blood-vessel-erased image was created using the method proposed by Nakagawa et al.[11]. Since there is high contrast between the cup and disc regions in the blue channel of a color image, a blue component was used for subsequent analysis. Thus, we classified the fundus images into the images of right and left eyes by comparing the pixel value in right and left side of an optic disc region, because the nose side of an optic disc was brighter than that of the ear. A profile was then obtained around center of gravity of the disc region extracted automatically (described in section 2.1) in the vertical direction. The blood vessel's region un-erased affected the vertical profile. Moreover, the contrast of ear side of the optic disc was high in the blue component image. Thus, twenty profiles were obtained from the gravity of the disc region to the ear side, and profiles were then averaged so that the result would not depend on the specific line selected. This profile was smoothed as shown in Fig. 4 (b), and impulses if there was any were removed, in order to reduce the effect of noise and remaining blood vessels.

Subsequently, the region under the threshold value of profile (as shown in Fig. 4 (c)) was determined as the skirt region. The threshold value was determined experimentally. The edge of the skirt region and mountain one was determined as the cup edge. Finally, C/D ratio was measured by using the cup edge and disc region, and the case with over 0.6 C/D ratio was determined as glaucoma.

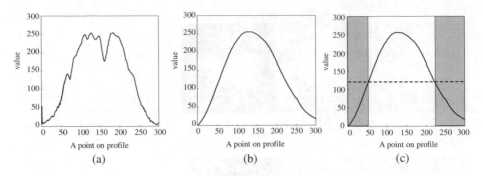

Fig. 4. An analysis of profile. (a) Original profile. (b) Smoothed profile. (c) Determination of cup edge. Gray regions show Rim regions, and a dashed line shows a thresholding value of determination of cup edge. The minimum distance between two gray region was cup diameter.

3 Results

3.1 Extraction of Optic Disc

The proposed method for extracting the optic disc was evaluated based on manual outlines drawn by an ophthalmologist. Fifty fundus images, including twenty-five glaucoma images, were included in this test. Comparison of results by the proposed method and our previous method is shown in Table 1. The method was useful in the extraction of disc regions with a concordance rate of 86%, which expresses the area of intersection as a percentage of the area of union. Our method correctly extracted on average 96% of the disc regions determined by the ophthalmologist, whereas it over-extracted 11% of the disc regions. In contrast, the concordance rate using our previous method [11] was 63% when the extraction rate was 83% and the over-extraction rate was 28%. An example result is shown in Fig. 5. Although the previous method mis-extracted slightly in the top of the disc region, it could extract the approximate disc region. Compared with the previous method, this method can be useful for detecting the optic disc in glaucoma examinations.

Table 1. Comparison between proposed method and previous method

	Concordance rate	Extraction rate	Over extraction rate
Proposed method	86%	96%	11%
Previous method [11]	63%	83%	28%

3.2 Determination of Cup-to-Disc Ratio

Seventy-nine images, including twenty-five glaucoma images, were used to evaluate the proposed method for determination of glaucoma. The proposed method obtained a sensitivity of 80% (20/25) and specificity of 85% (46/54). And, an ophthalmologist's sensitivity was 80% (20/25) when his specificity was 93% (51/54). In addition, we analyzed the result using ROC, as shown in Fig. 6. An AUC of a proposed method was 0.87 and an ophthalmologist's one was 0.96. Therefore, the performance of our

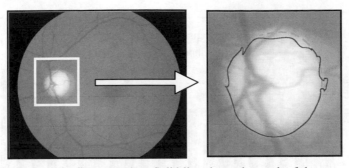

Fig. 5. Result of an optic disc extraction. Solid line shows the result of the proposed method. Dashed line shows a sketch by an ophthalmologist. This concordance rate was 92%, the extraction rate 97%, and the over-extraction rate 5.1%.

method approached to that of an ophthalmologist. On the other hand, the average C/D ratios by using the proposed method were 0.46 and 0.60 for abnormal and normal cases, respectively, while those by ophthalmologist were 0.57 and 0.83, respectively. A comparison of C/D ratios from an ophthalmologist and the proposed method in ten random images is shown in Fig. 7. Although C/D ratios of an ophthalmologist and a proposed method were similar in normal cases, the differences of an ophthalmologist and a proposed method were large in glaucoma cases. Thus, our results were slightly different from an ophthalmologist's value, but our method tended to show the small values in the normal cases and show the large values in the glaucoma cases. Such a problem will be improved by addition of detailed analysis of the vertical profile on the disc.

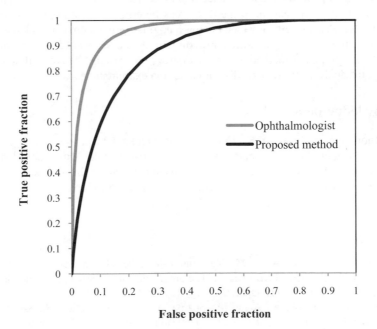

Fig. 6. Two ROC curves are shown. A gray curve shows ROC curve of an ophthalmologist. A black curve shows ROC curve of a proposed method.

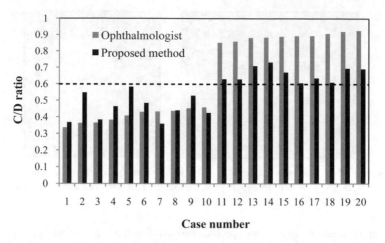

Fig. 7. Comparison of C/D ratio measured by an ophthalmologist and the proposed method. Case 1-10 are normal cases, and case 11-20 are glaucoma cases. Horizontal dashed line shows a threshold value for determination of glaucoma. A threshold value of determination of glaucoma was set 0.6.

4   Conclusion

We proposed a method for extracting the optic disc in retinal fundus images. The proposed method achieved a concordance rate of 86%, extraction rate of 96%, and over-extraction rate of 11% relative to the disc areas determined by an ophthalmologist. In this study, we also presented a method for recognizing glaucoma by calculating C/D ratio. The method correctly identified 80% of glaucoma cases and 85% of normal cases. Although the proposed method is not error-free, the results indicated that it can be useful for the analysis of the optic disc in glaucoma examinations.

Acknowledgement

The authors are grateful to K. Fukuta from Denso Corp., K. Ishida from Gifu Prefecture General Medical Center, A. Aoyama from Ogaki Municipal Hospital, T. Nakagawa from Kowa Company Ltd., Y. Hayashi, K. Sugio from Tak Co. Ltd., and the co-workers from the hospitals, the companies, and Gifu University involved in the "Big CAD projects" at the Knowledge Cluster Gifu-Ogaki.

References

1. Iwase, A., Suzuki, Y., Araie, M., Yamamoto, T., Abe, H., Shirato, S., Kuwayama, Y., Mishima, H., Shimizu, H., Tomita, G.: The prevalence of primary open-angle glaucoma in Japanese: The Tajimi study. Ophthalmology 111, 1641–1648 (2004)
2. Yamamoto, T., Iwase, A., Araie, M., Suzuki, Y., Abe, H., Shirato, S., Kuwayama, Y., Mishima, H., Shimizu, H., Tomita, G.: The Tajimi study report 2 prevalence of primary angle closure and secondary glaucoma in a Japanese population. Ophthalmology 112, 1661–1669 (2005)

3. Quigley, H., Broman, A.: The number of people with glaucoma worldwide in 2010 and 2020. British J. Ophthalmology 90, 262–267 (2006)
4. Hatanaka, Y., Hara, T., Fujita, H., Aoyama, M., Uchida, H., Yamamoto, Y.: Automatic distribution and shape analysis of blood vessels on retinal images. In: Proc. SPIE, vol. 5370, pp. 1621–1628. SPIE, Bellingham (2004)
5. Takahashi, R., Hatanaka, Y., Nakagawa, T., Hayashi, Y., Aoyama, A., Mizukusa, Y., Fujita, A., Kakogawa, M., Hara, T., Fujita, H.: Automated analysis of blood vessel intersections in retinal images for diagnosis of hypertension. Medical Imaging Technology 24, 270–276 (2006)
6. Hayashi, T., Nakagawa, T., Hatanaka, Y., Hayashi, Y., Aoyama, A., Mizukusa, Y., Fujita, A., Kakogawa, M., Hara, T., Fujita, H.: An artery-vein classification using top-hat image and detection of arteriolar narrowing on retinal images. IEICE Technical Report, vol. 107, pp. 127—132. IEICE, Tokyo (2007)
7. Hatanaka, Y., Nakagawa, T., Hayashi, Y., Fujita, A., Kakogawa, M., Kawase, K., Hara, T., Fujita, H.: CAD scheme to detect hemorrhages and exudates in ocular fundus images. In: Proc. SPIE, vol. 6514, pp. 65142M-1–65142M-8 (2007)
8. Hatanaka, Y., Nakagawa, T., Hayashi, Y., Kakogawa, M., Sawada, A., Kawase, K., Hara, T., Fujita, H.: Improvement of automatic hemorrhages detection methods using brightness correction on fundus images. In: Proc. SPIE, vol. 6915, pp. 69153E-1–69153E-10. SPIE, Bellingham (2008)
9. Mizutani, A., Muramatsu, C., Hatanaka, Y., Suemori, S., Hara, T., Fujita, H.: Automated microaneurysms detection method based on double ring filter in retinal fundus images. In: Proc. SPIE, vol. 7260, pp. 72601N-1–72601N-8. SPIE, Bellingham (2009)
10. Muramatsu, C., Hayashi, Y., Sawada, A., Hatanaka, Y., Yamamoto, T., Fujita, H.: Detection of retinal nerve fiber layer defects on retinal fundus images for early diagnosis of glaucoma. J. Biomedical Optics 15, 016021-1–7 (2010)
11. Nakagawa, T., Hayashi, Y., Hatanaka, Y., Aoyama, A., Mizukusa, Y., Fujita, A., Kakogawa, M., Hara, T., Fujita, H., Yamamoto, T.: Recognition of optic nerve head using blood-vessel-erased image and its application to production of simulated stereogram in computer-aided diagnosis system for retinal images. IEICE Transactions on Information and Systems J89-D, 2491–2501 (2006)
12. Nakagawa, T., Suzuki, T., Hayashi, Y., Mizukusa, Y., Hatanaka, Y., Ishida, K., Hara, T., Fujita, H., Yamamoto, T.: Quantitative depth analysis of optic nerve head using stereo retinal fundus image pair. J. Biomedical Optics 13, 064026-1-10 (2008)
13. Muramatsu, C., Nakagawa, T., Sawada, A., Hatanaka, Y., Hara, T., Yamamoto, T., Fujita, H.: Determination of cup and disc ratio of optical nerve head for diagnosis of glaucoma on stereo retinal fundus image pairs. In: Proc. SPIE, vol. 7260, pp. 72603L-1–72603L-8. SPIE, Bellingham (2009)
14. Lalonde, M., Beaulieu, M., Gagnon, L.: Fast and robust optic disc detection using pyramidal decomposition and Hausdorff-based template matching. IEEE Transactions on Medical Imaging 20, 1193–1200 (2001)
15. Li, H., Chutatape, O.: Automated feature extraction in color retinal images by a model based approach. IEEE Transactions on Biomed. Engineering 51, 246–254 (2004)
16. Chrastek, R., Wolf, M., Donath, K., Niemann, H., Paulus, D., Hothorn, T., Lausen, B., Lammer, R., Mardin, C.Y., Michelson, G.: Automated segmentation of the optic nerve head for diagnosis of glaucoma. Medical Image Analysis 9, 297–314 (2005)
17. Merickel, M.B., Wu, X., Sonka, M., Abramoff, M.D.: Optimal segmentation of the optic nerve head from stereo retinal images. In: Proc. SPIE, vol. 6143, pp. 61433B-1–61433B-8. SPIE, Bellingham (2006)

18. Xu, J., Ishikawa, H., Wollstein, G., Bilonick, R.A., Sung, K.R., Kagemann, L., Townsend, K.A., Schuman1, J.S.: Automated assessment of the optic nerve head on stereo disc photographs. Investigative Ophthalmology & Visual Science 49, 2512–2517 (2008)
19. Abramoff, M.D., Alward, W.L.M., Greenlee, E.C., Shuba, L., Kim, C.Y., Fingert, J.H., Kwon, Y.H.: Automated segmentation of the optic disc from stereo color photographs using physiologically plausible features. Investigative Ophthalmology & Visual Science 48, 1665–1673 (2007)
20. Corona, E., Mitra, S., Wilson, M., Krile, T., Kwon, Y.H., Soliz, P.: Digital stereo image analyzer for generating automated 3-D measures of optic disc deformatin in glaucoma. IEEE Transactions on Medical Imaging 21, 1244–1253 (2002)
21. Mikelberg, F.S., Parfitt, C.M., Swindale, N.V., Graham, S.L., Drance, S.M., Gosine, R.: Ability of the Heidelberg Retina Tomograph to detect early glaucomatous visual field loss. J. Glaucoma 4, 242–247 (1995)
22. Wong, D.W.K., Liu, J., Lim, J.H., Li, H., Wong, T.Y.: Automated detection of kinks from blood vessels for optic cup segmentation in retinal images. In: Proc. SPIE, vol. 7260, pp. 72601J-1–72601J-8. SPIE, Bellingham (2009)
23. Parker, J.R.: Algorithms for image processing and computer vision. Wiley Computer Publishing, New York (1997)
24. Canny, J.: A computational approach to edge detection. IEEE Transactions on Pattern Analysis and Machine Intelligence 8, 679–698 (1986)

Tongue Image Identification System on Congestion of Fungiform Papillae (CFP)

Bo Huang and Naimin Li

Bio-computing Research Center, Department of Computer Science and Technology,
Shenzhen Graduate School of Harbin Institute of Technology, Shenzhen, China
cstongue@gmail.com

Abstract. Tongue diagnosis is a unique and important diagnostic method in Traditional Chinese Medicine (TCM). It is used to observe abnormal changes in the tongue for identifying diseases. However, due to its qualitative, subjective and experience-based nature, traditional tongue diagnosis has very limited applications in clinical medicine. To date, no work has sought to automatically recognize a distinctive diagnostic and textural feature of the tongue, Congestion of Fungiform Papillae (CFP) on the middle of tongue surface. In this paper, we present a novel computerized tongue inspection system for identifying the presence or absence of CFP images. We first define and partition a region of interest (ROI) for texture acquisition. After preprocessing for reflective points, we apply the Gabor filter banks and Bayesian Network to identify the texture blocks. Finally, a fusion strategy is employed to model the relationship between the image classes and the presence of CFP. The results are promising.

Keywords: Bayesian Network, Congestion of Fungiform Papillae (CFP), Gabor filter banks, tongue texture.

1 Introduction

Computerized tongue diagnosis allows image acquisition of tongue in a simple, inexpensive, and non-invasive way which is a prerequisite for large scale screening. In a tongue image prescreening system, practitioners make decisions on the basis of physiological changes in the tongue, with local changes being taken to reflect pathological changes in the human body. One particular kind of local physiological change on the surface of the tongue, Congestion of Fungiform Papillae (CFP), is a more sensitive indicator of internal pathological changes and can serve as a useful guideline to differentiate health states and diseases [1]. CFP refers to the abnormal haematosis of mushroom shaped papillae, especially on the middle of the tongue surface. When a patient suffers from some diseases, say, hyperthyroidism, the fungiform papillae become congestive. This means that CFP can be found beyond the normal range of fungiform papillae, and become more prominent and noticeable (see Fig. 1). In contrast, a healthy tongue should be without any CFP signs, the normal fungiform papillae are flat, not easily seen, and distributed evenly. The presence of CFP in different regions should predict various pathologic conditions [1]. Therefore, a system, which can decide whether or not any signs suspicious for CFP, can improve diagnostic efficiency.

D. Zhang and M. Sonka (Eds.): ICMB 2010, LNCS 6165, pp. 73–82, 2010.

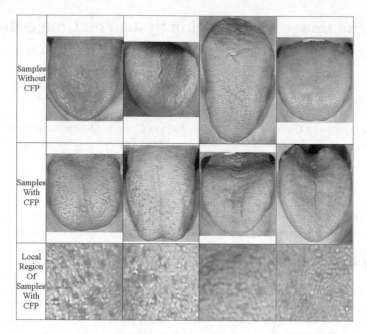

Fig. 1. Some samples

However, to date, no work has sought to automatically recognize the presence of CFP for computerized tongue diagnosis. However, our experience has determined that there are at least three basic challenges in this texture identification. First, watching the texture blocks carefully, we can find many reflective points. These reflective points are originated from the specular reflection of the saliva on the tongue surface. They will have a severe effect on the identification accuracy. The second difficulty is to render the judgment of a tongue texture by experts as a well-defined and measurable machine readable representation. Human judgment of texture characteristics depends on the subjective description of textural information. It is challenging because these subjective factors must be measured and discriminated in some way. The third difficulty is the changes in the tongue texture, which do not begin and complete in 24 hours, but may occur and persist over many months, reflecting the course of the disease. Such persistent variation produces diagnostic uncertainty, which is the ambiguity that belongs to one or more classes and then, only to some degree. This uncertainty should also be formalized within any framework.

The method in this paper is intended to be a first step toward such a prescreening system. This paper only concerns the presence of CFP on the middle of the tongue surface, which are among the first unequivocal signs of tongue image. The rest of this paper is organized as follows: section 2 describes the details of the whole prescreening system. Section 3 reports our experimental results and comparison. Section 4 offers our conclusion.

2 System Description

This section first describes the acquisition device to ensure the quality and quantity of tongue images. Secondly, the system then automatically extracts the geometric contour of tongue area from the image. Thirdly, ROI partition ensures accurate positioning of pathological area. Fourthly, reflective points are detected and removed. Fifthly, we employ Gabor textural measurements and Bayesian Network to identify the presence or absence of CFP. Finally, a fuzzy fusion framework is utilized to classify each tongue image into two classes.

2.1 Image Acquisition

The goal of image acquisition module is to reduce the environmental effects and improve the quality of tongue images, which is essential to maintain the accuracy of diagnostic results. The device (see Fig. 2) uses a video camera with a three-CCD kernel to capture tongue images. By taking a separate reading of RGB values for each pixel, three-CCD cameras achieve much better precision than a single-CCD. A lighting system, composed of two cold-light type of halogen lamps with a color temperature of 5000k, ensured that it presented the colors in a true and stable manner. The lamp type and optical light path both have elaborate designs and optimal selections. A much more humanistic design; ergonomics (also human engineering or human factors engineering) takes into account the capabilities and limitations of individuals to ensure that the equipment suits each individual.

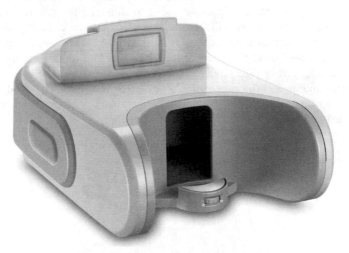

Fig. 2. Tongue image acquisition device.

2.2 Automatic Contour Extraction

The system then automatically extracts the geometric contour from the tongue image, which usually includes the lips, part of the face, or the teeth. We began by segmenting the tongue area from its surroundings using a model derived from a bi-elliptical parameterization of the bi-elliptical deformable template (BEDT) [2], a structure composed of two ellipses with a common center. BEDT makes the algorithm more robust to noise, which in our case is usually caused by pathological details. BEDT produces a rough segmentation through an optimization in the ellipse parameter space using a gradient descent method. The BEDT is then sampled to form a deformable contour, known as the BEDC [2]. To further improve the performance, we replaced the traditional internal force in the BEDC with an elliptical template force which is capable of accurate local control. We then obtained a series of points, using BEDC and a few manual readjustments, and defined the contour of the tongue image. Based on this set of edge points, it is then possible to use the seed filling algorithm to define every pixel in a tongue image as being either in or out of the boundary of the tongue area.

2.3 ROI Partition

The region of interest (ROI) for identification of CFP is the middle of the tongue surface. From a holistic point of view [1], any local part of the human body is representative of the whole body. The tongue is no exception. People in ancient times had already established the doctrine that different parts of the tongue correspond to various internal organs. This theory is very valuable for clinical diagnosis. In a tongue image prescreening system, practitioners make decisions on the basis of physiological changes in the tongue, with local changes being taken to reflect pathological changes in the human body.

The clinical significance of CFP first depends on its absence or presence. Based on the suggestion of clinical practitioners, the texture blocks partitioned from the middle of tongue surface are thus taken as the primary ROI. The tongue body can be considered as an approximate ellipse, we use it to illustrate the progress (see Fig. 3). As shown in Fig. 3, the dotted ellipse is a shrunken tongue contour, which has an identical top point with original contour (solid ellipse). Then some square texture blocks are partitioned along the boundary of the shrunken tongue contour. The first block of each side (left or right) touches the point p_{tip}, which is the crossing point of the shrunken contour and the perpendicular bisector in tongue tip. And each block share an angle point with another block on the shrunken contour. Then some blocks are partitioned until they cross the horizontal center line.

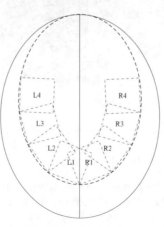

Fig. 3. ROI extraction

2.4　Reflective Point Removal

Observing the images in Fig. 1, we can see many reflective points. These reflective points are originated from the specular reflection of the saliva on the tongue surface. These will have a severe effect on the classification accuracy. As CFP has a number of varying features and the CFP-defined tongue image is difficult to capture, we have chosen to use an adaptive rather than a fixed threshold:

1) Initialization: Let *RatioThres* be the reflective point ratio threshold by manual statistics, *LumThres* be the luminance of the point with highest luminance, and *Steplen=1~5* be the corresponding step length in the iterative procedure;
2) Calculate *MinThres* of the luminance of the reflective points.

$$
MinThres = \begin{cases} laver + \dfrac{laver - l\min}{l\max - l\min} \times (l\max - laver) \times 2 \times Iper, Iper \leq 0.5 \\[3mm] l\max - \dfrac{l\max - laver}{l\max - l\min} \times (l\max - laver) \times 2 \times (1 - Iper), otherwise \end{cases} \tag{1}
$$

where *lmax* is the max luminance value, *lmin* is the min luminance value *l*, and *laver* is the mean luminance value of this image. *Iper* is the percentage of reflective points.

3) Calculate the ratio R of the pixel set whose $L > LumThres$;
4) If $R > RatioThres$, go to (7);
5) $LumThres > LumThres - Steplen$;
6) If $LumThres > MinThres$, go to (3);
7) Label the pixel set whose $L > LumThres$ into reflective point. This method is more adaptive the value range is broader as a result of interaction and co-adjustment between thresholds R and L.
8) The detected reflective points are filled in a bilinear interpolation function of the ROI texture block. This step will greatly reduce the impact of reflective points and facilitate the identification of CFP. A bilinear interpolator can be obtained by applying the following equation:

$$
f(i+u,j+v)=(1-u)(1-v)f(i,j)+(1-u)vf(i,j+1)+u(1-v)f(i+1,j)+uvf(i+1,j+1) \tag{2}
$$

where $f(i,j)$, $f(i,j+1)$, $f(i+1,j)$ and $f(i+1,j+1)$ are the neighbors of $f(i+u,j+v)$.

2.5　Texture Identification

We utilize texture identification to render the human judgment as a well-defined and measurable machine readable representation. Human judgment of a tongue texture depends on the subjective description of textural information, such as "red prickle", "red star" and so on [1]. These descriptions are extracted from the visual feeling of experienced experts. The responses of the visual cortex can be simulated by using 2D Gabor functions with image decomposition. These local spatial band pass banks of Gabor filter have good spatial localization, orientation and frequency selectivity. This means that Gabor based features will have a certain robustness when the CFP vary in size, shape, number, color, density and other factors. To represent the texture as a well-defined and measurable machine readable representation, we first apply a Gabor transform to filter the original tongue texture and then use the mean and standard deviation to represent the texture in terms of their orientation, periodicity, and spatial correlations in the tongue image. These two types of textural attributes are too complex to be represented by one or several filtered responses. In the Gabor filter bank theory, S is the number of scales, and K is the number of orientations in the multi-resolution decomposition. The filters are convolved with the signal, resulting in a so-called Gabor space. The texture analysis depends on using suitable numbers for direction and scale. In this paper, the optimal accuracy is achieved by using eight scales and four directions.

Bayesian Network is a probabilistic graphical model that provides a compact representation of the probabilistic relationships for a group of variables $(y_1, y_2, \ldots y_n)$. Formally, the nodes of Bayesian Networks represent variables and its arcs describe conditional independencies between the variables. If there is an arc from node A to another node B, A is called a parent of B, and B is a child of A. The set of parent nodes of a node y_i is denoted by *parent(yᵢ)*. The joint distribution of the node values can be written as:

$$p(y_1, y_2, \cdots, y_n) = \prod_{i=1}^{n} p(y_i \mid parents(y_i))$$ (3)

In general terms, this equation states that each node has an associated conditional probability table. We directly input the feature vector into Bayesian Network for training and testing. Then the class of each texture block is obtained.

2.6 Fuzzy Classification Framework

We proposed a fuzzy fusion framework to classify each tongue image into two classes: "samples with CFP" and "samples without CFP". There are two reasons for utilizing this fuzzy method. First, although Gabor filter banks and Bayesian Network converts human judgment into well-defined and measurable machine representation, each Bayesian Network classifier can only distinguish a texture block. Secondly, a pathological change does not appear in just one day, but may in fact, take place over a period of many months. As we observed, the distribution of CFP can change over time, which makes the classification difficult. The fuzzy membership theory is premised on the observation that many phenomena cannot be discreetly categorized as members of one class or another, but rather, share features with other phenomena so that they may be said to belong to one or more classes, and only to some degree.

Our framework attempts to deal with this categorical ambiguity by using numerical weightings to describe these various degrees of belonging. Bayesian Network module outputs a class for each texture block. The global classification decision can be considered as the degree that a tongue belongs to "samples with CFP" and "samples without CFP". After identification of each partitioned texture block, we can fusion the decisions in clinical medicine. Based on the clinical cases [1], CFP is less common in the region of tongue root, and CFP is relatively common in the tongue tip. This means that the texture block decision near tongue root should possess more weight in the fusion. Thus, the inverse of distance to tongue root is a natural weight of fusion. Suppose each side of middle part can be partioned N texture blocks, the weight of the kth block is

$$P_i = i / \sum_{k=1}^{N} j \qquad (4)$$

Then the class of the given image is

$$IM_{class} = (\sum_{i=1}^{N} P_i C_i + \sum_{j=1}^{M} P_j C_j) / 2 \qquad (5)$$

where C_i is the class of the ith blocks on the left side, C_j is the class of the jth blocks on the right side. Each IM_{class} is located in [0,1] and can be considered as the categorical ambiguity that results from diagnostic uncertainty.

3   Experiment Results

3.1   Accuracy Evaluation

In many previous approaches, the texture identification is usually determined by detailed visual discrimination based on the experience and knowledge of practitioners. Therefore, there may be much uncertainty and imprecision in the diagnostic results. In our system, this subjectivity is removed by labeling a CFP texture simply as either present or absent of CFP, with the labeling being conducted in consultation with three physicians who have more than twenty years of experience. They identified every sample and their opinion of labeling was highly consistent. The labels given to the texture blocks are compared with the results obtained by Bayesian Network. These form the basis for evaluating all of the experiments.

Accuracy was determined against the receiver operating characteristic (ROC) curve and equal error rate (EER). The ROC curve is a graphical plot to select possibly optimal models independently from the class distribution when its discrimination threshold is varied. Bayesian Network produces probability values which represent the degree to which an instance belongs to a class. Setting a threshold value will determine a point in the ROC space. The area under the ROC curve (AUC) is often preferred because of its useful mathematical properties as a non-parametric statistic. This area is often simply known as the discrimination power of the system. We also used EER, the accuracy where both accept and reject errors are equal, to evaluate the performance. More AUC and less EER mean better performance.

To estimate the accuracy of classifiers, we use a 5-fold cross validation for assessing how the classification results will generalize to an independent data set. Cross validation is a statistical practice of partitioning the original datasets into approximately equal sized subsets. In these five subsets, each subsets contains roughly the same proportions of the two types of class labels (either present or absent of CFP). Then each subset is used as a testing set for a classifier trained on the remaining subsets. The empirical accuracy is defined by the average of the accuracies of these subset classifications.

3.2   Effect of Reflective Points Removing

The top part of Fig. 4 shows the histogram distribution of all pixels of more than 6000 texture blocks in the RGB channels. For every channel, except for a smooth ridge in

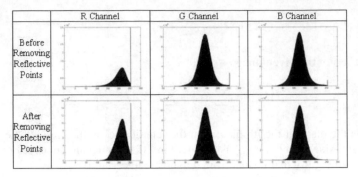

Fig. 4. Distribution of the RGB channel

the middle, there is a sharp peak near the maximum of the spectral intensity value. This peak is the distribution of reflective points. It can be concluded that the intensive values of these points exceed the thresholds. As shown in the lower part of Fig. 4, after removal of the reflective point, the peaks of G and B channel disappear and the number of reflective points in the R channel is also down sharply. Clearly, preprocessing removes the reflective points and keeps the CFP points down.

3.3 Classification and Comparison

We developed a prototype implementation of our system and applied it to a total of 857 samples. The chosen samples included 501 tongue samples with the presence of CFP and 356 samples without CFP. Each image is associated with a person. The healthy volunteers were chosen from students at Harbin Institute of Technology, mainly in the age group of 17-23. All of the patients were in-patients from six departments at the Harbin Hospital and included every age group.

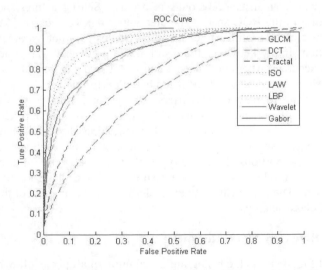

Fig. 5. ROC curve of the two-class classification of blocks

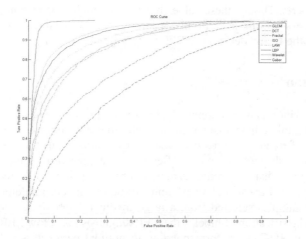

Fig. 6. ROC curve two-class classification of images

The use of many tongue samples supports the representativeness of the CFP texture blocks, including present and absent classes. From the 857 image samples, 6,002 texture blocks are partitioned, including 2,523 blocks with the presence of CFP and 3479 samples without CFP. Experiments were carried out by the Gabor filter and Bayesian Network. The two-class classification of blocks produced an EER of 0.0924 and AUC of 0.9638 and the EER and AUC of the two-class classification of the image samples are 0.0734 and 0.9771 respectively.

There are three kinds of widely-used texture analysis methods: filter, statistics and model-based. We compared the performance of our method with following methods: discrete cosine transform (DCT) filter [3], isotropic filter [4], wavelet filter [5], co-occurrence statistics [6], LBP statistic [7], Laws' statistic [8], and fractal model [9].

Fig. 5 and Fig. 6 show the ROC curves of the block classification and the image classification for the eight texture analysis methods, with ours being the Gabor filter bank-based method. Our method obtains the best performance. The ability of the other methods to cover the great range of the two textures is, for model-based methods, the result of a lack of a well-fitted model, and, for the statistics-based methods, their inability to detect texture using multiple orientations and scales. Moreover, the other three classical filters are not better than the Gabor filter in the multi resolution analysis. Obviously, these methods are not sufficient to represent the complexity of various distributions of CFP while the Gabor filter has the capability of doing so.

3.4 Application in Clinical Medicine

By identification of each partitioned texture block, we can fusion the decisions in clinical medicine. After normalization, this fusion has been applied the chronic observation of several patients' health state. The images in Fig. 1 are eight samples. Their class values, IM_{class}, are 0.031, 0.027, 0.053, 0.011, 0.930, 0.779, 0.713 and 0.691 respectively. These values are consistent with the visual feeling of three experience experts. Though these values have not enough ground truth, they represent some seriousness of disease. In this process, the seriousness of disease varies gradually, and the fusion decision of CFP texture blocks as well.

4 Conclusion

This paper describes a novel automated CFP texture prescreening system. This system can capture digital tongue images in a stable lighting environment, automatically segment areas of tongue textures from their surroundings, partition ROI from middle of tongue surface, detect and remove reflective points, perform a quantitative analysis of the pathological changes on the surface of a tongue using Gabor features, provide identification using Bayesian Network and . Experiments are implemented on a total of 857 tongue samples selected from a large medical database of tongue images. In the two-class classification of blocks, the CFP diagnosis has a low EER of just 9.24%, and the AUC of ROC curve corresponding to optimal parameters is 0.9638. And in image classification, the EER and AUC are 0.0734 and 0.9721 respectively. The experimental results demonstrate that the computer-aided classification of CFP textures will contribute toward a more standardized and accurate methodology for the use of CFP in the diagnosis of health state.

References

1. Li, N., Zhang, D., Wang, K.: Tongue Diagnostics. Shed-Yuan Publishing, Peking (2006)
2. Pang, B., Zhang, D., Wang, K.: The bi-Elliptical deformable contour and its application to automated tongue segmentation in Chinese medicine. IEEE Trans. Med. Imag. 24, 946–956 (2005)
3. Rao, K.R., Yip, P.: Discrete Cosine Transform: Algorithms, Advantages, Applications. Academic Press, Boston (1990)
4. Coggins, J.M., Jain, A.K.: A spatial filtering approach to texture analysis. Pattern Recognition Letters 3, 195–203 (1985)
5. Mallat, S.G.: Theory for multiresolution signal decomposition: the wavelet representation. IEEE Trans. Pattern Anal. Mach. Intell. 11, 674–693 (1989)
6. Haralick, R.M., Shapiro, L.G.: Computer and Robot Vision, vol. 1. Addison-Wesley, Reading (1992)
7. Ojala, T., Pietikainen, M., Maenpaa, T.: Multiresolution gray-scale and rotation invariant texture classification with local binary patterns. IEEE Trans. Pattern Anal. Mach. Intell. 24, 971–987 (2002)
8. Laws, K.I.: Rapid texture identification. In: Proc. SPIE Conf. Image Processing for Missile Guidance, vol. 238, pp. 376–380 (1980)
9. Edgar, G.A.: Measure, Topology, and Fractal Geometry. Springer UTM, Heidelberg (1995)

Newborn Footprint Recognition Using Band-Limited Phase-Only Correlation

Wei Jia[1], Rong-Xiang Hu[1,2], Jie Gui[1,2], and Ying-Ke Lei[1,2,3]

[1] Hefei Institute of Intelligent Machines, CAS, Hefei, China
[2] Department of Automation, University of Science and Technology of China
[3] Department of Information, Electronic Engineering Institute, Hefei, China
icg.jiawei@gmail.com

Abstract. Newborn and infant personal authentication is a critical issue for hospital, birthing centers and other institutions where multiple births occur, which has not been well studied in the past. In this paper, we propose a novel online newborn personal authentication system for this issue based on footprint recognition. Compared with traditional offline footprinting scheme, the proposed system can capture digital footprint images with high quality. We also develop a preprocessing method for orientation and scale normalization. In this way, a coordinate system is defined to align the images and a region of interest (ROI) is cropped. In recognition stage, Band Limited Phase-Only Correlation (BLPOC) based method is exploited for feature extraction and matching. A newborn footprint database is established to examine the performance of the proposed system, and the promising experimental results demonstrated the effectiveness of proposed system.

Keywords: biometric, newborn, infant, footprint recognition.

1 Introduction

In information and network society, there are many occasions in which the personal authentication is required, e.g., access control to a building, information security in a computer or mobile-phone, visitor management, and electronic payment etc. There is no doubt that biometric is one of the most important and effective solutions for this task. Generally, biometric is a field of technology that uses automated methods for identifying or verifying a person based on a physiological or behavioral trait [1].

In real applications, the traits that are commonly measured in different systems are the face, fingerprints, hand geometry, palmprint, handwriting, iris, and voice etc [1]. Recently, some interesting biometrics systems have been developed by exploiting new traits including hand vein and finger-knuckle-print, etc [9~13]. However, most biometric systems mentioned above are developed for adults. How to design a biometric system for newborns and infants has not been well studied in the past. Therefore, few literatures and products about this issue can be found from scientific document databases and markets. According to the following reasons, it is very necessary to study biometric techniques for newborn and infant personal authentication: (1) All over the

D. Zhang and M. Sonka (Eds.): ICMB 2010, LNCS 6165, pp. 83–93, 2010.
© Springer-Verlag Berlin Heidelberg 2010

world, there are about 80~90 million newborns every year, and total population of infants and young children at the age of 0~5 is about 400~500 million. Thus, the research of biometrics should not ignore this special group with large population. Otherwise, the whole architecture of biometrics technique is incomplete; (2) Identification of infants at birth is a critical issue for hospitals, birthing centers and other institutions where multiple births occur. Biometric is a good and feasible choice to deal with this task considering several factors, i.e., easy usage, fast processing and low-cost etc; (3) There are many crimes involving baby-switching or the abduction of infants. As a low-cost and fast solution, biometrics has been regarded as a promising tool to help the police find those abducted infants.

Actually, in many countries, footprint recognition has been used for newborn personal authentication for a long time. Usually the footprints are collected with ink spread on the foot with a cylinder or a paper, and then printed on the newborn's medical record, along with the mother's fingerprint [2].

Here, we will explain why other human traits such as face, iris, fingerprint and palmprint have not been used for newborn personal authentication. Up to now, face recognition with high accuracy is still a difficult task even for adults due to illumination, pose, expression and other varieties. Particularly, the face of newborn may have a drastic change within several days after birth. Thus, face recognition is not recommended for newborn personal authentication. Meanwhile, the use of the iris as identification feature is also a difficult method for newborns, especially the premature, because they hardly open their eyes, they do not have the ability of looking into a scanning device, and touching their eyelids to collect an image could hurt them. Besides, the iris pattern only stabilizes after the child's second year [2]. Although, fingerprint recognition has been widely and successfully used, it is not feasible for newborns. The main reason is that the finer of newborn is too small. As a result, newborn's fingerprint can't be clearly captured. Palmprint recognition is not suitable for newborns yet since it is often difficult to let a newborn open his hand. Although the DNA examination is proven to be efficient in the univocal identification of individuals, but it comes at high cost and can't be used in real time applications, demanding sophisticated laboratory procedures [2]. Compared to other techniques, footprint recognition is very attractive for newborn personal authentication, since it is a non invasive method, of easy applicability, high availability, very low cost, wide acceptance and has effectively been used for more than 100 years [2].

Obviously, traditional footprint recognition using inked footprint is offline. Although capturing offline newborn's footprint has been exploited in many countries, there exists a big debate on the effectiveness of offline footprint recognition caused by the image quality of offline footprint. In fact, there is no innovation for offline newborn's footprint acquisition in the past 100 years, and nearly most of offline footprint images are illegible due to the following reasons: (1) Use of inadequate materials (ink, paper, cylinder); (2) Untrained personal for footprint acquisition; (3) Baby's skin covered with an oily substance; (4) Reduced thickness of the newborn epidermis easily deforming the ridges upon contact and filling the valleys between ridges with ink; (5) Reduced size of the newborns ridges, which are three to five times smaller than that of adults [2]. Fig. 1 depicts two inked newborn footprint images. It can be seen that some thin lines can't be clearly observed, and some thick lines

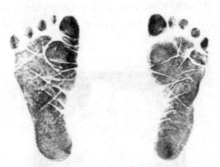

Fig. 1. Inked footprint images

become white areas. Due to bad image quality of offline footprinting, it is nearly impossible to obtain desirable recognition rates. Several researchers have conducted experiments to evaluate the recognition performance of offline footprinting. Unfortunately, they drawn a conclusion that newborn' offline footprinting can't be used for identification purposes, and then the acquisition of footprints in hospital should be abandoned because it only generates unnecessary work and costs [2].

Besides of bad image quality, newborn offline footprinting has other disadvantages. Up to now, most offline footprint images are stored in the papers, therefore it is difficult to form image database. At the same time, fast image retrieval can't be conducted, and offline footprint image can't be used in Electronic Medical Record. Meanwhile, offline footprint recognition is performed manually, which is very time-consuming.

As stated previously, offline footprinting can't satisfy the demanding for fast and reliable newborn personal authentication. The online system based on digital image acquisition and processing is becoming a promising choice for this task. Generally, an online system captures footprint images using a digital capture sensor that can be connected to a computer for fast processing. In this paper, we propose an online newborn footprint recognition system based on low-resolution imaging. The resolution of image used for recognition is about 80 dpi.

2   Image Acquisition and Preprocessing

In our system, the first step is image acquisition. We captured the newborns' footprint images using a digital camera, whose type is Cannon Powershot SX110 IS. The image capturing work was done in Anhui Province Hospital, which is one the biggest hospital in Anhui province, China. When capturing images, two persons are needed. One person is one of the authors of this paper, whose task is to take pictures using camera. The other person is a nurse of hospital, whose tasks are to pacify the newborn and hold the foot.

In order to facilitate image segmentation, we used a black cloth to wrap the ankle. Fig. 2 depicts a captured color footprint image, whose cloth background has been removed by image segmentation method. It can be seen that the quality of online footprint image is much better than that of inked footprint image. So it is possible to achieve promising recognition rates using online system. The size of raw images captured by camera is 3456×2592 pixels, which is too large to be fast processed. We resized the raw image into a smaller one, whose size is 691×518 pixels, and converted it from color space to gray space.

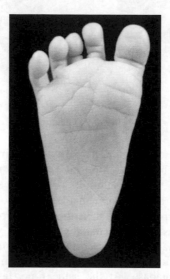

Fig. 2. A captured digital footprint image

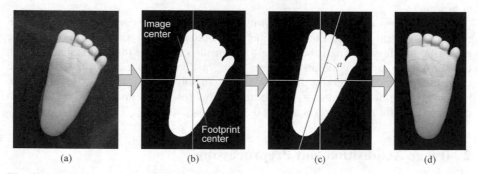

Fig. 3. Orientation normalization. (a) Original gray image $G(x,y)$, (b) Binary image $B(x,y)$, and the center of footprint, (c) estimating the angle of major axes of footprint, α, from image $C(x,y)$, (d) orientation normalized

All the images were collected in one session during the first two days following birth. After we explained some knowledge about the importance and significance about Newborn's Biometrics to newborn's parents, they consented that we can capture footprint images once and use these images for scientific research purpose. However, we failed to collect the images again from a same newborn since most parents declined our request of second image capturing. The main reason is that, in China, a newborn is so important for a family that the parents are unwilling to let other persons, especially strangers, to touch their baby once again. So they were very impatient for our second request. At last, we regretted to abandon the second image collection.

In image acquisition stage, a crucial problem is to select an opportune time to capture images. If a newborn is hungry or crying, he/she will ceaselessly move his/her hands, feet, and whole body. In this time, it is difficult to capture footprint images with desirable quality. On the contrary, if a newborn is calm or sleeping, the task of image capturing will become easy. In this paper, all images were captured when newborns were calm or sleeping.

After image acquisition, the next task is image preprocessing. In our image acquisition system, orientation and scale changes between different images captured from a same foot are inevitable caused by unconstrained image acquisition. Thus, several tasks should be done in preprocessing stage, i.e., orientation normalization, scale normalization and ROI extraction.

In [17], Nakajima et al. proposed a footprint recognition scheme utilizing pressure-sensing, and described an orientation normalization method for footprint pressure images, which can also be used in our system. The main steps (see Fig. 3) of orientation normalization are given as follows:

Step 1: Conduct image segmentation to extract footprint area from gray image G(x,y). Since the background of G(x,y) is a black cloth (see Fig. 3(a)), it is easy to perform image segmentation. A threshold, T, is used to convert the segmented image to a binary image, B(x,y), as shown in Fig. 3(b).

Step 2: Calculate the center (Center_x, Center_y) of footprint in B(x,y) by the following formula:

$$\begin{cases} Center\_x = \frac{\sum_x x \sum_y B(x,y)}{\sum_{x,y} B(x,y)} \\ Center\_y = \frac{\sum_y y \sum_x B(x,y)}{\sum_{x,y} B(x,y)} \end{cases} \tag{1}$$

And then, translate the footprint in B(x,y) in order to let the center of footprint and the center of image coincide with each other, as shown in Fig. 3(b). The formed binary image is named as C(x,y).

Step 3: Calculate the orientation of footprint area in C(x,y). First, we obtain the 2×2 covariance matrix summation of C(x,y):

$$M = \begin{bmatrix} \sum_{x,y} C(x,y)x^2 & \sum_{x,y} C(x,y)xy \\ \sum_{x,y} C(x,y)xy & \sum_{x,y} C(x,y)y^2 \end{bmatrix} \tag{2}$$

And then, perform eigen-decomposition on M to obtain the principal components, i.e., $V = [v1, v2]$, where v1 and v2 are 1×2 eigen-vectors corresponding to the largest and smallest eigen-value, respectively. The full matrix of V can be written as:

$$V = [v1, v2] = \begin{bmatrix} v_{1,1} & v_{2,1} \\ v_{1,2} & v_{2,2} \end{bmatrix} \tag{3}$$

Assuming that the footprint is an oval shape, the angle, α, of its major axes can be estimated using v1 (see Fig. 3(c)):

$$\alpha = -\operatorname{atan}\left(\frac{v_{1,1}}{v_{1,2}}\right) / \left(\frac{\pi}{180}\right) \tag{4}$$

After estimating α, we can rotate the major axes of footprint to vertical orientation to obtain normalized binary image.

Step 4: Perform the translation processing in step 2 and rotation processing in step 3 using the same parameters on gray image, we can obtain orientation normalized gray image, as shown in Fig. 3(d).

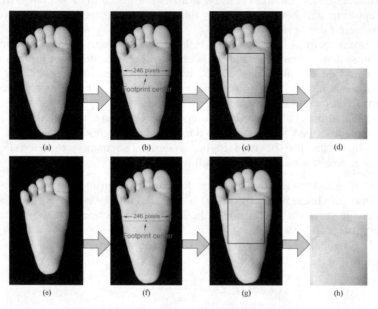

Fig. 4. Scale normalization and ROI extraction

We can perform the scale normalization at the vertical or horizontal direction. Vertical based scale normalization means that all footprints should have the same height in the position of crossing foot center. However, toes of the foot may have different poses. Thus, vertical based scale normalization may be not very reliable. In this paper, we conducted the scale normalization at the horizontal direction. Here, WH denotes the horizontal width of foot crossing foot center. For horizontal based scale normalization, all footprints were normalized to have the same WH, i.e., 246 pixels, which is the mean of all footprints' WH. Fig. 4(a) and (e) are two images captured from a same newborn's foot. Obviously, there exist scale changes between them. Their normalization images are Fig. 4(b) and (f), respectively. It can be observed that the scale varieties have been well corrected. After scale normalization, the center part (ROI) of normalized image was cropped for feature extraction (see Fig. 4(c), (d), (g), and (h)). The size of ROI image is 220×180.

3   Fundamentals of Band-Limited Phase-Only Correlation

From Fig. 2, it can be seen that newborns' footprint image at low resolution contain rich features such as line, direction and texture, etc. Particularly, the pattern of footprint is similar to the one of palmprint. In this regard, many methods proposed for palmprint recognition can also be used for footprint recognition [3~8] [18]. Among all kinds of methods, Band-Limited Phase-Only Correlation (BLPOC) is one of effective and efficient methods, which has been successfully applied to iris [16], palmprint [14, 15] and finger-knuckle-print [13] recognition. In our system, it was also exploited for newborn's footprint recognition.

Firstly, the definition of POC is described as follows:

Consider two $N_1 \times N_2$ images, $f(n_1, n_2)$, and $g(n_1, n_2)$. Let $F(k_1, k_2)$ and $G(k_1, k_2)$ denote the 2D DFTs of the two images. Here, $F(k_1, k_2)$ is given by:

$$F(k_1, k_2) = \sum_{n_1=0}^{N_1} \sum_{n_2=0}^{N_2} f(n_1, n_2) e^{-j2\pi(\frac{n_1 k_1}{N_1} + \frac{n_2 k_2}{N_2})}$$

$$= A_F(k_1, k_2) e^{j\theta_F(k_1, k_2)} \tag{5}$$

where $A_F(k_1, k_2)$ is amplitude and $\theta_F(k_1, k_2)$ is phase. $G(k_1, k_2)$ can be defined in the same way. The cross-phase spectrum $R_{FG}(k_1, k_2)$ is given by:

$$R_{FG}(k_1, k_2) = \frac{F(k_1, k_2)\overline{G(k_1, k_2)}}{|F(k_1, k_2)\overline{G(k_1, k_2)}|} = e^{j\theta(k_1, k_2)} \tag{6}$$

where $\overline{G(k_1, k_2)}$ is the complex conjugate of $G(k_1, k_2)$ and $\theta(k_1, k_2)$ denotes the phase difference $\theta_F(k_1, k_2)$-$\theta_G(k_1, k_2)$. The POC function $r_{fg}(n_1, n_2)$ is the 2D Inverse DFT (2D IDFT) of $R_{FG}(k_1, k_2)$ and is given by:

$$r_{fg}(n_1, n_2) = \frac{1}{N_1 N_2} \sum_{k_1 k_2} e^{j\theta(k_1, k_2)} e^{j2\pi(\frac{n_1 k_1}{N_1} + \frac{n_2 k_2}{N_2})} \tag{7}$$

From formulas (6) and (7), we can see that original POC exploits all components of image's 2D DFT to generate the out plane. In [14], ITO et al., found that BLPOC can achieve better recognition performance by removing the high frequency components and only using the inherent frequency band for matching.

Here we denote the center area of $\theta_F(k_1, k_2)$ and $\theta_G(k_1, k_2)$ as $\theta_F(k_1, k_2)_{BL}$ and $\theta_G(k_1, k_2)_{BL}$, whose size is $J_1 \times J_2$. Thus, the BLPOC function is given by:

$$r_{fg}(n_1, n_2)_{BL} = \frac{1}{J_1 J_2} \sum_{k_1 k_2} e^{j(\theta_F(k_1, k_2)_{BL} - \theta_G(k_1, k_2)_{BL})} e^{j2\pi(\frac{n_1 k_1}{J_1} + \frac{n_2 k_2}{J_2})} \tag{8}$$

4 Experiments

A newborn's footprint database was established. In total, the database contains 1968 images from 101 newborns' feet. That is, about 19~20 images were collected from each foot. In each class, we use the first three footprint images for training and leave the remaining footprint images for test. Therefore, the numbers of images for training and test are 303 and 1635, respectively. When a test image A matches with three training images belonging to a same class i, three matching scores will be generated. The largest one will be selected as the final matching score between A and class i. In experiments, the nearest neighbor rule (1NN) is used for classification. The EER is adopted for evaluate the verification performance, and the accuracy recognition rate (ARR) is exploited to evaluate the identification performance. The experiments were conducted on a personal computer with an Intel Duo T7500 processor (2.20GHz) and 2.0G RAM configured with Microsoft Vista and Matlab 7.0.

In [14] and [15], the *peak* was adopted for similarity measure. In this paper, we will also investigate the recognition performance of measures *peak-to-correlation energy* (PCE), and *peak-to-sidelobe ratio* (PSR). As the name suggests, *peak* is the maximum peak value in the correlation out plane (COP). PCE and PSR are defined by:

$$PCE = \frac{peak - mean_{COP}}{std_{COP}}, \quad PSR = \frac{peak - mean_{sidelobe}}{std_{sidelobe}} \quad (9)$$

where $mean_{COP}$ is the average of the COP, std_{COP} is the standard deviation of the COP, $mean_{sidelobe}$ is the average of the sidelobe region surrounding the peak (21×21 pixels with a 5×5 excluded zone around the peak), and $std_{sidelobe}$ is the standard deviation of the sidelobe region values.

Two experiments were conducted, which are identification and verification, respectively. In these experiments, determining suitable values of J_1 and J_2 is a key problem that should be solved firstly. For convenience, we let J_1 equal to J_2. That is to say, the selected center area of the 2D DFT spectrum is a square, whose size is $J_1\times J_1$. Furthermore, in order to choose the best J_1, we conduct the tests exploiting different values of J_1. Here, the value of J_1 is set to an even number, and changes from 92 to 22 with an interval of -2. Since the size of sidelobe region in PSR is 21, the smallest value of J_1 is set to 22. Fig.5(a) shows ARRs obtained from identification experiments using three similarity measures corresponding to different J_1. Fig.5(b) depicts the EERs obtained from verification experiments using three similarity measures. It can be easily seen that using PSR can generate the lowest EER and highest ARR. Table 1 lists the lowest EER and highest ARR obtained by *peak*, PCE and PSR, respectively. We can know the results obtained by PSR are best.

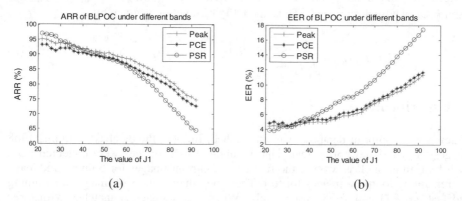

(a)   (b)

Fig.5. (a) the Accuracy Recognition Rate (ARR) of BLPOC under different bands (J_1), (b) the Equal Error Rate (ARR) of BLPOC under different bands (J_1)

Table 1. Highest ARR of lowest EER using different similarity measures and corresponding J_1

	Peak	PCE	PSR
Highest ARR	95.05% (22)	93.3% (22)	97% (22)
Lowest EER	4.34% (30)	4.52% (30)	3.82% (24)

From Table 1, we can see that the highest ARR is 97% and the lowest EER is 3.82%, which is promising recognition performance. When the value of J_1 is 22 and PSR is used for similarity measure, the best verification results can be obtained. Under these parameters, Fig. 6(a) depicts the corresponding FRR (False Rejection Rate) and FAR (False Acceptance Rate) curves. The distance distributions of genuine matching and imposter matching obtained in this experiment are plotted in Fig. 6(b). And the ROC curve is given in Fig. 10. Experimental results indicate that the proposed system has a good capability for newborn's personal authentication.

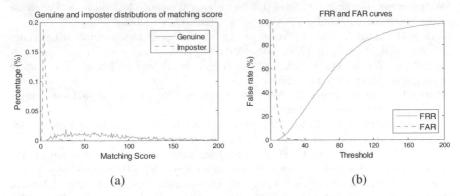

(a)   (b)

Fig.6. (a) the genuine and imposter distributions of matching score while the band width is 22 and the similarity measure is PSR, (b) the FRR and FAR curves of BLPOC while the band width is 22 and the similarity measure is PSR.

5 Conclusion

In the past decades, the research on biometrics has made a significant progress. However, in this field, the research on newborn personal authentication has been ignored nearly. Although recording inked newborns' footprint has been done in many countries for about 100 years, it can't be used for real newborn personal authentication due to its bad image quality and needing manual processing. In this paper, we proposed a novel online newborn recognition system for this issue. We developed preprocessing methods for orientation and scale normalization of footprints, and proposed an effective recognition method based on BLPOC. Compared with traditional offline footprinting scheme, the proposed system can capture footprint images with high quality, has fast processing speed, and more importantly, achieves very promising recognition performance. So, our work has important significance in the research of newborn personal authentication. In the future, we will develop other new algorithms to further improve the recognition performance.

Acknowledgment

This work was supported by the grants of the National Science Foundation of China, Nos. 60705007 & 60805021 & 60975005, the grant from the National Basic Research Program of China (973 Program), No.2007CB311002, the grants from the National High Technology Research and Development Program of China (863 Program), Nos. 2007AA01Z167, the grant of Postdoc Foundation of China, No. 200801231, and the Knowledge Innovation Program of the Chinese Academy of Sciences.

References

1. Jain, A.K., Ross, A., Prabhakar, S.: An Introduction to Biometric Recognition. IEEE Transactions on Circuits and Systems for Video Technology, Special Issue on Image and Video based Biometrics 14(1), 4–20 (2004)
2. Weingaertner, D., Bello, O., Silva, L.: Newborn's Biometric Identification: Can It Be Done? In: Proceedings of the VISAPP, vol. (1), pp. 200–205 (2008)
3. Zhang, D., Kong, A., You, J., Wong, M.: Online Palmprint Identification. IEEE Trans. Pattern. Anal. Mach. Intell. 25(9), 1041–1050 (2003)
4. Kong, A., Zhang, D., Kamel, M.: A Survey of Palmprint Recognition. Pattern Recognition 42(7), 1408–1418 (2009)
5. Huang, D.S., Jia, W., Zhang, D.: Palmprint Verification Based on Principal Lines. Pattern Recognition 41(4), 1316–1328 (2008)
6. Jia, W., Huang, D.S., Zhang, D.: Palmprint Verification Based on Robust Line Orientation Code. Pattern Recognition 41(5), 1504–1513 (2008)
7. Yue, F., Zuo, W.M., Zhang, D.: FCM-based Orientation Selection for Competitive Code-based Palmprint Recognition. Pattern Recognition 42(11), 2841–2849 (2009)
8. Hennings-Yeomans, P., et al.: Palmprint Classification Using Multiple Advanced Correlation Filters and Palm-specific Segmentation. IEEE Trans. on Information Forensics and Security 2(3), 613–622 (2007)
9. Kumar, A., Prathyusha, K.V.: Personal Authentication Using Hand Vein Trangulation and Knuckle Shape. IEEE Transactions on Image Processing 38, 2127–2136 (2009)
10. Kumar, A., Ravikanth, C.: Personal Authentication Using Finger Knuckle Surface. IEEE Transactions on Information Forensics & Security 4(1), 98–110 (2009)
11. Zhang, L., Zhang, L., Zhang, D., Zhu, H.: Online Finger-knuckle-print Verification for Personal Authentication. Pattern Recognition 3(7), 2560–2571 (2010)
12. Zhang, L., Zhang, L., Zhang, D.: Finger-knuckle-print: a New Biometric Identifier. In: Proceedings of the IEEE International Conference on Image Processing, ICIP (2009)
13. Zhang, L., Zhang, L., Zhang, D.: Finger-knuckle-print Verification Based on Band-limited Phase-only Correlation. In: Proceedings of the 13th International Conference on Computer Analysis of images and patterns, pp. 141–148 (2009)
14. ITO, K., Aoki, T., Nakajima, H.: A Palmprint Recognition Algorithm using Phase-Only Correlation. IEICE Transactions on Fundamentals 91(4), 1023–1030 (2008)
15. Litsuka, S., ITO, K., Aoki, T.: A Practical Palmprint Recognition Algorithm Using Phase Information. In: Proc. of ICPR 2008, pp. 8–11 (2008)

16. Miyazawa, K., Ito, K., Aoki, T., Kobayashi, K., Nakajima, H.: An Effective Approach for Iris Recognition Using Phase-Based Image Matching. IEEE Transactions on Pattern Analysis and Machine Intelligence 30(10), 1741–1756 (2008)
17. Nakajima, K., Mizukami, Y., Tanaka, K., Tamura, T.: Footprint-based Personal Recognition. IEEE Transactions on Biomedical Engineering 47(11) (2000)
18. Guo, Z.H., Zhang, D., Zhang, L., Zuo, W.M.: Palmprint Verification Using Binary Orientation Co-occurrence Vector. Pattern Recognition Letters 30(13), 1219–1227 (2009)

Radii Solaris Extraction through Primitive Modelling

Wang Junhui, Ma Lin, Wang Kuanquan, and Li Naimin

School of Computer Science and Technology
Harbin Institute of Technology, 150001
Harbin, P.R. China
wangjh0426@163.com, malin_li@hit.edu.cn

Abstract. On Iris, features such as lacunae, crypts, lines are closely related with diseases and health status of human body. As one of such features, Radii Solaris plays an important role in automatic diagnosis through iridology. In this paper, we present a novel Radii Solaris extraction method based on primitive definition. Firstly, Radii Solaris primitives are defined and standard templates are obtained from training iris data. Secondly, Radii Solaris is extracted from iris images according to standard template and primitive. Experimental results show that this method can obtain Radii Solaris' primitive and Radii Solaris' templates preferably which provide basis for future work, and get good recognition results.

Keywords: Iridology; Radii Solaris Detection; primitive definition; template matching.

1 Introduction

Both Traditional Chinese Medicine's and Western Medicine's Iridology believe that iris tightly correlates with body appa[1]ratus: every apparatus has its exact mapping locations on iris, and iris changes directly reflect health condition of human body [1]. Iris abnormality mainly embodies color change, local patch, fibrin crypt and the change of fibrin density. The advantage of Iridology is non-contact, non-pain and easy to dissemination, such advantages make it stands in the trend of development of future medicine and preventive medicine [2]. At present, the researches of Iris Diagnosis are still in underway phase. Because Western Medicine's Iridology based on experimental observation and statistical analysis, so its chart can locate exactly and its standard is unambiguous and stable. But Western Medicine is impossible to unify relation of those living apparatus in clinic. Though Traditional Chinese Medicine's Iridology defines the mapping as the all and the one, but the accuracy of Traditional Chinese Medicine's chart is poor and lack of unified, rigorous and quantitative standard [3].

Here we focus on the most important iris feature --- Radii Solaris: we define Radii Solaris primitives and obtain standard template from training iris samples. Firstly, we

---

[1] This work was supported in part by the Specialty Leader Support Foundation of Harbin (2003AFLXJ006), the Foundation of Heilongjiang Province for Scholars Returned from Abroad (LC04C17).

D. Zhang and M. Sonka (Eds.): ICMB 2010, LNCS 6165, pp. 94–103, 2010.

obtain coordinate of Radii Solaris' edge. Secondly, we calculate feature attributes from the coordinate. Thirdly, we extract edges of iris, divide edge image into 12 sub-areas and bestow every sub-area as feature templates. At last, feature templates are clustered by K-Means method. According to clustering results, we get primitive definition and standard template.

Radii Salaries are extracted from iris image according to primitive definition and template matching. First of all, according to template matching we calculate cursory position of Radii Solaris in iris image. And then, we calculate precious position of Radii Solaris by primitive matching.

The rest of the paper is organized as follows. Section 2 introduces iris image preprocess briefly. Section 3 describes primitive definition of Radii Solaris. In section 4, Radii Solaris is extracted from iris image according to former primitive definition. Finally, conclusion is drawn in section 5.

2 Preprocess Iris Image

Because of the effect of illumination and difference of instrument in the process of image acquisition, iris images involve a lot of noises. At the same time, eyelid and eyelash may cover up parts of iris, and reduce corresponding apparatus contributions to iris diagnosis. So it is necessary to carry out preprocessing before iris partition is executed. The preprocessing of iris image mainly includes iris location and iris image normalization.

2.1 Iris Location

The purpose of iris location is getting rid of the information (eyelid, eyelash and pupil) which is useless for iris diagnosis, in other words, is locating inner and outer limbus of iris. Accurate iris location is the precondition of feature extraction. Iris location includes iris rough location and iris precious location.

2.1.1 Rough Location of Iris

The purpose of iris rough location is estimating center and radius of corresponding circle of inner and outer limbus. In this paper, pupil partition method is used for rough location of inner and outer limbus. Our method estimates gray level threshold using gray level distributing information of iris image, and then separates pupil from iris image. At last, the traditional Daugman algorithm is used for rough location.

The result of iris rough location is given in Figure 1.

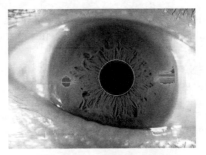

Fig. 1. Iris rough location

2.1.2 Precious Iris Location Based on Transfiguration Template

We locate outer limbus of iris by calculating maximum of gray integral difference of points which are round circumference. A transfiguration circle template whose center and radius are alterable is defined as Equation 1.

$$\max(r, x_0, y_0) \left| G_\sigma(r) * \frac{\partial}{\partial r} \oint_{(r, x_0, y_0)} \frac{I(x, y)}{2\pi r} ds \right| \tag{1}$$

Where: * denotes convolution operation, smoothness function is denoted by $G\sigma(r)$, the gray level of pixel (x,y) is defined as $I(x,y)$.

The process of search are as follow: when center (x_0, y_0) and radius r of pupil are known, radius r is increased gradually and coordinate values x_0, y_0 are changed into a small range until the maximum is discovered[4]. In actual engineering, because of interference of eyelid and eyelash, a part of iris is covered, especially at the top and bottom of iris. So the search of outer limbus always limit in two sector regions whose angle range is $-\pi/4 \sim \pi/4$ and $3\pi/4 \sim 5\pi/4$.

The result of iris precious location is depicted in Figure 2.

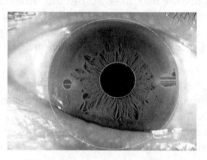

Fig. 2. Iris precious location

2.2 Iris Image Normalization

Because of the effect of illumination and camera angle, iris' size and angle have a great difference which decreases recognition accuracy of Radii Solaris greatly. The purpose of iris image normalization is mapping iris into same size and same corresponding area, thereby this step eliminate the effect of translation and rotation to texture detection [5].

Iris image normalization is mapping Cartesian coordinates into polar coordinates of iris image. The conversion can be fulfilled by following formula:

$$\begin{cases} x(r,\theta) = (1-r) * x_i(\theta) + r * x_o(\theta) \\ y(r,\theta) = (1-r) * y_i(\theta) + r * y_o(\theta) \end{cases} \tag{2}$$

Where the following condition must be satisfied: $r \in [0, 1]$, $\theta \in [0, 2\pi]$.

The result of iris image normalization is given in Figure 3.

Fig. 3. Iris image normalization

3   Primitive Definitions of Radii Solaris

In this paper, we focus on Radii Solaris extraction. Our aim is increasing recognition accuracy through the aid of prior knowledge about Radii Solaris. To obtain such knowledge, we define Radii Solaris primitive and calculate standard primitive templates as follows. Firstly, we pick out 20 iris images from our own database as training samples. Secondly, we calculate feature attributes according to coordinate of Radii Solaris' edge. Thirdly, we extract edges of iris image using canny operator, divide edge images into 12 clockwise sub-areas and each segment correspond to feature template of sub-area. Fourthly, K-Means clustering method is used for classing above feature template. We take clustering centers as primitive definition.

3.1   Coordinate Calculation of Edge of Radii Solaris

3.1.1   Filtering and Binarization
We need to reduce noise from original image before further processing. In this paper, Gauss function [6] is used for image filtering. Satisfactory filtering results may be achieved at relatively low computational cost. Gauss function is given by:

$$G_\sigma(r) = \frac{1}{\sqrt{2\pi}\sigma} e^{-\frac{(r-r_0)^2}{2\sigma^2}} \tag{3}$$

For the sake of obtaining coordinate of Radii Solaris' edges easily, we carry out binarization process to iris image.

The result of iris image process is given in Figure 4.

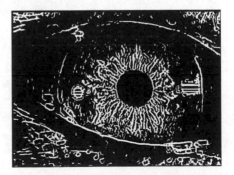

Fig. 4. Iris image after filtering and binarization

3.1.2 Input Assistant Points

The purpose of assistant points is to aid the edge extraction of Radii Solaris. They need to provide following information: position of gaps and direction of Radii Solaris' edge. On one hand, to confirm the position of gaps, we need to input assistant points at the endpoints of gaps; on the other hand, to confirm the direction of Radii Solaris' edge, we need to input assistant points at the points whose curvatures are more than $\pi/2$. According to assistant points, we can fill gaps and continuous regions between assistant points.

3.1.3 Fill Continuous Regions between Assistant Points

Continuous region is whole points of edge between two assistant points. In this paper, the points which satisfy following conditions are defined as edge points:

   a)   *The gray level of the point is 255;*
   b)   *There is at least one adjacent point whose gray level is 0.*

In this paper, we fill continuous regions between assistant points as follows. Firstly, according to those two assistant points, we calculate rough trend of this continuous region (eight azimuths: north, northeast, east, southeast, south, southwest, west, and northwest). Secondly, we extract edge of Radii Solaris by the azimuth.

3.1.4 Fillup the Gap

There are two reasons for gap. On the one hand, because of Gauss filtering some edges become illegibility or even disappearance; on the other hand, binarization makes some edges discontinuity. So we need to fill those gaps in the process of Radii Solaris' edge extraction. In this paper, we take beeline between two assistant points as edges of gaps.

3.2 Primitive Definition of Radii Solaris

Radii Solaris is a strip and through center of outer limbus generally. Radii Solaris are depicted in Figure 5. Primitive definition of Radii Solaris includes length, width, curvature, and angle. Primitive definition is calculated as follows. First of all, we calculate end-points' coordinate of Radii Solaris' long axis. And then, we calculate feature such as length, width, curvature, and angle.

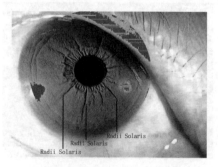

Fig. 5. Radii Solaris Examples

3.2.1 Calculate End-points' Coordinates of Radii Solaris' Long Axis

Figure 6 is an image about Radii Solaris. In this section, we will take this Radii Solaris as an example to calculate end-points' coordinates of long axis. We will take A as starting point to explain. We take midpoints of A, D and B, C as end-points of long axis. The steps are as follows:

   a) Take midpoint of A, D as end-point of long axis which is near pupil.

   b) Calculate coordinates of B and C. Firstly, we calculate distances between whole edge's points and A. Secondly, we take two points of longest distance as B and C.

   c) Take midpoint of B, C as end-point of long axis which is away from pupil.

Fig. 6. Radii Solaris' assistant points for definition

3.2.2 Calculate Feature Attribute of Radii Solaris

We calculate length of Radii Solaris as follows. According to end-points' coordinates of Radii Solaris' long axis, we can get length of long axis and take it as length of Radii Solaris.

For the width of Radii Solaris, we firstly calculate width at every point, and then we take average of all widths as width of Radii Solaris.

Considering the curvature of Radii Solaris, we firstly calculate secants of angle make up of all three adjacent points. And then, we take average of secants as curvature of Radii Solaris.

We calculate angle of Radii Solaris as follows. First of all, we calculate slope of long axis and intersection of long axis and pupil. And then, we calculate slope of tangent which crosses the intersection. At last, we calculate angle according to those two slopes.

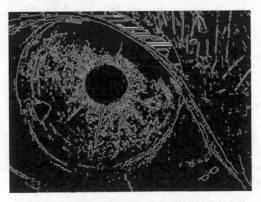

Fig. 7. Iris image after edge extraction

3.3 Edge Extraction

We extract edges of iris image using canny operator [7]. The edge image will be used for feature templates. The result of edge extraction is given in Figure 7.

3.4 The Extraction of Feature Template

We choose some parts of edge image as feature template. From Iridology chart which is shown in figure 8, we can figure out that 12 sub-areas of iris correspond with different apparatus and systems. The disease reflected by Radii Solaris is different along with different sub-area. So we divide edge image into 12 sub-areas clockwise, each segment correspond to feature template of sub-area.

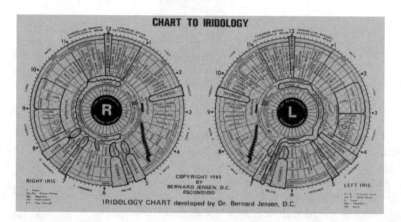

Fig. 8. Jensen's chart to Iridology

3.5 K-Means Clustering

In this paper, K-Means clustering method is used for classing above feature template. K-Means is a clustering arithmetic based on compartmentalizes [8]. It makes target function maximum according to incomplete search in the self-contained space.

In this paper, class number k equals to 2, in other words template with Radii Solaris and template without Radii Solaris. All feature templates make up of data objects. We take clustering centers as primitive definition. We take templates belong to different classes as standard templates.

3.6 Experimental Results

In this paper, we pick out 50 iris images from our own database. According to experiment, we get Radii Solaris primitive definition which is given by Table 1. In this table, angle stands for angle which is made by Radii Solaris' long axis and tangent of inner limbus.

Table 1. Radii Solaris Primitive Definition

Length (pixel)	Width (pixel)	Curvature	Angle
84	3	0.42	1.43

4 Radii Solaris Extraction Method Based on Template Matching and Primitive

In brief, Radii Solaris extraction method is as follows. First of all, we calculate rough positions of Radii Solaris using template matching. And then, we search continuous lines at above positions and label lines which satisfy primitive definition as Radii Solaris.

4.1 Radii Solaris Extraction Method Based on Template Matching

In this section, we will detect Radii Solaris using template matching roughly. We adopt weighted template to extract Radii Solaris from iris image. Its essential ideas are as follows.

We take sum of eight points which are adjacent to one point as weighed value [9].

$$W(i,j) = \sum_{k=i-1,l=j-1}^{k=i+1,l=j+1} f(k,l) \tag{4}$$

Where $W(i,j)$ denotes weighted value at point (i,j); H expresses height of template; W expresses width of template.

Because the weights of Radii Solaris' edge points are bigger than other points, so the key role of Radii Solaris' edges can be expressed better in the process of template matching.

4.2 Radii Solaris Extraction Method Based on Primitive

In this section, we will detect Radii Solaris accurately using primitive definition. The main idea of Radii Solaris extraction is labeling lines which satisfy primitive definition as Radii Solaris.

The algorithm of Radii Solaris extraction method based on primitive describes as follows:

Input: iris image and primitive definition
Output: Radii Solaris recognition results

- a) *Filtering and Binarization Process*
- b) *Thinning grayscale edge*
- c) *Detect continuous lines from edge image*
- d) *Label lines which satisfy primitive definition as Radii Solaris.*

4.3 Experimental Results and Analysis

In Figure 9, red lines show all lacunae. From Figure 9(b), we can see that Radii Solaris are detected accurately from lacunae according to primitive definition. But result has a shortage: some short or anomalous Radii Solaris will be wiped off after using primitive definition.

There are two reasons why the shortage happens: (1) When we detect lacunae, some lacunae rupture because of noises. Take lacunae of sub-area 7 for an example, we can see that the detected lines are shorter than actual lacunae. Because of lacunae rupture, their lengths don't satisfy primitive definition, so they are wiped off. (2) If Radii Solaris is present in eyelid or light spot, our program will be hard to detect it.

We pick out 50 iris images from our own database to calculate recognition ratio. According to experiment, the recognition ratio is about 73.2%.

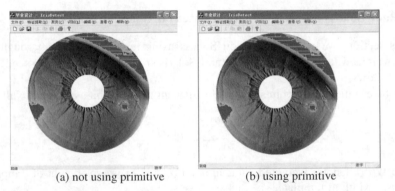

(a) not using primitive   (b) using primitive

Fig. 9. Recognition results comparison for the effect of primitives

5   Conclusion

In this paper, we present a novel Radii Solaris extraction method based on primitive definition. First of all, we define Radii Solaris primitive and obtain standard template from training iris samples. And then, we extract Radii Solaris from iris image according to standard template and primitive.

We select lots of iris images from iris image database and detect Radii Solaris of these iris images. Experimental results show that our method can detect Radii Solaris accurately. We can divide Radii Solaris from other lines precisely using primitive definition. But the recognition rate of Radii Solaris is low for some iris images of low quality, and the recognition result is dissatisfactory for some Radii Solaris of inconspicuous feature. For the sake of improving detection precision of Radii Solaris, we should adopt more effective edge detection operator.

References

1. Navratil, F.: Iridology - For your eyes only. C.V.Mosby, 81–107 (2001)
2. Smolnik, J.: Iridology - The windows to the soul, mind and body. Iridology, 5–6 (2000)
3. Csekey, A.: Iridology: Case Study. Iridology, 1–2 (2000)

4. Daugman, J.: Recognizing people by their iris patterns. Information Security Technical Report, pp. 33–39 (1998)
5. Ma, L., Tan, T., Wang, Y., Zhang, D.: Personal identification based on iris texture analysis. IEEE Transactions on Pattern Analysis and Machine Intelligence, 1519–1533 (2003)
6. Foggia, P., Sansone, C., Tortorella, F., Vento, M.: Combining statistical and structural approaches for handwritten character description. Image and Vision Computing, 701–711
7. Wang, X., Jin, J.: An edge detection algorithm based on improved Canny operator. Intelligent Systems Design and Applications, 623–628 (2007)
8. Ding, C., He, X.: K-means Clustering via Principal Component Analysis. In: International Confess of Machine Learning (ICML), pp. 225–232 (2004)
9. Wang, C.: Edge Detection Using Template Matching. Duck University (1985)

A DIAMOND Method for Classifying Biological Data

Han-Lin Li[1], Yao-Huei Huang[1], and Ming-Hsien Chen[2]

[1] Institute of Information Management, National Chiao Tung University, Hsinchu, Taiwan
{hlli0410,yaohuei.huang}@gmail.com
[2] Department of Information Management, Tatung University, Taipei, Taiwan
mhchen@ttu.edu.tw

Abstract. This study proposes an effective method called DIAMOND to classify biological and medical data. Given a set of objects with some classes, DIAMOND separates the objects into different cubes, where each cube is assigned to a class. Via the union of these cubes, we utilize mixed integer programs to induce classification rules with better rates of accuracy, support and compactness. Two practical data sets, one of HSV patient results and the other of Iris flower, are tested to illustrate the advantages of DIAMOND over some current methods.

Keywords: DIAMOND; Integer Program; Classification rules.

1 Introduction

Given a biological data set with several objects, where each object owns some attributes and belongs to a specific class, the rules of classifying these objects are the combination of attributes which can describe well the features of a specific class. Three criteria of evaluating the quality of a rule, as described in Li and Chen [1], are listed (i)Accuracy rate: The rule fitting a class had better not cover the objects of other classes. (ii)Support rate: The rule fitting a class had better being supported by large number of objects of the same class. (iii)Compact rate: The rule is specified by less number of attributes.

Decision tree methods, Support vector hyper-plane methods and Integer programming hyper-plane methods are three of well-known classification methods; which are reviewed as follows:

Decision tree methods. Decision tree methods ([2], [3] and [4]) are heuristics in nature and are closer to the techniques of statistical inference approaches. These methods split recursively the data along a single variable into hyper-rectangular regions. Backward propagation is preformed to prevent over-fitting of the data sets. Attributes leading to substantial entropy reduction are included as condition attributes to partition the data. Its main shortcoming is its fundamentally greedy approach, which may only find a feasible solution, instead of finding an optimal solution with respect to the maximal rates of accuracy and coverage.

Support vector hyper-plane methods. Support vector hyper-plane methods ([5], [6] and [7]) separate points of different classes by a single hyper-plane, where the optimal

D. Zhang and M. Sonka (Eds.): ICMB 2010, LNCS 6165, pp. 104–114, 2010.

separating hyper-plane is modeled as a convex quadratic programming problem. Since the number of variables is required to equal to the number of training data, the training becomes quite slow for a large number of training data.

Integer program hyper-plane methods. Recently, Bertsimas and Shioda [8] use a mixed-integer optimization method to solve the classical statistical problems of classification and regression. Their method separates data points into different regions by various hyper-planes. Each region is assigned a class in the classification. By solving the mixed-integer program, the rules with high rate of accuracy can be induced. However, this approach may generate too many polyhedral regions which reduce the rate of compact in the induced rules. By utilizing integer programming techniques, Li and Chen [1] develop a multiple criteria method to induce classification rules. Their method clusters data points into polyhedral regions by solving integer programs. Thus to induce the rules with high rate of accuracy, is highly supported, and is highly compact. However, since their approach is based on the concept of the separating hyper-planes, it may also need to generate large number of hyper-planes in order to generate classification rules, especially for the data with large number of attributes.

This study therefore proposes a novel method called DIAMOND to improve current classification techniques. For a data set with objects of various classes, DIAMOND method groups these objects into some sets of hyper-cubes. Each object is assigned to a cube belongs to a class via solving mixed 0-1 programs iteratively. Thus to ensure the most of objects are assigned to a proper set of cubes, restricted the condition that the number of total cubes are minimized.

Comparing with Decision tree methods and two hyper-plane methods mentioned above, the features of DIAMOND method are listed as follows: (i) Decision tree methods are heuristic approach which can only induce feasible rules. DIAMON method is an optimization approach which can find the optimal rules with high rates of accuracy, support and compact. In addition, Decision tree methods split the data along a single variable into hyper-rectangular regions, which may generate large number of branches. While DIAMOND method clusters the data based on multiple variables into cubes, where the number of cubes can be pre-specified. (ii) Hyper-plane methods need to use numerous hyper-planes to separate objects of different classes. When the objects of various classes are scattered into indistinct groups, Hyper-plane methods are computationally ineffective for the cases with large number of attributes. DIAMOND method is more powerful to classifying objects which are clustered into various groups, or to classify objects with large number of attributes.

In order to examine the efficiency of DIAMON method, two practical data sets, one of HSV patients and the other of Iris flowers, are tested. The results illustrate clearly the advantage of DIAMOND method over some of current Decision-tree methods and Separating hyper-plane methods.

This study is organized as follows. Section 2 uses an example to illustrate the basic idea of DIAMOND method. Section 3 is the formulation of optimization program for the proposed model. Numerical experiments are reported in Section 4.

2   Basic Ideas of DIAMOND Method

This section uses an example to express the basic idea of DIAMOND method.

Table 1. Data sets of Example 1

Object	a_1	a_2	c	Symbol	Object	a_1	a_2	c	Symbol
x_1	6	8	1	○	x_9	22	15	2	△
x_2	12	20	1	○	x_{10}	30	11	2	△
x_3	13	8	1	○	x_{11}	33.5	7.5	2	△
x_4	18	12.5	1	○	x_{12}	24.5	3.5	3	×
x_5	21	19	1	○	x_{13}	26.5	8	3	×
x_6	23.5	14.5	1	○	x_{14}	23.5	7.5	3	×
x_7	17.5	17.5	2	△	x_{15}	6	30	3	×
x_8	22	17	2	△					

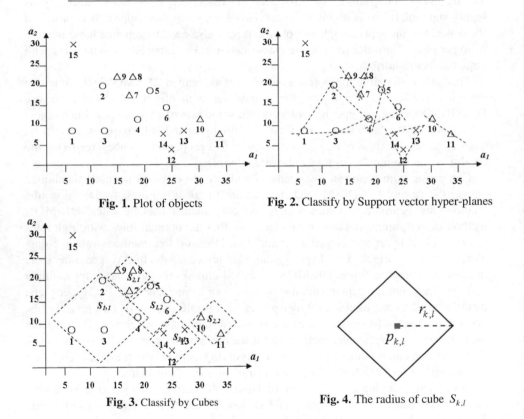

Fig. 1. Plot of objects

Fig. 2. Classify by Support vector hyper-planes

Fig. 3. Classify by Cubes

Fig. 4. The radius of cube $S_{k,l}$

Example 1. Consider a data set T in Table 1 which has 15 objects $(x_1,...,x_{15})$, two attributes (a_1,a_2), and an index of classes c. The data set T is expressed as $T = \{x_i(a_{i,1},a_{i,2};c_i)\,|\,i=1,...,15\}$. The domain values of c is $\{1,2,3\}$. Since there are only two attributes, we can plot these 15 objects on a plane (see Fig. 1). If we use the hyperplane methods to discriminate these three classes then there may need many hyper-planes (see Fig. 2). It becomes more complicated to combine these hyper-planes to form the regions for the objects of a specific class.

Instead of using "hyper-planes", DIAMOND using "cubes" (shaped like diamonds) to classifying these objects, where a rule is expressed by the union of cubes belong to a class. We intend to use least number of cubes to classify these objects, subjected to the constraint that a cube covers the objects of a single class as many as possible. Examining Fig. 1 to know that a good way to classify these 15 objects is to cluster them by 5 cubes (see Fig. 3), where Cube $S_{1,1}$ contains (x_1, x_2, x_3, x_4) ; Cube $S_{1,2}$ contains (x_4, x_5, x_6) ; Cube $S_{2,1}$ contains (x_7, x_8, x_9) ; Cube $S_{3,1}$ contains (x_{12}, x_{13}, x_{14}) ; Cube $S_{2,2}$ contains (x_{10}, x_{11}) ; x_{15} is not covered by any cube. Here we denote $S_{k,l}$, $p_{k,l}$ and $r_{k,l}$ respectively as cube, centroid, and radius of the l'th cube for class k . The radius of a cube is the distance between its centroid point and a corner point (Fig. 4). The attribute values of $p_{k,l}$ are denoted as $(b_{k,l,1}, b_{k,l,2})$. The situation that an object $x_i(a_{i,1}, a_{i,2}; c_i)$ is "covered" by a cube $S_{k,l}$ is expressed as

$$\left| a_{i,1} - b_{k,l,1} \right| + \left| a_{i,2} - b_{k,l,2} \right| \le r_{k,l} \tag{1}$$

The situation that an object x_i is "not covered" by a cube $S_{k,l}$ is expressed as

$$\left| a_{i,1} - b_{k,l,1} \right| + \left| a_{i,2} - b_{k,l,2} \right| > r_{k,l} \tag{2}$$

Here we ask that each cube should cover at least two objects. Since object x_{15} is not covered by any cubes, it is regarded as an outlier. A rule for class 1 can then be expressed as

"If an object x_i is covered by a cube $S_{1,1}$ or $S_{1,2}$ then x_i belongs to class 1", rewritten in a compact way as

R_1 : if x_i is covered by $S_{1,1} \cup S_{1,2}$ then $c_i = 1$.

Mathematically, R_1 can be rewritten as

R_1 : if $\left| a_{i,1} - b_{1,1,1} \right| + \left| a_{i,2} - b_{1,1,2} \right| \le r_{1,1}$ or $\left| a_{i,1} - b_{1,2,1} \right| + \left| a_{i,2} - b_{1,2,2} \right| \le r_{1,2}$ then x_i is covered by $S_{1,1} \cup S_{1,2}$.

Fig. 1 shows that, the objects $x_1, ..., x_6$ are covered by R_1 . Similarly, we can express Rule 2 for classifying class 2 and Rule 3 for classifying class 3 as follows:

R_2 : if x_i is covered by $S_{2,1} \cup S_{2,2}$, then $c_i = 2$.

R_3 : if x_i is covered by $S_{3,1}$, then $c_i = 3$.

Notably that cubes $S_{1,1}$ and $S_{1,2}$ overlap with each other.

By referring to Li and Chen [1], the rates of accuracy, support, and compact in R_1 , R_2 and R_3 can be specified below. Which are used to measures the quality of a rule.

$$AR(R_k) = \frac{\text{number of objects covered correctly by } R_k}{\text{number of objects covered by } R_k} \tag{3}$$

For instance, $AR(R_1) = \dfrac{6}{6} = 1$. An object x_i is called covered correctly by R_k, if $c_i = k$.

$$SR(R_k) = \frac{\text{number of objects covered correctly by } R_k}{\text{number of objects of the class } k} \tag{4}$$

For instance, $SR(R_1) = SR(R_2) = 1$ but $SR(R_3) = 0.67$.

$$CR = \frac{\text{number of classes}}{\text{total number of cubes}} \tag{5}$$

Take above example for instance, where there are 3 classes but 5 cubes therefore $CR(R_1, R_2, R_3) = \dfrac{3}{5}$.

3   Proposed DIAMOND Method and Algorithm

Consider a data set T with n objects, where each object has m attributes $\{a_1, \ldots, a_m\}$ and belongs to one of g classes, expressed as $T = \{x_i(a_{i,1}, \ldots, a_{i,m}; c_i) \mid i = 1, \ldots, n\}$ where $c_i \in \{1, \ldots, g\}$. Denote the number of objects belong to k'th class as $n(k)$, $1 \le k \le g$.

Notation 1. An object x_i in T is specified as $x_i = (a_{i,1}, a_{i,2}, \ldots, a_{i,m}; c_i)$, where $a_{i,j}$ is the value of j'th attribute for i'th object, and c_i is the class where the i'th object belong to, $c_i \in \{1, \ldots, g\}$.

Notation 2. A rule R_k is used to classify the objects of k'th class which is specifies by the union of a set of q_k cubes, expressed as $R_k = S_{k,1} \cup S_{k,2} \cup \ldots \cup S_{k,q_k}$.

Notation 3. A l'th cube for k'th class, denoted as $S_{k,l}$, is specified by its centroid and radius, expressed as $S_{k,l} = (b_{k,l,1}, \ldots, b_{k,l,m}; r_{k,l})$, where $b_{k,l,j}$ is its centroid's value at j'th dimension (i.e., attribute), and $r_{k,l}$ is its radius.

Remark 1. The total number of cubes is $\displaystyle\sum_{k=1}^{g} q_k$.

Remark 2. An object $x_i = (a_{i,1}, \ldots, a_{i,m}; c_i)$ is covered by $S_{k,l} = (b_{k,l,1}, \ldots, b_{k,l,m}; r_{k,l})$ if

$$\sum_{j=1}^{m} \left| a_{i,j} - b_{k,l,j} \right| \le r_{k,l} \tag{6}$$

An object x_i is not covered by a cube $S_{k,l} = (b_{k,l,1}, \ldots, b_{k,l,m}; r_{k,l})$ if and only if

$$\sum_{j=1}^{m} \left| a_{i,j} - b_{k,l,j} \right| > r_{k,l} \tag{7}$$

Remark 3. Given a cube $S_{k,l}$ and two objects; $x_i(a_{i,1},...,a_{i,m};c_i)$ where $c_i = k$ and $x_{i'}(a_{i',1},...,a_{i',m};c_{i'})$ where $c_{i'} \neq k$, denote $u_{k,l,i}$ and $v_{k,l,i'}$ respectively as two binary variables specified as (i) $u_{k,l,i} = 1$ if object x_i is covered by $S_{k,l}$; and $u_{k,l,i} = 0$ otherwise. (ii) $v_{k,l,i'} = 1$ if object $x_{i'}$ is covered by $S_{k,l}$; and $v_{k,l,i'} = 0$ otherwise.

That means if an object x_i is covered correctly by a cube $S_{k,l}$ of the same class then $u_{k,l,i} = 1$. But if the object $x_{i'}$ is covered by a cube $S_{k,l}$ which is not the same class then $v_{k,l,i'} = 1$.

Definition 1. The accuracy rate of rule R_k denoted as $AR(R_k)$, is specified below:

$$AR(R_k) = \frac{\|R_k\| - \sum_{i'=1}^{n(k')}\sum_{l=1}^{q_k} v_{k,l,i'}}{\|R_k\|} \quad (8)$$

where $\|R_k\|$ means the number of total objects covered by R_k.

Definition 2. The support rate of a rule R_k, denoted as $SR(R_k)$, is specified below:

$$SR(R_k) = \frac{\sum_{i=1}^{n(k)}\sum_{l=1}^{q_k} u_{k,l,i}}{n(k)} \quad (9)$$

Definition 3. The compact rate of a set of rules $R_1,...,R_g$ is expressed below:

$$CR(R_1,...,R_g) = g / \sum_{k=1}^{g} q_k \quad (10)$$

DIAMOND model generates a set of diamonds (cubes) to induce a rule which maximizes the support rate subjected to the constraint that the accuracy rate is higher than a threshold value. We also design an iterative algorithm to keep the rate of compact as higher as possible. The proposed model is formulated below:

Model 1 (Nonlinear DIAMOND model)

$$\text{Maximize} \sum_{l=1}^{q_k}\sum_{i=1}^{n(k)} u_{k,l,i} \quad (11)$$

For a cube $S_{k,l}$, following constraints need to be satisfied:

$$\sum_{j=1}^{m} |a_{i,j} - b_{k,l,j}| \leq r_{k,l} + M\left(1 - u_{k,l,i}\right) \ \forall x_i \text{ where } c_i = k, \quad (12)$$

$$\sum_{j=1}^{m} |a_{i',j} - b_{k,l,j}| > r_{k,l} - M v_{k,l,i'} \ \forall x_{i'} \text{ where } c_{i'} \neq k, \quad (13)$$

$$AR(R_k) = \frac{\|R_k\| - \sum_{i'=1}^{n(k')} \sum_{l=1}^{q_k} v_{k,l,i'}}{\|R_k\|} \geq \text{Threshold value}, \tag{14}$$

where $M = \max\{a_{i,j} \ \forall i = 1,\dots,n \text{ and } j = 1,\dots,m\}$, $b_{k,l,j} \geq 0$, $r_{k,l} \geq 0$,

$$u_{k,l,i}, v_{k,l,i'} \in \{0,1\}; a_{i,j} \text{ and } a_{i',j} \text{ are constants.} \tag{15}$$

The objective function (11) is to maximize the support rate. Constraints (12) and (13) come from (6) and (7). Constraint (14) is to ensure that the accuracy rate should higher than a threshold value. Constraint (12) implies that if a cube $S_{k,l}$ covers an object x_i of same class then $u_{k,l,i} = 1$; and otherwise $u_{k,l,i} = 0$. Constraint (13) implies that if a cube $S_{k,l}$ does not cover an object $x_{i'}$ of another class then $v_{k,l,i'} = 0$; and otherwise $v_{k,l,i'} = 1$.

Inequalities (12) and (13) are nonlinear, which need to be linearized. The related techniques in linearizing Model 1 are expressed by three propositions listed in **Appendix**. Model 1 can then be reformulated as a linear mixed-binary program below:

Model 2 (Linearized DIAMOND model)
Maximize (11)
subject to (14) and (15),

$$\sum_{j=1}^{m} a_{i,j} - b_{k,l,j} + 2e_{k,l,i,j} \leq r_{k,l} + M\left(1 - u_{k,l,i}\right),$$

$$a_{i,j} - b_{k,l,j} + e_{i,k,l,j} \geq 0,$$

$$\sum_{j=1}^{m} \left(a_{i',j} - b_{k,l,j} - 2a_{i',j}\lambda_{k,l,i',j} + 2z_{k,l,i',j}\right) > r_{k,l} - Mv_{k,l,i'},$$

$$a_{i',j} - b_{k,l,j} - 2a_{i',j}\lambda_{k,l,i',j} + 2z_{k,l,i',j} \geq 0,$$

$$\overline{b}_j\left(\lambda_{k,l,i',j} - 1\right) + b_{k,l,j} \leq z_{k,l,i',j} \leq b_{k,l,j} + \overline{b}_j\left(1 - \lambda_{k,l,i',j}\right),$$

$$0 \leq z_{k,l,i',j} \leq \overline{b}_j\lambda_{k,l,i',j},$$

$$\lambda_{k,l,i,j} \leq \lambda_{k,l,i',j} \text{ for all } i \text{ and } i' \text{ where } a_{i,j} > a_{i',j}, \text{ and } M \text{ is specified before.}$$

The solution algorithm is listed below, which intends to find the rules where the rate of compact is as high as possible.

Step 0: Initialization: Let $k = 1$ and $l = 1$, specify the threshold value in (14).
Step 1: Solve Model 2 to obtain the l'th cube of class k. Remove the objects covered by $S_{k,l}$ from the data set.
Step 2: Let $l = l + 1$, resolve Model 2 until all objects in class k are assigned to the cubes of same class.
Step 3: Let $k = k + 1$ reiterates Step 1 until all objects are assigned.

4 Numerical Examples

Two data sets are tested in our experiments, one is the Iris flower introduced by Sir Ronald Aylmer Fisher (1936) [10] and the other is HSV (highly selective vagotomy) patients data set of F. Raszeja Mem. Hospital in Poland [11].

Table 2. Centroid points for Iris data set by DIAMOND method

Cube #	$b_{k,l,1}$	$b_{k,l,2}$	$b_{k,l,3}$	$b_{k,l,4}$	$r_{k,l}$
$S_{1,1}$	5.1	3.20	1.85	0.5	2.45
$S_{2,1}$	6.7	2.60	3.5	1.2	2.50
$S_{2,2}$	5.9	3.15	4	1.3	1.55
$S_{3,1}$	6.2	2.90	6.6	2.4	2.70
$S_{3,2}$	5.3	2.45	4.9	1.6	1.05

Iris flower data set

The Iris flower data set [10] contains 150 objects, each object described by four attributes and classified by three classes. By utilizing DIAMOND method, the induced classification rules are reported in Table 2. Table 2 lists 5 found cubes with centroid points and radiuses. Rule R_1 in Table 2 contains one cube $S_{1,1}$ which implies that "if $|a_{i,1} - 5.1| + |a_{i,2} - 3.2| + |a_{i,3} - 1.85| + |a_{i,4} - 0.5| \le 2.45$ then object x_i belongs to class 1". The accuracy rate of R_1 is 1 in DIAMOND method, which means none of object in class 2 or class 3 is covered by $S_{1,1}$. The support rate of R_1 for DIAMOND method is 1, which means all objects in class 1 are covered by $S_{1,1}$. The compact rate of rules R_1, R_2 and R_3 is computed as $CR(R_1, R_2, R_3) = 0.6$. Both Decision tree method [3] (see Fig. 5) and Polynomial hyper-plane support vector method [13] are used to deduce classification rules for the same data. The results are listed in Table 3.

Table 3. Comparing result for Iris data set (R_1, R_2, R_3)

Items	DIAMOND	Decision tree	Hyper-plane support vector
$AR(R_1, R_2, R_3)$	(1, 1, 1)	(1, 0.98, 0.98)	(1, 0.98, 0.96)
$SR(R_1, R_2, R_3)$	(1, 0.98, 0.98)	(1, 0.98, 0.98)	(1, 0.96, 0.98)
CR	0.6	0.5	0.1875

Table 3 lists proposed method with the other two methods and associates AR, SR and CR values. In addition, the CR value of these rules found by DIAMOND is higher than that found by the other two methods. The details can be searched at http://javen-classification. blogspot.com/.

HSV data set

HSV data set contains 122 patients ([1], [11], [12], [13]). The patients are classified into 4 classes, and each patient has 11 pre-operating attributes. In our experiment, we randomly take the 92 (75% for each class) patients as the training set and the rest of 30 patients as testing set. To maximize the support rate with respect to the constraint

Table 4. Centroid points for HSV data by DIAMOND method

$S_{k,l}$	$b_{k,l,1}$	$b_{k,l,2}$	$b_{k,l,3}$	$b_{k,l,4}$	$b_{k,l,5}$	$b_{k,l,6}$	$b_{k,l,7}$	$b_{k,l,8}$	$b_{k,l,9}$	$b_{k,l,10}$	$b_{k,l,11}$	$r_{k,l}$
$S_{1,1}$	0	63.0	14.0	2	14.7	86.5	180.0	13.8	23.3	627.0	61.8	565.5
$\vdots$							$\vdots$					$\vdots$
$S_{2,6}$	1	36.0	5.0	0	7.1	102.0	66.0	15.1	13.6	263.0	86.0	118.9
$S_{3,3}$	0	21.0	6.0	2	2.0	182.0	48.6	7.5	12.1	387.0	26.5	129.8
$\vdots$							$\vdots$					$\vdots$
$S_{4,5}$	1	71.0	8.0	1	4.4	50.0	80.2	3.8	5.7	223.0	13.2	126.0

*The details see http://javen-classification.blogspot.com/

that $AR \geq 0.9$ and to minimize the number of cubes, DIAMOND method generate 24 cubes iteratively. The centroids and radiuses of these cubes are listed in Table 4 and the results are listed in Table 5.

Table 5. Comparing result for HSV data set (R_1, R_2, R_3, R_4)

Items	DIAMOND	Decision tree	Hyper-plane support vector
$AR(R_1, R_2, R_3, R_4)$	(1, 1, 1, 0.91)	(0.93, 0.81, 0.7, 0.71)	(0.9, 1, 1, 0.9)
$SR(R_1, R_2, R_3, R_4)$	(0.96, 0.83, 0.78, 0.79)	(0.93, 0.72, 0.78, 0.71)	(0.9, 0.72, 0.67, 0.69)
CR	0.17	0.17	0.09

Decision tree method is applied to deduce rules for the same data set, which uses 24 branches as shown in Fig. 6. Similarly, Polynomial hyper-plane method [13] is applied to find rules for HSV data set, which has 45 hyper-planes. Table 5 also shows that DIAMOND method can find rules with higher (or equal) rates of *AR*, *SR* and *CR* comparing with other two methods. The details see http://javen-classification. blogspot.com/.

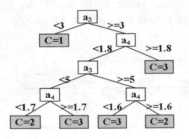

Fig. 5. Decision tree for Iris data set [10]

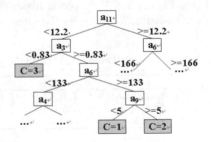

Fig. 6. Decision tree for HSV data set [11]

5 Conclusion

A method called DIAMOND is developed to classify objects with various classes. By solving a mixed 0-1 linear program, DIAMOND generates a set of cubes to cluster objects of the same class, where the accuracy rate (*AR*) is higher than a threshold value and the associated support rate is maximized. In addition, the compact rate (*SR*) for all rules found by DIAMOND is kept as high as possible via an iterative solution algorithm. Two commonly used data sets (Iris and HSV) are tested to illustrate that, comparing with a Decision tree method and a Polynomial hyper-plane support vector method. DIAMOND method can induce rules with higher *AR*, *SR* and *CR* values. One of possible research directions is to utilize an effective logarithmic method [14] to accelerate the speed of finding the classification rules thus to treat large data sets.

References

1. Li, H.L., Chen, M.H.: Induction of Multiple Criteria Optimal Classification Rules for Biological and Medical Data. Computers in Biology and Medicine 38(1), 42–52 (2008)
2. Breiman, L., Friedman, J., Olshen, R., Stone, C.: Classification and Regression Trees. Wadsworth International, Belmont (1984)
3. Quinlan, R.: C4.5: Programs for Machine Learning. Morgan Kaufman, San Mateo (1993)
4. Kim, H., Loh, W.Y.: CRUISE User Manual. Technical Report 989. Journal of the American Statistical Association 96, 589–604 (2001)
5. Vapnik, V.N.: The Nature of Statistical Learning Thoery. Springer, New York (1995)
6. Rifkin, R.: Everything old is new again: A Fresh Look at Historical Approaches in Machine Learning. Ph.D. thesis, Massachusetts Institute of Technology, Cambridge, MA (2002)
7. Katagiri, S., Abe, S.: Incremental Training of Support Vector Machines Using Hyperspheres. Pattern Recognition Letters 27, 1495–1504 (2006)
8. Bertsimas, D., Shioda, R.: Classification and Regression via Integer Optimization. Operations Research 55(2), 252–271 (2007)
9. Li, H.L.: An Efficient Method for Solving Linear Goal Programming Problems. Journal of Optimization Theory and Applications 9(2), 467–471 (1996)
10. Fisher, R.A.: The Use of Multiple Measurements in Taxonomic Problems. Annals of Eugenics 7, 179–188 (1936)
11. Slowinski, K.: Rough classification of HSV patients. In: Slowinski, R. (ed.) Intelligent Decision Support-Handbook of Applications and Advances of the Rough Sets Theory, pp. 77–94. Kluwer, Dordrecht (1992)
12. Dunn, D.C., Thomas, W.E.G., Hunter, J.O.: An Evaluation of Highly Selective Vagotomy in the Treatment of Chronic Ulcer. Surg. Gynecol. Obstet 150, 145–151 (1980)
13. Chang, C.C., Lin, C.J.: LIBSVM: a library for support vector machines, Software available at, http://www.csie.ntu.edu.tw/~cjlin/libsvm/index.html
14. Li, H.L., Lu, H.C.: Global Optimization for Generalized Geometric Programs with Mixed Free-sign Variables. Operations Research 57(3), 701–713 (2009)

Appendix

Proposition 1. Inequality (12) is linearized as follows referring to Li [9]:

$$\sum_{j=1}^{m}\left(a_{i,j} - b_{k,l,j} + 2e_{k,l,i,j}\right) \leq r_{k,l} + M\left(1 - u_{k,l,i}\right), \tag{16}$$

$$a_{i,j} - b_{k,l,j} + e_{k,l,i,j} \geq 0, \tag{17}$$

where $e_{k,l,i,j} \geq 0$.

Proposition 2. Inequality (13) can be linearized as follows:

$$\sum_{j=1}^{m}\left|a_{i',j}-b_{k,l,j}\right| = \sum_{j=1}^{m}\left(1-2\lambda_{k,l,i',j}\right)\left(a_{i',j}-b_{k,l,j}\right) \tag{18}$$

$$= \sum_{j=1}^{m}\left(a_{i',j}-b_{k,l,j}-2a_{i',j}\lambda_{k,l,i',j}+2z_{k,l,i',j}\right) > r_{k,l}-Mv_{k,l,i'}$$

where

$$a_{i',j}-b_{k,l,j}-2a_{i',j}\lambda_{k,l,i',j}+2z_{k,l,i',j} \geq 0 \tag{19}$$

$$\bar{b}_{j}\left(\lambda_{k,l,i',j}-1\right)+b_{k,l,j} \leq z_{k,l,i',j} \leq b_{k,l,j}+\bar{b}_{j}\left(1-\lambda_{k,l,i',j}\right), \tag{20}$$

$$0 \leq z_{k,l,i',j} \leq \bar{b}_{j}\lambda_{k,l,i',j}, \tag{21}$$

$\bar{b}_{j}$ is constant, $\bar{b}_{j} = \max\left\{a_{i',j}; \forall i' \notin i\right\}$ and $\lambda_{k,l,i',j} \in \{0,1\}$.

Proposition 3. By referring to (18), consider the two inequalities

(i) $\left|a_{i,j}-b_{k,l,j}\right| = a_{i,j}-b_{k,l,j}-2a_{i,j}\lambda_{k,l,i,j}+2z_{k,l,i,j} \geq$ constant,

(ii) $\left|a_{i',j}-b_{k,l,j}\right| = a_{i',j}-b_{k,l,j}-2a_{i',j}\lambda_{k,l,i',j}+2z_{k,l,i',j} \geq$ constant',

where $\lambda_{k,l,i,j}, \lambda_{k,l,i',j} \in \{0,1\}$ and $i' \neq i$. If $a_{i,j} > a_{i',j}$ then $\lambda_{k,l,i,j} \leq \lambda_{k,l,i',j}$.

Proposition 3 is useful to tight the relationship between $\lambda_{k,l,i,j}$ and $\lambda_{k,l,i',j}$. If $\lambda_{k,l,i,j} = 1$ then $\lambda_{k,l,i',j} = 1$, and if $\lambda_{k,l,i,j} = 0$ then $\lambda_{k,l,i',j} = 0$.

The proofs of these propositions are available at http://javen-classification. blogspot.com/

Tongue Image Texture Segmentation Based on Gabor Filter Plus Normalized Cut

Jianfeng Li[1], Jinhuan Shi[2], Hongzhi Zhang[1], Yanlai Li[1],
Naimin Li[3], and Changming Liu[1]

[1] School of Computer Science and Technology, Harbin Institute of Technology,
Harbin, China
[2] The General Hospital of the Second Artillery of P.L.A, Beijing, China
[3] The No.211 Hospital of P.L.A., Harbin, China
lijeff@hit.edu.cn

Abstract. Texture information of tongue image is one of the most important pathological features utilized in practical Tongue Diagnosis because it can reveal the severeness and change tendency of the illness. A texture segmentation method based on Gabor filter plus Normalized Cut is proposed in this paper. This method synthesizes the information of location, color and texture feature to be the weight for Normalized Cut, thus can make satisfactroy segmentation according to texture of tongue image. The experiments show that the overall rate of correctness for this method exceeds 81%.

Keywords: Tongue diagnosis; Texture Segmentation; Gabor Filter; Normalized Cut.

1 Introduction

Tongue Diagnosis is well-recognized as one of the most important diagnostic methods of the Traditional Chinese Medicine which has a history of more than 5,000 years. It has been practically applied to make diagnosis of disease by observing abnormal changes of the tongue by physicians allover the world, especially in China and the East Asia [1]. Nowadays, the development of computer science and technology has made it possible to design an automated tongue diagnosis system based on technologies such as image processing and pattern recognition [2]. To achieve this goal, successful extraction of various tongue pathological features are indispensable.

Textural feature is an important feature of tongue image through the process of tongue diagnosis, and is also an important part of pathological feature on the surface of the tongue. Successful extraction of texture information is highly valued in computerized tongue diagnosis system because textural feature can reveal the severeness and change tendency of the illness by observing the patient's tongue. Due to the complicity of the tongue image texture itself and the limit of texture-analyzing method, little research has been carried out in the field of tongue texture analysis.

The department of Computer Science of Hong Kong Baptist University has used Gabor Wavelet Opponent Color Features (GWOCF) for tongue texture diagnosis. They divided the whole tongue image into small Tongue Texture Blocks of size

D. Zhang and M. Sonka (Eds.): ICMB 2010, LNCS 6165, pp. 115–125, 2010.
© Springer-Verlag Berlin Heidelberg 2010

36×36, and proposed Gabor-filter-based method combined with linear discriminant function to analysis tongue texture. The Institute of Image Processing and Pattern Recognition of Shanghai Jiaotong University has used 2D Gabor wavelet energy distribution function to make the segmentation of tongue and analyze tongue's crackle. The Department of Computer Science and Engineering of Harbin Institute of Technology has used a set of 2D Gabor filter banks to extract and represent textural features, then apply the Linear Discriminant Analysis (LDA) to identify the data [3]-[5]. Up to now, all researches about tongue texture analysis were conducted on the basis of tongue image block, whose analysis result is vitally dependent on the choice of location of the blocks. And their tongue image segmentations were only limited in the area of color threshold and clustering, did not take the effect of different kinds tongue coating and tongue substance into consideration. These drawbacks have cumbered the correct analysis of tongue texture because of the specialty of tongue texture comparing to other general texture features.

In this paper, a series of texture analysis methods based on Normalized Cut and Gabor filters are proposed aiming to making segmentation of tongue image according to its different texture features. This method can be helpful to analysis various thickness of tongue coating and tongue substance. Experiments have been carried out with calibrated parameters, and it shows satisfactory result can be achieved by using this method to segment texture of tongue image.

The organization of the paper is given as follows: Section 2 describes preprocessing methods including Gabor filters and boundary enlargement. Section 3 presents our extraction method in detail. Section 4 is the experiments and discussions. Conclusion and future work comprise Section 5.

2   Preprocessing and Gabor Filter

2.1   Gabor Filter and Its Application in Preprocessing

The Gabor Transform is a strong tool in the field of image processing to analyze texture, with its impulse response defined by a harmonic function multiplied by a Gaussian function. A 2D Gabor filter is a Gaussian kernel function modulated by a sinusoidal plane wave in the spatial domain. The kernel function of a Gabor filter is defined as:

$$\Psi(X) = \frac{k_v^2}{\sigma^2} \exp(-\frac{k_v^2 X^2}{2\sigma^2})\{\exp(iKX) - \exp(-\frac{\sigma^2}{2})\}. \tag{1}$$

$\exp(iKX)$ is an oscillation function whose real part consist of cosine function and imaginary part consist of sine function, and $\exp(-\frac{k_v^2 X^2}{2\sigma^2})$ is a Gauss function. It restricted the scope of oscillation function and make oscillation function available only locally. $\exp(-\sigma^2/2)$ is component of direct current, can keep the filter away from falling under the influence of the amount of direct current, and are also called direct current compensation. This component can help to evade the influence of the absolute value of image gray level, thus make the filter insensitive to illumination intensity.

The kernel function of 2D Gabor filter is a compound function which consist of two components: real component and imaginary component, namely,

$$G_k(x, y) = G_r(x, y) + G_i(x, y).$$ (2)

Real component can be represented by

$$G_r(x, y) = \frac{k_v^2}{\sigma^2} \exp(-\frac{k_v^2(x^2 + y^2)}{2\sigma^2})$$
$$*[\cos(k_v \cos(\varphi_u) \cdot x + k_v \sin(\varphi_u) \cdot y) - \exp(-\frac{\sigma^2}{2})]$$ (3)

And imaginary part can be described by

$$G_i(x, y) = \frac{k_v^2}{\sigma^2} \exp(-\frac{k_v^2(x^2 + y^2)}{2\sigma^2})$$
$$*[\sin(k_v \cos(\varphi_u) \cdot x + k_v \sin(\varphi_u) \cdot y) - \exp(-\frac{\sigma^2}{2})]$$ (4)

Figure 1 is an example of Gabor Filter result of a tongue image.

    (a)   (b)   (c)   (d)

Fig. 1. An example of Gabor Filter result (a) the origin tongue image, (b) The amplitude of filtered image, (c)Real part (d)Imaginary part

From the result we can seen that the real part and imaginary part are more difficult to reveal the texture information of tongue than the amplitude part. Therefore, we choose amplitude part to extract texture information.

2.2 Filling the Enlarged Pixels

It can be easily seen from Fig.1(a) that there exist a abnormal stripe-shaped area whose energy value is much higher than its nearby area in the boundary region. This phenomenon shows the strong ability of Gabor filter in edge detection, but it is a great interference to our texture analysis and must be removed in order to get appropriate texture information of the boundary region. The method to erase this affect is to enlarge the tongue boundary and fill the enlarged pixels, then exert the Gabor filter to the enlarged image and remove the filled pixels from the result image. The ameliorated result can thus be acquired without the interference of the boundary effect.

Usually, the filling methods are composed of filling the enlarged pixels with the mean value of the nearby image block, or filling with the value of the nearest pixel of the pixel to be enlarged. But these methods are not suitable to texture analysis because they make the image too smooth to get the original texture information. In order to get the precise texture information of the boundary area, a new algorithm of boundary enlargement is proposed, and is described as follow:

Input: Original tongue image.

Output: filtered image of the original tongue image without any abnormal stripe-shaped area in the boundary region.

Step 1. Denote the input tongue surface image as I, and calculate its center (x_c, y_c). x_c equals to the average value of vertical axis coordinates of all the pixels that belong to I, and equals to y_c the average value of all the horizontal axis coordinates.

Step 2. Acquire the outermost pixel of the tongue image, define them as boundary line.

Step 3. Exert a morphological dilation operation on the tongue image, the difference pixels of the dilated image and the original image are the enlarged pixels that need to be filled.

Step 4. Assume (x_t, y_t) to represent the target pixel that need to be filled, as shown in Fig.2, draw a line to connect the target pixel and the center pixel, calculate the intersection pixel of this line and the boundary edge of the image. Then we calculate the symmetric point of (x_t, y_t) with respect to the intersection pixel, denoted as f, and use the gray level of (x_t, y_t) to fill the pixel of (x_r, y_r). If the coordinates of (x_r, y_r) do not exactly fall into the central point of the pixel, we use bilinear interpolation value of the gray level of the 4-neighbourhood pixels to fill in this pixel. The thorough process is described below.

For a symmetric pixel (x_r, y_r), assume its coordinates in float-point form to be $(i+u, j+v)$. i,j are the integer part of the float point coordinates and u,v stand for the decimal part of the float point coordinates. The gray level value of this pixel $f(i+u, j+v)$, which equal to (x_r, y_r), can be expressed by the values of the four neighboring pixels of the original image (i, j) , $(i, j+1)$, $(i+1, j)$, $(i+1, j+1)$ by the following equation:

$$f(i+u, j+v) = (1-u)(1-v)f(i, j) + (1-u)vf(i, j+1)$$
$$+u(1-v)f(i+1, j) + uvf(i+1, j+1)$$

(5)

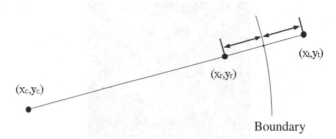

(x_t, y_t)

(x_r, y_r)

(x_c, y_c)

Boundary

Fig. 2. Illustraion of Boundary enlargement algorithm

Step 5. The enlargement scope is set to be no less than half of the Gabor filter window. If Step 4 has not reached the satisfied enlargement scope, the algorithm goes back to morphological dilation operation in Step 3.

Step 6. Process the enlarged image with Gabor filter, and make edge segmentation to the filtered image according to the size of the original tongue image, then we get the filtered image of the original tongue image without any abnormal stripe-shaped area in the boundary region.

An illustration of the effect of tongue image after boundary enlargement is shown in below figure. It can be easily seen that the enlarged region manifests the similar texture of boundary region, which can contribute much in getting correct Gabor filter result.

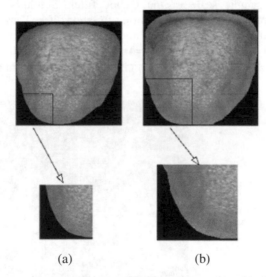

(a)   (b)

Fig. 3. (a) Original tongue image and its magnified boundary region. (b) effect after boundary enlargement.

Because the shape of tongue is very close to elliptical, there is bound to be the only one intersection point of the boundary line and the line between center point and filling pixel. The result of the Gabor filter after boundary enlargement with scope 10 is shown in below. Being compared to Fig 1(b), the result image has no relatively high-energy stripe-shaped area near the boundary.

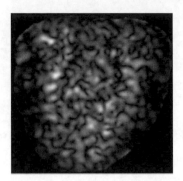

Fig. 4. The amplitude of Gabor filtered tongue image after boundary enlargement

Since the texture of tongue image is relatively slender, the choice of Gabor filter must correspond to this characteristic. There are two parameters k_v and φ_u which determine the direction and the scale of the filter.

$$k_v = 2^{\frac{v+3}{4}}, v = 0, \cdots, M-1, \quad \varphi_u = u\frac{\pi}{N}, u = 0. \cdots, N-1. \tag{6}$$

Selecting different subscript v can describe different width of Gauss Window, thus can control the sample scale of the Gabor filter. Similarly, selecting different subscript u can control the sample direction of Gabor filter. Thus, M×N Gabor filters can be determined to filter image by the above equations. Here we chose Gabor filter with 5 scales and 4 directions (M=5, N=4) to process the small image block of size 17×17 from the original tongue image, and the result is shown in Fig.5.

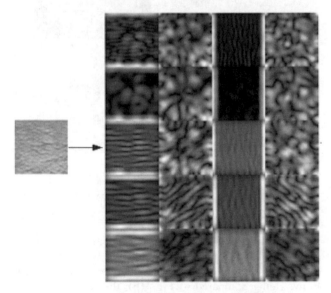

Fig. 5. Gabor filter result of tongue image block

3 Normalized Cut

From the graph theory point of view, the process of tongue texture segmentation equals to find the optimal cut to divide the whole graph (the tongue image) into several sub-graph, each of which have their own similar attribute within their sub-graph.

Normalized Cut (Ncut) is an improved method to find the minimal cut of the graph which can eliminate the segmentation bias of traditional methods [6]. Normalized Cut has a global criterion, and can maximize the similarity within the same group and minimize the similarity between different groups [7]. Normalized Cut treat the sum of weights of all edges as numerator, calculate the proportion of weights of these edges to the weights of the set of all edges, namely

$$E(u) = Ncut(G_1, G_2) = \frac{cut(G_1, G_2)}{assoc(G_1, G)} + \frac{cut(G_2, G_1)}{assoc(G_2, G)}.$$

$$cut(G_1, G_2) = \sum_{u \in G_1, v \in G_2} w(u, v), assoc(G_1, G) = \sum_{v_i \in E_1(G_1), v_j \in E(G)} w(v_i, v_j)$$

(7)

Minimizing $E(u)$ can be achieved by some approximation methods, which are described in detail in [8].

Usually, spatial Euclidean distance and RGB space Euclidean distance of pixels are taken into consideration when building up the weight matrix. Although this method referred to color information, it ignored the texture information of the image. In order to describe the texture information of the tongue image, we introduced a group of Gabor filter result into the weight matrix calculating equation, which is listed below.

$$w_{ij} = \begin{cases} \exp(-\dfrac{dist_{space}(i, j)}{\alpha}) \times \exp(-\dfrac{dist_{color}(i, j)}{\beta}) \\ \times \exp(-\dfrac{\chi^2(h_i, h_j)}{\gamma}) \qquad dist(i, j) \le r \\ \\ 0 \qquad\qquad\qquad otherwise \end{cases}$$

(8)

$dist_{space}(i, j)$ stands for the spatial Euclidean distance of pixel i and pixel j, and $dist_{color}(i, j)$ represents chessboard distance of RGB value between pixel i and pixel j. α, β, γ are adjustment parameters. $\chi^2(h_i, h_j)$ stands for χ^2 probability statistics of Gabor filter result image amplitude histogram of the 17×17 window centered at pixel i and pixel j.

$$\chi^2(h_i, h_j) = \frac{1}{2} \sum_{k=1}^{K} \frac{[h_i(k) - h_j(k)]^2}{h_i(k) + h_j(k)}.$$

(9)

In the above equation, $h_i(k)$ stands for the histogram of the Gabor filter result image of the 17×17 window centered at pixel. Here we use 4-direction Gabor Filter with $k_v = 1$. Then we use the following algorithm to finish texture segmentation.

Input: tongue image.

Output: different sub-graphs as segmentation result

Step 1. Denote the input tongue image as I, compute its three components of RGB I_r, I_g, I_b, and then get the gray level image I_{gray} of the image I by averaging its three components.

Step 2. Exert 4-direction Gabor Filter on the above four images, and calculate corresponding attribute of each pixel.

Step 3. Make primary segmentation to image I. Treat every pixels of the same class as the same vertex of the graph model, with its attribute determined by the average attribute of all pixels from the class which it belongs.

Step 4. Compute weight matrix according to equation (8), build up the graph model of the image I.

Step 5. Segment the image once with Normalized Cut method, and divide the image into two sub-graphs.

Step 6. If the number of sub-graph is less than 32, continue the process of Step5.

One result image processed by the above algorithm is shown in Fig.6. Each class is represented by region of different gray level.

Fig. 6. Segmentation result of Normalized Cut

It is easily seen from Fig.6 that there exist many tiny regions which consist of only dozens of pixels. These small regions are too small to be entitled with value in tongue texture segmentation, and can increase the expenditure of the algorithm as well as execute time. Therefore, it would be better to erase these small regions and merge them into neighboring larger region. We set a merging threshold of small region. If the number of pixel of a region is smaller than the merging threshold, this region is merged into its neighboring region whose average spatial Euclidean distance of its pixels is most close to that of the merged region. After a lot of experiments to test the effectiveness of the merging threshold, we finally chose 2100 to be optimal threshold.

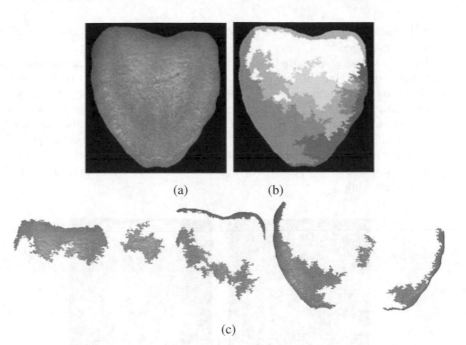

Fig. 7. (a) Original tongue image, (b)result of texture segmentation, (c) sub-graphs representing each class

An example to show the practicability of our proposed method is shown in Fig.7. From the result we can see that different texture of tongue coating and tongue substances are clearly recognized and classified, and the dividing of each sub-graph of class accord with perceptual observation.

4   Experiments and Discussions

We had selected 170 tongue images which consists of different types of textures that can be observed in tongue, and exert our tongue texture segmentation method on these images. Result of each image is examined by professional Traditional Chinese Medicine doctor to judge whether our segmentation is correct or not, the statistics is list in Table 1.

Table 1. Correctness rate of proposed method

Type	Strict Correct	Moderate Correct	Unsatisfied
Samples	79	59	32
Percentage	46.5%	34.7%	18.8%

If evaluation result judged by the professional doctor of the classification effectiveness of a filtered tongue image is not good enough to be labeled as "Strictly

Correct" and not bad enough to be considered as "Unsatisfied", this case fall into the class of "Unsatisfied".

Generally speaking, proposed method can acquires more regular segmented region with better perceiving effectiveness, and can even get better performance in discriminating very similar classes. But, because we use the average property of a region as its region property, the effectiveness of classifying detail is not good enough. An example of unsatisfied segmentation is shown in Fig.8, from which we can see thin tongue coating at the tip of tongue tip and thick tongue coating at the center part of the tongue were inappropriately classified into the same class. This is the reason why we get 18.8% unsatisfied samples. Fortifying the differentiae ability in detail is the key point of future work.

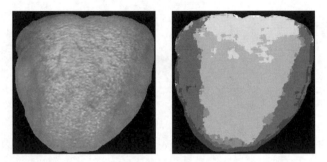

Fig. 8. An example of unsatisfied segmentation

5 Conclusion

In this paper, a novel tongue image texture segmentation method based on Gabor filter plus Normalized Cut is proposed. After synthesizing the information of location, color and textural feature, the proposed method make satisfactory texture segmentation on tongue image. In other words, the tongue image can be segmented into several parts according to their different texture, which is very useful to observe tongue image's pathological feature more clearly. The experiments show that the overall rate of correctness for this method exceeds 81%.

Acknowledgments. This work is partially supported by the NSFC funds of China under Contract Nos. 60902099 and 60871033, and the Education Ministry Young Teacher Foundation of China under No. 200802131025. The authors would like to thank Wangmeng Zuo and Hao Zhu and Dongxue Wang for their kindly help.

References

1. Li, N.M., et al.: The contemporary investigations of computerized tongue diagnosis. In: The Handbook of Chinese Tongue Diagnosis, pp. 1315–1317. Shed-Yuan Publishing, Peking (1994)
2. Zhang, D.: Automated Biometrics: Technologies and Systems. Kluwer Academic Publishers, Dordrecht (2000)

3. Yuen, P.C., Kuang, Z.Y., Wuand, W., Wu, Y.T.: Tongue texture analysis using Gabor wavelet opponent for tongue diagnosis in Traditional Chinese Medicine. Series in Machine Perception and Artificial Intelligence 40, 179–188 (1999)
4. Huang, B., Li, N.: Fungiform Papillae Hyperplasia (Fph) Identification by Tongue Texture Analysis. Int. J. Image Graphics 9(3), 479–494 (2009)
5. Pang, B., Wang, K.: Time-adaptive Snakes for Tongue Segmentation. In: Proc. of ICIG 2000, Tianjin (2000)
6. Shi, J., Malik, J.: Normalized Cuts and Image Segmentation. In: Proc. of IEEE Conf. Computer Vision and Pattern Recognition, pp. 731–737 (1997)
7. Carballido-Gamio, J., et al.: Normalized Cuts in 3D for Spinal MRI Segmentation. IEEE Trans. Medical Imaging 23(1), 36–44 (2004)
8. Shi, J., Malik, J.: Normalized Cuts and Image Segmentation. IEEE Trans. on Pattern Analysis and Machine Intelligence 22(8), 888–905 (2000)

Chaos Synchronization Detector Combining Radial Basis Network for Estimation of Lower Limb Peripheral Vascular Occlusive Disease

Chia-Hung Lin[1,*], Yung-Fu Chen[2,*], Yi-Chun Du[3],
Jian-Xing Wu[3], and Tainsong Chen[3,**]

[1]Department of Electrical Engineering, Kao-Yuan University, Lu-Chu Hsiang,
Kaohsiung 821, Taiwan
eech153@cc.kyu.edu.tw
[2]Department of Health Services Administration, China Medical University,
Taichung 40402, Taiwan
yungfu@mail.cmu.edu.tw
[3]Institute of Biomedical Engineering, National Cheng Kung University,
No. 1 University Road, Tainan 70101, Taiwan
Tel.: 886-6-2757575, Ext.: 63425; Fax: 886-6-2760677
terryfresh@gmail.com, chenta@mail.bme.ncku.edu.tw

Abstract. Early detection of lower limb peripheral vascular occlusive disease (PVOD) is important to prevent patients from getting disabled claudication, ischemic rest pain and gangrene. This paper proposes a method for the estimation of lower limb PVOD using chaos synchronization (CS) detector with synchronous photoplethysmography (PPG) signal recorded from the big toes of both right and left feet for 21 subjects. The pulse transit time of PPG increases with diseased severity and the normalized amplitudes decreases in vascular disease. Synchronous PPG pulses acquired at the right and left big toes gradually become asynchronous as the disease progresses. A CS detector is used to track bilateral similarity or asymmetry of PPG signals, and to construct various butterfly motion patterns. Artificial neural network (ANN) was used as a classifier to classify and assess the PVOD severity. The results demonstrated that the proposed method has great efficiency and high accuracy in PVOD estimation.

1 Introduction

Photoplethysmography (PPG) is an optical measurement technique that can be used to non-invasively detect blood volume changes in a vascular bed of tissue. It requires a light source (laser or light emitting diode) to illuminate the skin using reflection mode and a photo-detector to measure the variation of light intensity induced by the change of blood volume. The technique is compact with low cost, and can be designed as a portable instrument for clinical applications. A PPG signal consists of alternating current (AC) and direct current (DC) components that reveal physiological

---

[*] Co-first authors who contributed equally to this work.
[**] Corresponding author.

D. Zhang and M. Sonka (Eds.): ICMB 2010, LNCS 6165, pp. 126–136, 2010.
© Springer-Verlag Berlin Heidelberg 2010

information, such as cardiac synchronous changes in the blood volume within each heartbeat, thermoregulation, respiration, and vasomotor activity [1-2]. PPG signals can be acquired by applying multi-channel data measurement at the ear lobes, index fingers and thumbs of hands, and big toes of feet. These measurements are further analyzed with digital signal processing (DSP) and pulse waveform analysis [3]. It was shown that this optical technique is a promising solution for monitoring peripheral vascular occlusive disease (PVOD).

PVOD is a widespread chronic vascular disease, which is associated with an increased risk of cardiovascular disease and nerve damage. If treatments are not adequately controlled, it may cause ischemic rest pain, poor wound healing, and disabling claudicating for diabetes mellitus. PPG characteristics for lower limb PVOD estimation have been described using various features, including timing delay, amplitude reduction, and shape distortion [4-5]. Bilateral differences were studied by comparing the patient's big toe data with the normative pulse ranges. In clinical investigations, the transit time and shape distortion increase with disease severity, and the calibrated amplitude decreases in vascular diseases. These phenomena are attributed to increased resistance and decreased blood in the peripheral arteries [6]-[8]. PPG pulses are synchronously induced at the right and left sites with each heartbeat. When PVOD becomes severe progressively, the bilateral pulses gradually become delayed and asynchronous.

In order to develop an assistant tool, chaos synchronization (CS) detector is proposed to assess the grade of lower limb PVOD. The CS has been extensively studied in some applications, such as secure communication, adaptive control system, and information processing [9-11]. A master system and a slave system oscillating in a synchronized manner are needed to design a CS detector. Through a coupling variable, the response of the slave system can automatically track the response of the master system with the same dynamical pattern in a proportional scale. The proposed CS detector is used to estimate the error states using a discrete-time chaotic system. The error dynamic equation is defined as the difference between the slave system and the master system, and is used to construct the butterfly motion patterns from time-series signals. The probabilistic neural network (PNN) was applied to develop a classifier for PVOD estimation with subjects being classified into three groups, i.e. normal and lower-grade (LG) and higher-grade (HG) disease. The particle swarm optimization (PSO), an evolutionary optimization algorithm, was used to automatically adjust the targets and the best network parameters in a dynamic modeling environment. PPG signals obtained from 21 subjects were used to train a PNN model and then signals of 6 subjects for verifying its predictive performance.

2 The Design Mode of CS Detector

Chaos Synchronization (CS) Description: Biomedical signals are seemingly random or periodic in time and have been ascribed to noise or interactions between various physiological functions. A dynamic system has been proposed to provide a mathematical model in nonlinear time series analysis, which behavior changes continuously in time are described by a coupled set of first-order autonomous ordinary differential equations [11], as shown in the following equation:

$$\frac{dx(t)}{dt} = f(x(t), \mu) \tag{1}$$

where the vector $x(t)$ indicates the dynamical variables, the vector μ represents parameters, and the vector field f denotes the dynamical rules governing the behavior of dynamical variables. Equation (1) represents a non-autonomous system in R^n. A deterministic chaos is a nonlinear dynamical system (NDS) and has some remarkable characteristics, such as excessive sensitivity to initial conditions, broad spectrums of Fourier transform, and fractal properties of the motion in phase space.

An NDS with numerous applications in engineering is the Duffing-Holmes system [12], which can be delineated as:

$$\frac{d^2 z}{dt^2} + \mu_1 z + \mu_2 \frac{dz}{dt} + z^3 = p\Delta f \tag{2}$$

where μ_1, μ_2, and p are parameters, and Δf is an uncertain term representing the unmodeled dynamics or structural variation of the NDS. In general, Δf is assumed to be bounded with $0 \leq |\Delta f| < \alpha$. This second-order non-autonomous equation can be represented as the first-order system by defining dynamic variables $x_1 = z$ and $x_2 = dz/dt$.

$$\begin{cases} dx_1 / dt = x_2 \\ dx_2 / dt = -\mu_1 x_1 - \mu_2 x_2 - x_1^3 + p\Delta f \end{cases} \tag{3}$$

Eq. (3) can also be described as the behavior changes at discrete time intervals, namely discrete NDS.

For lower limb PVOD estimation, the bilateral differences (right to left) of PPG transit time in healthy subjects (bilateral similarity) and unhealthy subjects (bilateral asymmetry) represents different grades of disease [6-8]. The chaos synchronization system (CSS) is applied to quantify the chaotic motion for PVOD estimation. A typical configuration of CSS consists of a master system and a slave system. For a data sequence of n points, $i=1, 2, 3, \ldots, n$, two n-dimensional chaotic systems with the term $\Delta f = 0$ are expressed as follows:

$$\textbf{Master:} \begin{cases} \dot{x}_1[i] = x_2[i+1] \\ \dot{x}_2[i+1] = -\mu_1 x_1[i] - \mu_2 x_2[i+1] - (x_1[i])^3 \end{cases} \tag{4}$$

$$\textbf{Slave:} \begin{cases} \dot{y}_1[i] = y_2[i+1] \\ \dot{y}_2[i+1] = -\mu_1 y_1[i] - \mu_2 y_2[i+1] - (y_1[i])^3 \end{cases} \tag{5}$$

where $x_1[i]$ and $x_2[i+1]$, $i=1, 2, 3, \ldots, n-1$, indicate the data sequence of the right-side PPG signals, $y_1[i]$ and $y_2[i+1]$ denote the data sequence of the left-side PPG signals. Let the error states be $e_1[i] = x_1[i] - y_1[i]$ and $e_2[i] = x_2[i+1] - y_2[i+1]$. Subtracting Eq. (5) from Eq. (4) yields the synchronization error dynamics, as shown below:

$$\begin{cases} \dot{e}_1[i] = e_2[i] \\ \dot{e}_2[i+1] = -\mu_1 e_1[i] - \mu_2 e_2[i] - (x_1[i])^3 + (y_1[i])^3 \end{cases} \tag{6}$$

The error state can be written in a matrix form as shown in the following equation:

$$\begin{bmatrix} \dot{e}_1 \\ \dot{e}_2 \end{bmatrix} = \begin{bmatrix} 0 & 1 \\ -\mu_1 & -\mu_2 \end{bmatrix} \begin{bmatrix} e_1 \\ e_2 \end{bmatrix} = \mathbf{A} \begin{bmatrix} e_1 \\ e_2 \end{bmatrix} \tag{7}$$

Since the eigenvalues of matrix $\mathbf{A}$ are roots of a 2nd degree polynomial having real or complex coefficients. The polynomial $p_A(\lambda)=\lambda^2-(trace(\mathbf{A}))\lambda+det(\mathbf{A})$ is called the characteristic equation of $\mathbf{A}$. The design parameters μ_1 and μ_2 can be determined to result in real eigenvalues $\lambda(\mathbf{A})<0$, as shown below:

$$\lambda = -\frac{1}{2}\mu_2 \pm \frac{1}{2}\sqrt{{\mu_2}^2 - 4\mu_1}, \ \ \mu_2 > \sqrt{{\mu_2}^2 - 4\mu_1} \tag{8}$$

By selecting $\mu_2=\beta\mu_1$, $\mu_1>0$, $\beta>0$, the asymptomatic stability of (7) is surely guaranteed. The scaling factors μ_1 and μ_2 are nonzero constants, which control the chaotic motion of CSS. The oscillation path is a limit cycle attractor. The master and slave systems are synchronized such that $e_1[i]=0$, $e_2[i]=0$, and $(x_1[i])^3-(y_1[i])^3=0$.

Equation (6) can be realized to track the bilateral similarity or asymmetry of PPG signals, namely CS detector. The dynamic error equation E_i, $i=1, 2, 3, \ldots, n-1$, is defined as the difference between the slaves $y_1[i]$ and $y_2[i+1]$, and that of the references $x_1[i]$ and $x_2[i+1]$, as delineated below:

$$E_i=-\mu_1 e_1[i]-\mu_2 e_2[i]-[(x_1[i])^3-(y_1[i])^3] \tag{9}$$

In view of equation (9), data sequence $E = [E_1, E_2, E_3, \ldots, E_{n-1}]\in R^{n-1}$ changes with time. A CS detector for PVOD estimation is designed such that the resulting tracking error is zero or within a bounded range.

Chaotic Motion of Bilateral PPG Signals: Lower Limb PVOD estimation is an important index for providing information to prevent arteriosclerosis, leg pain, and arterial disease. If peripheral resistance increases or peripheral vascular is occlusive, PPG pulses are asynchronously induced at the right and left legs. Bilateral pulses gradually become asynchronous with disease severity. The proposed CS detector could use this information to assess the grade of PVOD. Figure 1 shows the PPG waveforms by measuring simultaneously at the right and left toes from normal subjects and patients with PVOD. The horizontal axis indicates the sampling number and the vertical axis is the normalized amplitude.

The dynamic errors E_i, $i=1, 2, 3, \ldots, n-1$, can be obtained from equation (9). As shown in Fig. 2, using CS detector with $\beta=2$ and $\mu_1=1, 2, 3$, the phase plot of error state e_2 and dynamic error E reflects the chaotic behaviors for normal condition (Nor), lower-grade (LG) disease, and higher-grade (HG) disease. Butterfly chaotic motions lie in the second quadrant (systolic rising edge) and fourth quadrant (dicrotic notch edge). Trajectories coming close to a stable point need not actually approach this point. The origin point is seen as asymptotically stable that there is a circle C surrounding the stable point such that every trajectory is inside C. The choice of parameters β and μ_1 control the chaotic motion, and its trajectory increases as the parameters μ_1 and μ_2 increase. According to equations (6) and (9), we select $\beta=2$, $\mu_1=1$, and $\mu_1=2$ resulting in a stable CS detector (real eigenvalues and $\lambda(\mathbf{A})<0$).

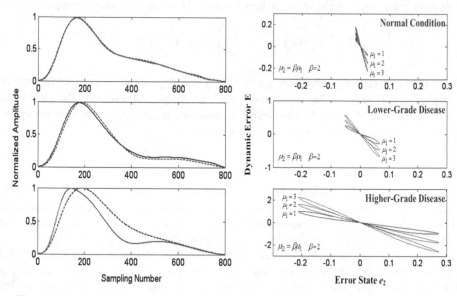

Fig. 1. PPG waveforms at the toes of right and left feet for normal subjects and patients with PVOD

Fig. 2. Butterfly chaotic motion of the CS detector: phase plot at e_2-E plane

Therefore, we used the error states and dynamic errors to construct various butterfly motion patterns. These patterns have various morphologies and limited motion paths that can be used for discriminating different PVOD grades. Their trajectories also increase as the diseases gradually become serious. The butterfly patterns were systematically calculated for training the PNN-based classifier.

3 Classifier Design

RBF-Based Classifier: Radial basis network (RBN) can be designed in two different ways, generalized regression neural network (GRNN) and probabilistic neural network (PNN) [13-14]. The first network is often used for linear or non-linear function approximation, while the second network can be applied in control system, identification, and classification. A PNN is guaranteed to converge to a Bayesian classifier if it is given enough training data, and the number of radial basis neurons is proportional to the complexity of the problem. In a continuous modeling system, an adaptive PNN is promising in designing a classifier for PVOD estimation.

A general structure of the PNN consists of four layers, including input layer, pattern layer, summation layer, and output layer, that has the parallelism distributed process, learning, and pattern recognition ability. Let the input pattern $\mathbf{E}=[E_1, E_2, E_3, ..., E_{n-1}]$ connect to the nodes (n=799 in this study) in the input layer, and its Euclidean distance (ED) to the recorded patterns are computed in each pattern node:

$$ED(k) = \sqrt{\sum_{i=1}^{n-1}(E_i - w_{ki}{}^{IH})^2} \tag{10}$$

where the weights $w_{ki}{}^{IH}$ are created by training patterns $\mathbf{E}(k)=[E_1(k), E_2(k), E_3(k), \ldots, E_{n-1}(k)]\in N$, $k=1, 2, 3, \ldots, K$. The output of pattern node H_k can be computed by radial basis function shown in the following equation:

$$H_k = \exp[-ED(k)^2/(2\sigma_k^2)] \tag{11}$$

where σ_k is the smoothing parameter with $\sigma_1=\sigma_2=\cdots=\sigma_k=\sigma$. If $ED(k)$ approaches zero, H_k will be a maximum value, meaning the input pattern is similar to the kth training pattern. The output O_j with $j=1, 2, \ldots, m$ and $0 \le O_j \le 1$ can be computed by

$$O_j = (\sum_{k=1}^{K} w_{jk}{}^{HS} H_k)/\sum_{k=1}^{K} H_k = S_j / \sum_{k=1}^{K} H_k \tag{12}$$

where S_j is the output of summation node, and each predicted output $w_{jk}{}^{HS}\in\{0,1\}$ is a function of the corresponding output O_j associated with the stored pattern E(k).

Adaptive mechanism may be suitable to apply in a continuous modeling system, such as tuning smoothing parameter and adjusting targets automatically. The optimization techniques, such as gradient descent method, are used to minimize the objective function defined as the mean squared error function (MSEF):

$$MSEF = \frac{1}{K}\sum_{k=1}^{K}\sum_{j=1}^{m}[T_j(k) - O_j(k)]^2 \tag{13}$$

where $T_j(k)\in\{0,1\}$ is the desired output for the kth training pattern. The optimal parameter σ is obtained by minimizing the $MSEF$. However, $MSEF$ is a non-linear function and can't guarantee convergence to a global optimal solution.

The particle swarm optimization (PSO) algorithm is an evolutionary optimization technique, which has been proven to be efficient in solving optimization problems with high dimensionality, non-linearity, and non-differentiability [15-16]. It guides searches using multiple particles rather than individuals and can avoid trapping at a local minimum. Particles modify their search points around multi-dimensional search space until unchanged positions have been achieved. Each position is adjusted by dynamically altering the velocity of each particle according to its flying experience and the flying experiences of the other particles. Suppose σ_g^p is the current position of agent g at the pth iteration with $g=1, 2, 3, \ldots, G$ and G is the population size. In addition, σ_{opt} indicates the global optimum in the group, and $\sigma_{opt(g)}$ is the individual optimum. The modification of parameter σ can be represented by the velocity $\Delta\sigma_g^p$. The mathematical representation of PSO is given by [17-18]:

Velocity: $\Delta\sigma_g^{p+1} = \omega\Delta\sigma_g^p + c_1 rand_1(\sigma_{opt(g)} - \sigma_g^p) + c_2 rand_2(\sigma_{opt} - \sigma_g^p)$ (14)

Position: $\sigma_g^{p+1} = \sigma_g^p + \Delta\sigma_g^{p+1}$ (15)

where ω is the inertia weight that control the impact of the current velocity on the next velocity; c_1 and c_2 are the positive acceleration parameters that pull each particle toward the best positions; and $rand_1$ and $rand_2$ are the uniformly random numbers between 0 and 1. The second part of equation (14) is the "*cognitive component*", which encourages the particles to move toward their own best positions found so far. The third part is the "*social component*", which represents the collaborative effect of the particles to find the global optimal solution [18]. To efficiently converge to the global optimal solution, PSO with time-varying acceleration coefficients (TVAC) [19-20] are used to improve the performance of PSO. Fixed parameters c_1 and c_2 can be modified according to the following equations.

$$c_1 = (b_1 - a_1)p / p_{max} + a_1 \text{ and } c_2 = (b_2 - a_2)p / p_{max} + a_2 \qquad (16)$$

where a_1, b_1, a_2, and b_2 are constant, p_{max} is the maximum number of allowable iterations. With a large cognitive component and small social component, particles are allowed to move around the search space at the initial stage. By reducing the cognitive component and increasing the social component, the particles will converge to the global optimum at the end of search [19]. The values used in this study are selected by changing c_1 from 2.5 to 0.5 and c_2 from 0.5 to 2.5 according to numerical experiments.

Classifier Training: An optical measurement technique is used to acquire PPG signals operating in reflection mode. The measuring device consists of light sources (near infrared: 940nm, spectral bandwidth: 45nm, forward voltage: 1.2V), photo-detector, trans-impedance amplifier, filter circuits, amplifiers, and interface [7,8]. Each subject was asked to lie supine on a couch in a room with temperature controlled under 25±1°C. For bilateral measurement, two probes were placed at the right and left big toes, respectively. The PPG signals were captured at a sampling rate of 1 kHz for 15 minutes with an analog-to-digital (A/D) converter (NI DAQ Card, 16 channels, 1.25MS/s). Butterfly motion patterns were systematically calculated for 21 subjects (age: 24 to 65) who were divided into three groups, i.e. Nor, LG, and HG. Table 1 shows the preliminary PVOD assessments based on ankle-to-brachial pressure index (ABPI) [8] and the ranges of bilateral-timing differences for three groups. The subjects were first classified into two groups by using ABPI (ABPI≥0.9: Nor or LG diabetics, ABPI<0.9: HG diabetics) followed by the discrimination between the normal and LG groups based on the ranges of bilateral-timing differences.

Table 1. Subjects are classified into 3 groups, i.e. normal, LG, and HG

Subject Parameter		ABPI ≥ 0.9		ABPI < 0.9
		Normal (N=11)	LG (N=6)	HG (N=4)
PTT_f (ms)	Mean (Max-Min)	2.58 (0.3-7.4)	9.2 (5.1-23.7)	29.2 (23.6-34.8)
ΔPTT_P (ms)	Mean (Max-Min)	7.3 (0.4-22.3)	22.6 (14.3-56.5)	52 (46.2-57.8)
ΔRT (ms)	Mean (Max-Min)	6.84 (1.3-15.6)	14.6 (3.4-32.3)	23.4 (11.5-35.3)

Note: The values of ΔPTT_f, ΔPTT_p, and ΔRT are absolute values with (1) $\Delta PTT_f = |PTT_{Rf} - PTT_{Lf}|$, (2) $\Delta PTT_p = |PTT_{Rp} - PTT_{Lp}|$, and (3) $\Delta RT = |RT_R - RT_L|$, where suffix R and L indicated right and left feet.

Pattern data acquired from 21 subjects (11 normal subjects, 6 LG patients, and 4 HG patients) were used to train a PNN-based classifier with 799 input nodes, 21 pattern nodes, 4 summation nodes, and 3 output nodes. The weights between the input and pattern layers are expressed as w_{ki}^{IH} with k=1, 2, ..., 21 and i=1, 2, ..., 799. The weights between the pattern nodes and summation nodes are denoted as w_{jk}^{HS} with j=1, 2, 3. An additional node which adds the outputs of all the pattern nodes is provided, which is then summated with the outputs of individual summation node to form the output nodes. In this study, PSO with TVAC were used to classify the patterns into 3 groups. The parameters are given as follows: population size G=20, inertia weight ω=0, acceleration coefficients a_1=2.5 and b_1=0.5 (c_1=2.5→0.5), acceleration coefficients a_2=0.5 and b_2=2.5 (c_2=0.5→2.5), and number of iterations p_{max}=50. The optimal parameter σ_{opt} =0.0754 was obtained by minimizing the MSEF. It can rapidly converge for less than 20 learning iterations with an average time of 12.3972 seconds taken to classify the 21 training data. Fig. 3 shows smoothing parameters and mean squared errors for progressive iterations. For the same classifier and training data, PSO with constant acceleration coefficients (c_1=1 and c_2=1) is easier to trap local minimum, while PSO with TVAC is expected to obtain the optimal parameters. The number of population size was suggested to range from 20 to 60 [19]. As shown in Figure 4, only slight improvement of optimal parameters was achieved by increasing the population size to 40 and 60, hence a population size of 20 was chosen.

4   Experimental Results and Discussions

The proposed method was developed on a Pentium-IV 3.0GHz PC with 480MB RAM using Matlab software (MathWorks Inc.). The diagnostic performance was tested with data obtained from 21 subjects. The subjects were divided into three groups, including normal subjects, and LG and HG diabetic patients with PVOD determined based on ABPI and bilateral-timing differences (Table 1). According to pattern similarity for the same group, each trajectory of average butterfly patterns is limited in ellipse region C1, C2, and C3, Figures 5. It can be seen that each trajectory lies in the second and fourth quadrants. This type of analysis accentuates the subtle changes in butterfly patterns. Therefore, the bilateral similarity or asymmetry in systolic rising edge and dicrotic notch edge between the right and left sites has also been observed at the error state versus dynamic error plane. The butterfly patterns are similar for individual groups of normal subjects and PVOD patients. Owing to the butterfly motion pattern within limited region, the number of training data can be reduced.

The proposed method tested the detection accuracy of PNN model for 6 subjects divided into 3 groups, each includes 2 subjects. Table 2, compares the outcome determined by a physician and predicted by the bilateral timing difference method [7-8] and the proposed method. Ten cycles of time-domain PPG signals for a normal subject and a LG patient are shown in Figures 6(a) and 6(b) with their plots of error states versus dynamic errors shown in Figs. 5(a) and 5(b), respectively.

It is non-objective to evaluate lower limb PVOD by referring to ABPI only, arterial Doppler and magnetic resonance angiography are also valuable examinations. The ground truth classification of the 3 groups was decided by a professional physician based on data measured with instruments. However, the instruments are very expensive

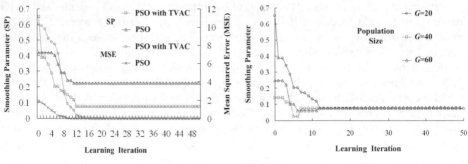

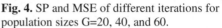

Fig. 3. SP and MSE of different iterations for PSO and PSO with TVAC

Fig. 4. SP and MSE of different iterations for population sizes G=20, 40, and 60.

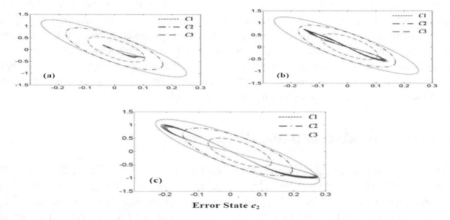

Fig. 5. Plots of dynamic errors versus error states for (a) normal subjects, and (b) LG and (c) HG patients, respectively

and it takes several measurements, which has limitation for routine screening in the primary care setting. For bilateral timing differences, each ECG R-peak was used as the timing reference for obtaining the parameters. The timing reference might be disturbed by heart-rate variation and asynchronous measurement in parameters extraction. In contrast, the proposed CS detector is simple but still can achieve high accuracy for PVOD estimation.

Table 2. PVOD estimation

Subject No.	Mean Values of Bilateral Timing Parameters (ms)			ABPI (R)	ABPI (L)	Physician Decision	Bilateral Timing Difference [7-8]	Proposed Method
	ΔPTT_f	ΔPTT_p	ΔRT					
1	2.7330	1.9970	3.1330	1.0650	1.1138	Nor	Nor	Nor
2	1.1730	5.8700	4.6950	1.1404	1.1333	Nor	Nor	Nor
3	15.321	17.286	1.9640	1.1788	1.2357	LG	LG	LG
4	23.667	33.333	32.833	1.0714	1.0446	LG	LG	LG
5	33.072	48.500	15.428	1.0817	0.8941	HG	HG	HG
6	23.606	57.700	35.000	1.0442	0.8945	HG	HG	HG

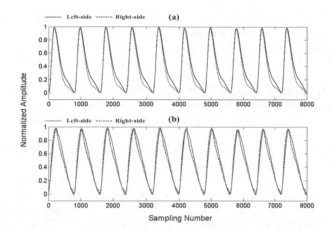

Fig. 6. Time-domain PPG signals of (a) normal subject and (b) LG patient

5 Conclusion

A simple CS detector has been proposed to detect PPG signals for lower limb PVOD estimation. The CS detector is used to track bilateral synchronous and asynchronous patterns of PPG signals, and to construct various butterfly patterns for discriminating normal subjects and patients with LG and HG. The proposed method can be used to prevent patients from further deterioration and complications, such as cardiovascular disease and nerve damage. Adequate treatment of diabetes may improve the risk of chronic complications. For the development of an automatic diagnostic tool, the proposed model can be further integrated with computer-based DSP and telemedicine techniques for healthcare applications at homes and elderly communities.

Acknowledgments. This work is supported in part by the National Science Council of Taiwan under contract numbers: NSC96-2221-E-006-002-MY3 and NSC98-2410-H-039-003-MY2. The authors would like to thank Dr. Chian-Ming Lee for providing his valuable suggestion and help on experiments.

References

1. Nitzan, M., Babchenko, A., Khonokh, B., Landau, D.: The variability of the photoplethysmographic signal- a potential method for the evaluation of the automatic nervous system. Physiological Measurement 19, 93–102 (1998)
2. Nitzan, M., Babchenko, A., Khonokh, B.: Very low frequency variability in arterial blood pressure and blood volume pulse. Med. Biol. Eng. Comp. 37, 54–58 (1999)
3. Allen, J.: Photoplethysmography and its application in clinical physiological measurement. Physiological Measurement 28, R1–R39 (2007)
4. Jago, J.R., Murray, A.: Repeatability of peripheral pulse measurement on ears, fingers, and toes using photoelectric plethysmography. Clin. Phys. Physiol. Meas. 9, 319–329 (1988)
5. Allen, J., Murray, A.: Variability of photoplethysmography peripheral pulse measurements at the ears, thumbs, and toes. IEE Proc. Meas. Technol. 147, 403–407 (2000)

6. Nitzan, M., Khonokh, B., Slovik, Y.: The difference in pulse transit time to the toe and figure measured by photoplethysmography. Physiological Measurement 23, 85–93 (2002)

7. Allen, J., Oates, C.P., Lees, T.A., Murray, A.: Photo-plethysmography detection of lower limb peripheral arterial occlusive disease: a comparison of pulse timing, amplitude and shape characteristics. Physiological Measurement 26, 811–821 (2005)

8. Allen, J., Overbeck, K., Nath, A.F., Murray, A., Stansby, G.: A prospective comparison of bilateral photo-plethysmography versus the ankle-brachial pressure index for detecting and quantifying lower limb peripheral arterial disease. Journal of Vascular Surgery 47, 794–802 (2008)

9. Chen, M., Zhou, D., Shang, Y.: A sliding mode observer based secure communication scheme. Chaos, Solitons & Fractals 25, 573–578 (2005)

10. Lin, J.S., Yan, J.J., Liao, T.L.: Chaotic synchronization via adaptive sliding mode observers subject to input nonlinearity. Chaos, Solitons & Fractals 24, 371–381 (2005)

11. Wagner, C.D., Mrowka, R., Nafz, B., Persson, P.B.: Complexity and chaos in blood pressure after baroreceptor denervation of conscious dogs. Am. J. Physiol. 269, 1760–1766 (1996)

12. Yau, H.T., Kuo, C.L., Yan, J.J.: Fuzzy sliding mode control for a class of chaos synchronization with uncertainties. International Journal of Nonlinear Sciences and Numerical Simulation 7(3), 333–338 (2006)

13. Chen, S., Cowan, C.F.N., Grant, P.M.: Orthogonal least squares learning algorithm for radial basis function networks. IEEE Transactions on Neural Networks 2(2), 302–309 (1991)

14. Lin, C.H., Du, Y.C., Chen, T.: Adaptive wavelet network for multiple cardiac arrhythmias recognition. Expert Systems with Applications 34(4), 2601–2611 (2008)

15. Bonabeau, E., Dorigo, M., Theraulaz, G.: Swarm intelligence: from natural to artificial systems. Oxford University Press, New York (1999)

16. Kennedy, J., Eberhart, R.: Swarm intelligence. Morgan Kaufmann, San Francisco (2001)

17. Kennedy, J., Eberhart, R.: Particle swarm optimization. In: Proc. of the 1995 IEEE Int. Conf. Neural Networks, pp. 1942–1948. IEEE Press, New York (1995)

18. Kennedy, J.: The particle swarm: social adaptation of knowledge. In: Proc. of the 1997 IEEE Int. Conf. Evolutionary Computation, pp. 303–308. IEEE Press, New York (1997)

19. Ratnaweera, A., Halgamuge, S.K., Watson, H.C.: Self-organizing hierarchical particle swarm optimizer with time-varying acceleration coefficients. IEEE Transactions on Evolutionary Computation 8(3), 240–255 (2004)

20. Lin, Y.L., Chang, W.D., Hsieh, J.G.: A particle swarm optimization approach to nonlinear rational filter modeling. Expert System with Applications 34, 1194–1199 (2008)

Classification of Wrist Pulse Blood Flow Signal Using Time Warp Edit Distance

Lei Liu, Wangmeng Zuo, Dongyu Zhang, Naimin Li, and Hongzhi Zhang

School of Computer Science and Technology, Harbin Institute of Technology,
Harbin, 150001, China
liulei8174@yahoo.com.cn

Abstract. The blood flow signals can be used to examine a person's health status and have been widely used in the study of the clinical diagnosis of cardiovascular diseases. According to the pulse diagnosis theory of traditional chinese, the pathological changes of certain organs could be reflected on the wrist pulse signals. In this paper, we use Doppler ultrasonic device to collect the wrist pulse blood flow signals from patients with pancreatitis (P), duodenal bulb ulcer (DBU), appendicitis (A) and acute appendicitis (AA) as well as healthy persons. After extracting the envelopes of ultrasonic pulse contour, the wrist pulse blood flow signals are pre-processed using wavelet transform. Finally, we adopted a recent time series matching method, time warp edit distance (TWED), on the pre-processed data for classification of wrist pulse blood flow signals. The proposed approach is tested on the wrist blood flow signal dataset, and achieves higher classification accuracy than several classical time series matching approaches, such as Euclidean distance (ED), dynamic time warping (DTW), and edit distance with real penalty (ERP).

Keywords: Wrist blood flow diagnosis; time series; time warp edit distance.

1 Introduction

Pulse diagnosis, one of the most important diagnostic methods in traditional Chinese medicine (TCM), has been used in disease examination and in guiding medicine selection for thousands of years. In pulse diagnosis, wrist pulse signal carries vital information and can reflect the pathological changes of the body condition. By feeling the pulse, although the practitioners could examine the person's health conditions, the diagnostic results, however, sincerely depend on the practitioner's subjective analysis and sometimes may be unreliable and inconsistent. Therefore, with the help of modern computer techniques, it is necessary to develop computerized pulse signal analysis techniques to make pulse diagnosis quantitative and objective [1, 2, 3, 4]. For example, Chen et al. [5, 6] developed several models to extract features from the wrist pulse blood flow signals, and use support vector machine (SVM) and fuzzy C-means (FCM) for classification. Using linear discriminant classifier, other researchers [7, 8] have also shown the effectiveness of identifying human sub-health status based on pulse signals. Most recently, Zhang et al. [9, 10] used the Hilbert-Huang transform and the wavelet method to extract pulse features including wavelet powers, wavelet packet powers, and other Doppler ultrasonic diagnostic parameters.

D. Zhang and M. Sonka (Eds.): ICMB 2010, LNCS 6165, pp. 137–144, 2010.

In this paper, we adopt a recently developed time series matching method, time warp edit distance (TWED) [11], for classification of the wrist pulse blood flow signals. In order to test the classification performance of TWED, experiments are carried out on the wrist blood flow signal dataset [12] which contains 100 healthy people, 54 patients with pancreatitis (P), 77 with duodenal bulb ulcer (DBU), 35 with appendicitis (A), and 54 with acute appendicitis (AA).

The reminder of the paper is organized as follows. Section 2 describes the pre-processing method of the Doppler ultrasonic signals. Section 3 introduces the time warp edit distance method. Section 4 provides the experimental results and Section 5 concludes the paper.

2 The Pre-Processing of the Wrist Pulse Blood Flow Signals

In our scheme, blood flow signals of the wrist radial artery are collected by a Doppler ultrasonic acquisition device. At the beginning of signal acquisition, operator uses

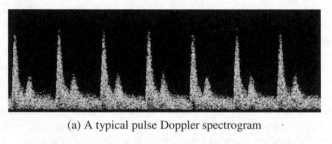

(a) A typical pulse Doppler spectrogram

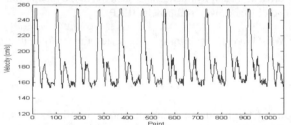

(b) The maximum velocity envelop of the wrist blood flow Signal

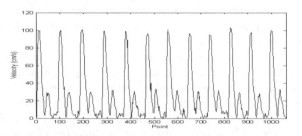

(c) The wrist blood flow signal after de-noising and drift removal

Fig. 1. The pre-processing of the wrist pulse blood flow signal. (a) A typical pulse Doppler spectrogram, (b) the maximum velocity envelop of the wrist blood flow signal, and (c) the wrist blood flow signal after de-noising and drift removal.

his/her finger to feel the fluctuation of pulse at the patient's styloid process of radius to figure out a rough area where the ultrasound probe is then put on and moved around carefully until the most significant signal is detected. Then, a stable signal segment is recorded and stored. As shown in Fig. 1 (a), the raw data acquired is represented in the form of Doppler spectrogram.

The wrist pulse blood flow signals are the extracted maximum velocity envelope of the spectrogram, as shown in Fig. 1 (b). The envelope signals may include the low-frequency drift and high-frequency noise in maximum velocity. Thus, the 7-level 'db6'wavelet transform [13] is used to reduce the low-frequency drift and high-frequency noise. The low frequency drift of the waveform is suppressed by reducing the 7th level wavelet approximation coefficients. Analogously, the high frequency noise can be removed by reducing the 1st level wavelet detail coefficients from the waveform. The result of drift and noise removal is shown in Fig. 1 (c).

3   The TWED Method

After the signals are pre-processed, the wrist pulse blood flow signals could be treated as discrete time series, and thus discrete time series similarity matching algorithms can be used to classify signals from healthy persons and from patients. By far, varieties of elastic matching methods, e.g., edit distance [14], dynamic time warping (DTW) [17], edit distance with real penalty (ERP) [16], have been developed and applied to various time series classification and clustering problems.

In this section, we suggested to use a recently developed time series matching method, TWED, for classification of wrist blood flow signals. The reason to choose TWED is based the following three reasons: first, TWED is a elastic metric, where this property might be valuable in designing effective and efficient pulse classification methods; second, TWED introduces a stiffness parameter to control the elasticity of the metric, and thus is more flexible for time series matching; finally, the effectiveness of TWED has been empirically proved in [11].

3.1   The Property of TWED Algorithm

Generally speaking, the TWED matching method has the following four properties:

- superior performance in dealing with local time shifting.
- satisfying the triangle inequality.
- including a *stiffness* parameter to control the elasticity of TWED.
- using time-stamp differences in comparing samples as part of the local matching costs.

TWED provides a similarity metric that is potentially useful in time series retrieval applications since it could benefit from the triangular inequality property to speed up the retrieval process while tuning the parameters of the elastic measure. Simultaneously, a parameter is highlighted in the TWED algorithm that controls a kind of *stiffness* of the elastic measure along the time axis that is placed between edit distance with infinite *stiffness* and DTW with no *stiffness*.

3.2 The TWED Algorithm

Give two time series A and B, TWED introduces three operations, $delete_A$, $delete_B$, and $match$, for the editing of two time series, and the discrete time series similarity between two time series is measured as the minimum cost sequence of editing operations need to transform one time series into another. In [11], one can see the editing operations of TWED are different from classical operations of edit distance [15]. Based on these three operations, the TWED algorithm adopted Eq. (1) to compute a sequence of editing operations at a cost for all pairs of time series. In Eq. (1), U denotes the set of finite time series: $U = \{A_1^p | p \in N\}$, where A_1^p denotes a time series with discrete time index varying between 1 and p. It should be noted that A_1^0 represents the empty time series (with null length). a_i' denotes the ith sample of time series A. We will consider that $a_i' \in S \times T$, where $S \subset R^d$ with $d \geq 1$ embeds the multidimensional space variables, and $T \subset R$ embeds the time-stamp variable. So that we can write $a_i' = (a_i, t_{a_i})$, where $a_i \in S$ and $t_{a_i} \in T$, with the condition that $t_{a_i} > t_{a_j}$ whenever $i > j$ (time stamp strictly increase in the sequence of samples). λ is a constant penalty. The similarity between time series B and time series A is recurrently calculated by,

$$\delta_{\lambda,\upsilon}(A_1^p, B_1^q) = \min \begin{cases} \delta_{\lambda,\upsilon}(A_1^{p-1}, B_1^q) + \Gamma(a_p' \to \Lambda) & delete_A \\ \delta_{\lambda,\upsilon}(A_1^{p-1}, B_1^{q-1}) + \Gamma(a_p' \to b_q') & match \\ \delta_{\lambda,\upsilon}(A_1^p, B_1^{q-1}) + \Gamma(\Lambda \to a_p') & delete_B \end{cases} \quad (1)$$

with

$$\Gamma(a_p' \to \Lambda) = d(a_p', a_{p-1}') + \lambda$$
$$\Gamma(a_p' \to b_q') = d(a_p', b_q') + d(a_{p-1}', b_{q-1}') .$$
$$\Gamma(\Lambda \to b_q') = d(b_q', b_{q-1}') + \lambda$$

The recursion is initialized as follows:

$$\delta_{\lambda,\upsilon}(A_1^0, B_1^0) = 0$$
$$\delta_{\lambda,\upsilon}(A_1^0, B_1^j) = \infty, \text{ for } j \geq 1,$$
$$\delta_{\lambda,\upsilon}(A_1^i, B_1^0) = \infty, \text{ for } j \geq 1$$

with $a_0' = b_0' = 0$ by convention.

TWED further introduces *stiffness* into the definition of $\delta_{\lambda,\upsilon}$ by choosing $d(a', b') = d_{LP}(a, b) + \upsilon * d_{LP}(t_a, t_b)$, where υ is a non-negative constant to characterize the *stiffness* of the elastic measures $\delta_{\lambda,\upsilon}$. Notice that $\upsilon > 0$ is required to let $\delta_{\lambda,\upsilon}$ be a distance. If $\upsilon = 0$, then $\delta_{\lambda,\upsilon}$ will be a distance on S but not on $S \times T$. In this equation d_{LP} is the l_p norm metrics [16]. The final formulation of $\delta_{\lambda,\upsilon}$ is shown as follows:

$$\delta_{\lambda,v}(A_I^p, B_I^q) = \min \begin{cases} \delta_{\lambda,v}(A_I^{p-1}, B_I^q) + \Gamma(a_p' \to \Lambda) & delete_A \\ \delta_{\lambda,v}(A_I^{p-1}, B_I^{q-1}) + \Gamma(a_p' \to b_q') & match \\ \delta_{\lambda,v}(A_I^p, B_I^{q-1}) + \Gamma(\Lambda \to b_q') & delete_B \end{cases} \quad (2)$$

with

$$\Gamma(a_p' \to \Lambda) = d_{LP}(a_p, a_{p-1}) + v \cdot (t_{a_p} - t_{a_{p-1}}) + \lambda$$

$$\Gamma(a_p' \to b_q') = d_{LP}(a_p, b_q) + d_{LP}(a_{p-1}, b_{q-1})$$

$$+ v \cdot (\left| t_{a_p} - t_{b_q} \right| + \left| t_{a_{p-1}} - t_{b_{q-1}} \right|),$$

$$\Gamma(\Lambda \to b_q') = d_{LP}(b_q, b_{q-1}) + v \cdot (t_{b_q} - t_{b_{q-1}}) + \lambda$$

Algorithm 1 (Time Warp Edit Distance with Sakoe-Chiba Band)

Input: A a time series A[1..n] with time index tA[1..n]

  B a time series B[1..m] with time index tB[1..m]

  v stiffness parameter to control the elasticity of the metric

  λ constant penalty for the insert or delete operations

  r the width of Sakoe-Chiba band

Output: TWED(A, B) the TWED distance between time series A and B

1: Initialization of the matrix DP[0..n, 0..m]

2: A[0] = 0, tA[0] = 0, B[0] = 0, tB[0] = 0

3: DP[0, 0] = 0

4: DP[i, 0] = ∞, i = 1,2,...,n

5: DP[0, j] = ∞, j = 1,2,...,m

6: **for** i = 1:n

7:   **for** j = max(1, $i-r$):min(m, $i+r$)

8:    $DP[i, j] = \min \begin{cases} DP[i-1, j] + d_{LP}(A[i], A[i-1]) + v \cdot (tA[i] - tA[i-1]) + \lambda, \\ DP[i-1, j-1] + d_{LP}(A[i], B[j]) + d_{LP}(A[i-1], B[j-1]) \\ \quad + v \cdot (\left| tA[i] - tB[j] \right| + \left| tA[i-1] - tB[j-1] \right|), \\ DP[i, j-1] + d_{LP}(B[j], B[j-1]) + v \cdot (tB[j] - tB[j-1]) + \lambda \end{cases}$

9: **end for**

10: **end for**

11: TWED(A, B) = DP[n, m]

To improve the efficiency of TWED, we further incorporate the Sakoe-Chiba band [18] to reduce the number of paths that need to be considered in the elastic matching. Finally, the detailed iterative implementation of the TWED with Sakoe-Chiba band algorithm which computes minimal global cost is summarized in Algorithm 1.

4 Experimental Results

In this section, the wrist pulse blood flow signals from the pulse signal dataset [14] are used to evaluate the classification accuracy of TWED. This dataset contains 320 wrist pulse blood flow signals, which includes healthy people and people with four different kinds of diseases, i.e., 100 healthy people, 54 patients with pancreatitis (P), 77 with duodenal bulb ulcer (DBU), 35 with appendicitis (A), and 54 with acute appendicitis (AA). All the samples are acquired from Harbin Binghua Hospital, and the type of disease of each sample is assigned according to the clinical diagnosis results. In our experiments, the dataset is divided into a test dataset and a training dataset before preprocessing. For the test dataset, 50 healthy people, 27 patients with pancreatitis, 40 with duodenal bulb ulcer, 20 with Appendicitis and 27 with acute appendicitis are chosen. The remained samples are used to construct the training dataset.

In the experiments, using the TWED matching method, the nearest neighbour classifier is adopted to classify the signals after de-noising and drift removal. For TWED, the value of the stiffness parameter v is selected from $\{10^{-5}, 10^{-4}, 10^{-3}, 10^{-2}, 10^{-1}, 1\}$ and the value of the constant penalty λ is selected from $\{0, 0.25, 0.5, 0.75, 1.0\}$. Using the training set, we determine the optimal parameter values by minimizing the classification errors with the leave-one-out strategy.

After determining the optimal parameter values, we test the classification performance of TWED on the test dataset. We carry out five classification experiments: classification of healthy people and people with pancreatitis, classification of healthy people and people with duodenal bulb ulcer, classification of healthy people and people with appendicitis, classification of healthy people and people with acute appendicitis, and classification of healthy people and people with diseases.

To further evaluate the classification performance of TWED, we also compare TWED with several other popular time series matching methods, i.e., Euclidean distance (ED), dynamic time warping (DTW) [17], and edit distance with real penalty

Table 1. Experimental results in distinguishing patients from healthy people

Sample Class	Sample number		Accuracy (%) ED	Accuracy (%) DTW [17]	Accuracy (%) ERP [16]	Accuracy (%) TWED
Healthy	50	77	61.04	76.62	76.62	90.91
Pancreatitis	27					
Healthy	50	90	61.11	73.33	73.33	77.78
DBU	40					
Healthy	50	70	62.86	74.29	77.14	77.14
Appendicitis	20					
Healthy	50	77	61.95	66.23	72.73	76.62
Acute Appendicitis	27					
Healthy	50	164	62.81	67.07	67.68	71.95
All kinds of diseases	114					

(ERP) [16]. The classification results of each classification tasks obtained using different matching methods are listed in Table 1. From Table 1, one can see that, compared with other time series matching methods, TWED is more effective for classification of the wrist blood flow signals.

5 Conclusions and Future Work

In this paper, we adopt an effective time series matching method, TWED, for classification of wrist pulse blood flow signals. Compared with other time series matching methods, TWED is an elastic metric which involves a stiffness parameter to control the elasticity of the metric and a second parameter to define a constant penalty for the *insert* and *delete* operations, and thus is expected to be effective for classification of wrist blood flow signals.

To evaluate the classification performance of TWED, experiments are carried out to classify healthy people from patients with pancreatitis, duodenal bulb ulcer, appendicitis, and acute appendicitis. Experimental results show that TWED is superior to other time series matching methods, e.g., the ED, DTW, and ERP.

For future work, we will develop more effective and efficient classification method with the help of the metric property of TWED, and will combine TWED with other time series modeling and analysis methods to further improve the classification performance.

Acknowledgment

This work is partially supported by the NSFC funds of China under Contract Nos. 60902099, 60872099, and 60871033, and the Education Ministry Young Teacher Foundation of China under No. 200802131025.

References

1. Lukman, S., He, Y., Hui, S.: Computational methods for traditional Chinese medicine: a survey. Computer Methods and Programs in Biomedicine 88, 283–294 (2007)
2. Zhu, L., Yan, J., Tang, Q., Li, Q.: Recent progress in computerization of TCM. Journal of Communication and Computer 3, 78–81 (2006)
3. Wang, K., Xu, L., Zhang, D., Shi, C.: TCPD based pulse monitoring and analyzing. In: Proceedings of the 1st ICMLC conference, Beijing (2002)
4. Wang, H., Cheng, Y.: A quantitative system for pulse diagnosis in Traditional Chinese Medicine. In: Proceedings of the 27th IEEE EMB conference, Shanghai (2005)
5. Chen, Y., Zhang, L., Zhang, D., Zhang, D.: Wrist Pulse Signal Diagnosis using Modified Gaussian Models and Fuzzy C-Means Classification. Medical Engineering & Physics 31, 1283–1289 (2009)
6. Chen, Y., Zhang, L., Zhang, D., Zhang, D.: Computerized Wrist Pulse Signal Diagnosis using Modified Auto-regressive Models. Journal of Medical Systems (September 2009)
7. Lu, W., Wang, Y., Wang, W.: Pulse analysis of patients with severe liver problems. IEEE Eng. in Med. and Biol. 18, 73–75 (1999)

8. Zhang, A., Yang, F.: Study on recognition of sub-health from pulse signal. In: Proceeding of ICNNB Conference, vol. 3, pp. 1516–1518 (2005)

9. Zhang, D., Zhang, L., Zhang, D., Zheng, Y.: Wavelet-based analysis of Doppler ultrasonic wrist-pulse signals. In: Proceeding of ICBBE Conference, Shanghai, vol. 2, pp. 589–543 (2008)

10. Zhang, D., Zuo, W., Zhang, D., Zhang, H., Li, N.: Wrist Blood Flow Signal-based Computerized Pulse Diagnosis Using Spatial and Spectrum Features. J. Biomedical Science and Engineering, 6 pages (2010)

11. Marteau, P.-F.: Time Warp Edit Distance with Stiffness Adjustment for Time Series Matching. IEEE Trans. Pattern Analysis and Machine Intelligence 31, 306–318 (2009)

12. Chen, Y., Zhang, L., Zhang, D., Zhang, D.Y.: The public Pulse Signal Dataset (2010), http://www4.comp.polyu.edu.hk/~cslzhang/Pulse_datasets.htm

13. Xu, L., Zhang, D., Wang, K.: Wavelet-based cascaded adaptive filter for removing baseline drift in pulse waveforms. IEEE Trans. Biomed. Eng. 52, 1973–1975 (2005)

14. Wagner, R.A., Fischer, M.J.: The String-to-String Correction Problem. J. ACM 21, 168–173 (1973)

15. Mäkinen, V.: Using Edit Distance in Point-Pattern Matching. In: Proc. Eighth Symp. String Processing and Information Retrieval (SPIRE 2001), pp. 153–161 (2001)

16. Chen, L., Ng, R.: On the Marriage of LP-Norm and Edit Distance. In: Proc. Int'l. Conf. Very Large Data Bases, pp. 792–801 (2004)

17. Keogh, E., Ratanamahatana, C.A.: Exact indexing of dynamic time warping. Knowledge and Information Systems 7, 358–386 (2005)

18. Sakoe, H., Chiba, S.: Dynamic programming algorithm optimization for spoken word recognition. IEEE Trans. Acoustics, Speech, and Signal Proc. 26, 43–49 (1978)

Computerized Pork Quality Evaluation System

Li Liu and Michael O. Ngadi

Department of Bioresource Engineering, McGill University, Macdonald Campus,
Ste-Anne-de-Bellevue,
Quebec, Canada H9X 3V9
Michael.ngadi@mcgill.ca, li.liu5@mail.mcgill.ca

Abstract. Pork quality assessment is important in the pork industry application. However, traditional pork quality assessment is conducted by experienced workers and thereby is subjective. In this paper, a computerized system scheme based on hyperspectral imaging technique is proposed for objective pork quality evaluation. This hyperspectral imaging technique used texture characteristics to develop an accurate system of pork quality evaluation. Hypercube, which is a set of spectral images over all wavelengths, were filtered by oriented Gabor filters to obtain texture characteristics. Spectral features were extracted from Gabor-filtered cube as well as from hypercube directly and then compressed by the principal component analysis (PCA) over the entire wavelengths (400-1000 nm) into 5 and 10 principal components (PCs). 'Hybrid' PCs were created by combining PCs from hypercube and from Gabor-filtered cube. Linear discriminant analysis (LDA) was employed to classify pork samples based on hybrid PCs as well as pure PCs. The cross-validation technique was applied on LDA to produce the unbiased statistical results. The overall average accuracy was 72% by 5 pure PCs and reached 84% by 5 hybrid PCs. The highest accuracy, 100% classification for all samples, was obtained when using 5 hybrid PCs. Thus, a statistical significant improvement was achieved using image texture features. Results showed that the proposed computerized system worked well on pork quality evaluation and has potential for on-line pork industry application.

Keywords: Computerized system, hyperspectral imaging, pork quality, texture, Gabor filter, PCA, LDA.

1 Introduction

Objective quality assessment of fresh pork is increasingly becoming vital in the pork processing industry due to globalization and segmentation of markets. This has led to an urgent need for efficient technologies to assess pork quality levels. Normally, pork quality can be classified into five categories based on color, texture/firmness and exudation/drip loss [1-3]. RFN meat, which is regarded as the ideal class, has normal color, texture and water-holding capacity (WHC). PSE meat has undesirable appearance and lacks firmness due to excessive drip loss, while DFD meat has firm and sticky surface with high WHC, very little or no drip loss, and very high pH [1].

D. Zhang and M. Sonka (Eds.): ICMB 2010, LNCS 6165, pp. 145–152, 2010.
© Springer-Verlag Berlin Heidelberg 2010

Both PSE and DFD are traditional quality defects in fresh pork. Over the years RSE and PFN have been recognized as major quality defects in Canada, which accounts for >13% in all defects compared to PSE (13%) and DFD (10%) [4]. Meat defects bring big economic loss which makes it more and more important in meat industry applications to develop an efficient and effective quality evaluation system for quick and objective meat defect assessment.

Recently, hyperspectral imaging technique has been applied for pork quality classification [5,6]. A hyperspectral image, also called *hypercube* [7], is a three-dimensional image with two spatial dimensions and one wavelength dimension. Each pixel in a hypercube contains the spectrum of the position over hundreds of contiguous wavebands. In a hyperspectral imaging system, spectral information of fresh pork samples was extracted by principle component analysis (PCA) and quality levels were classified by applying a feed-forward neural network model on the extracted spectral information. Experimental results show that PCA perfectly classified RFN samples and the feed-forward neural network yielded correct classification at 69% by 5 PCs and 85% by 10 PCs [6]. Although promising results were obtained by this study, some work still need to be done for improvement. In the hyperspectral imaging system, spectral features which are related with color and water content of pork were used for quality class classification without considering the image's texture feature. However, according to the pork quality standards, texture feature is as important for pork quality assessment as color and water content features. Therefore, the use of image texture feature for evaluation should be considered and consequently is expected to bring some improvement to pork quality evaluation system.

The two-dimensional (2-D) Gabor filter is a well known image texture detector. It is a linear filter which is a sinusoidal modulated Gaussian function in the spatial domain. The 2-D Gabor filter has the ability to achieve certain optimal joint localization properties in the spatial domain and in the spatial frequency domain [8]. This ability exhibits desirable characteristics of capturing salient visual properties such as spatial localization, orientation selectivity, and spatial frequency. Such characteristics make it an effective tool for texture analysis [9,10].

In this paper, we aimed to develop an improved pork quality evaluation system by combining texture features with the spectral features which characterize color and water content of pork. For this purpose, a hyperspectral imaging system which can utilize texture features was developed based on 2-D Gabor filters. The specific objectives were to extract texture features by applying the Gabor filter; to obtain spectral features as well as texture features for pork quality evaluation; and to determine the discriminating capability of various types of spectral features by employing the linear discriminant analysis (LDA).

2   Materials and Methods

2.1   Samples and Data Acquisition

In this study, we employed the data used in [6] to evaluate the pork quality levels. Fresh pork loins were collected from a local commercial packing plant and the quality levels of all samples were assessed by a trained employee. After shipping to the Instrumentation laboratory at McGill University, the loin samples were cut into 1 cm thick chops and forty fresh pork loin samples, including RFN (10), PSE (10), PFN (10) and RSE (10), were obtained for the experiment.

Hyperspectral images of the forty fresh pork loin samples were captured by a hyperspectral imaging system which consisted of a line-scan spectrograph (ImSpector, V10E, Spectra Imaging Ltd, Finland), a CMOS camera (BCi4-USB-M40LP, Vector International, Belgium), a DC illuminator (Fiber-Lite PL900-A, Dolan-Jenner Industries Inc, USA), a conveyor (Dorner 2200 series, Donner Mfg. Corp., USA) and a supporting frame as shown in Fig. 1 (a), as well as a data acquisition and preprocessing software (SpectraCube, Auto Vision Inc., CA, USA) as shown in Fig. 1 (b). This system captured hyperspectral images in a wavelength range of 400-1000 nm with a spectral resolution of 2.8 nm and a spot radius <9 µm.

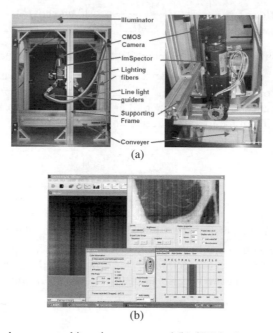

(a)

(b)

Fig. 1. (a) The hyperspectral imaging system and (b) GUI for image acquisition

2.2 Image Processing

The objective of image processing was to extract useful spectral features for the following data analysis. All operations in this section were performed using MATLAB 7.0 (The MathWorks, Inc., Mass., USA).

Image Preprocessing. For each sample, a hyperspectral image was captured by the hyperspectral imaging system and was stored in a RAW format. To correct spectral images from dark current of camera, the relative reflectance, I, of an image was calculated by

$$I = \frac{I_0 - B}{W - B},\tag{1}$$

where B was a dark image, W was a white image taken from a standard white reference, and I_0 was the original spectral image. In order to avoid the bias introduced by the selection of region of interest (ROI), the whole lion-eye area of each sample was selected as ROI by using a predefined mask.

Gabor Filter and Spectral Features. In order to extract useful and helpful texture features of pork samples, two-dimensional (2-D) Gabor filters were applied to the preprocessed hyperspectral images. An oriented Gabor filter G is a Gaussian function modulated by an oriented harmonic function and can be defined as follows [11]:

$$G(x, y; u, \sigma, \theta) = \frac{1}{2\pi\sigma^2}\exp\left\{-\frac{x^2 + y^2}{2\sigma^2}\right\}\cos[2\pi u(x\cos\theta + y\sin\theta)],\tag{2}$$

where (x, y) is the coordinate of point in 2-D space, u is the frequency of the sinusoidal wave, σ is the standard deviation of the Gaussian envelope and θ controls the orientation of the Gabor filter. To make Gabor filters more robust against brightness difference, a discrete Gabor filter was tuned to zero DC (direct current) with the application of the following formula:

$$\tilde{G} = G - \frac{\sum_{i=-n}^{n}\sum_{j=-n}^{n}G[i, j]}{(2n + 1)^2},\tag{3}$$

where $(2n + 1)^2$ is the size of the filter. The adjusted circular Gabor filter $\tilde{G}$ were used to filter the ROI of each sample's hypercube. The mean spectrum of the Gabor-filtered ROI and the mean spectrum of ROI were used as spectral features. For denoising both mean spectra were smoothed by a one-dimensional mean filter with 11 points wide.

2.3 Data Analysis

Spectral features obtained in Section 2.2 were first projected onto a lower dimensional linear space by using principal component analysis (PCA) to extract non redundant features and then the extracted features were classified by using typical supervised classification methods. All operations in this section were performed using MATLAB 7.0 (The MathWorks, Inc., Mass., USA).

Linear discriminant analysis, i.e. LDA [12], is a supervised technique which has been commonly applied for data classification and was used in this study for supervised classification. LDA best classifies objects into two or more classes by finding the linear combination of features. In our study, LDA was employed to classify pork samples into different quality levels. For each class, six samples were used as training set, while the rest of four samples were used as testing set for LDA.

The main problem of supervised classification is that the classification results vary with different training sets. To solve this problem, the cross-validation technique was used for LDA. Cross-validation is a statistical estimation method for evaluating how the model would generalize to independent data sets. The technique partitions the data set into two un-overlapped subsets. One is used for training and the other for testing. Normally, multiple rounds of cross-validation are performed by using different partitions to reduce variability. The validation results are averaged over the rounds. In this study, the cross-validation technique was applied based on an extensive investigation. For every class, a label from 1 to 10 was randomly assigned to each sample. The samples of different classes having the same label were regarded as the same group and thereby a total of 10 groups were obtained. Each group had 4 samples which were from 4 different classes. Selecting six groups as the training set and the remaining groups as the testing set, a total of 210 ($= C_{10}^6$) partitions were obtained. In the rest of this paper, all of the partitions were used to obtain the statistical classification results.

3 Results and Discussions

3.1 Spectral Features and PCA

The hyperspectral images of pork samples reported in [6] were used for quality level classification. Fig. 2 shows the procedure of image preprocessing and Gabor filter application. The ROI of each hypercube was obtained by using the predefined mask (Fig. 2b) and then all acquired ROI images (Fig. 2c) were filtered by four oriented

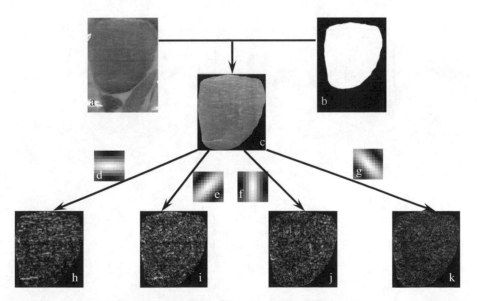

Fig. 2. The region of interest and its Gabor-filtered images. (a) Spectral image of a pork sample at wavelength 680 nm, (b) predefined ROI mask of the pork sample, (c) spectral ROI image of (a), (d-g) the oriented Gabor filters at four directions and (h-k) the corresponding Gabor-filtered ROI images using (d-g), respectively.

Gabor filters $\{\tilde{G}_0, \tilde{G}_1, \tilde{G}_2, \tilde{G}_3\}$ (Fig. 2d-2g), where $u = 0.1$, $\sigma = 5$, $n = 11$, and $\theta = \{0, \pi/4, \pi/2, 3\pi/4\}$. The corresponding Gabor-filtered ROI images were shown in Fig. 2h-2k.

Typical curves of spectral features for different pork quality levels are shown in Fig. 3, including mean spectral curves of ROI, denoted by **MS**, as shown in Fig. 3(a) and mean response curves of Gabor-filtered ROI, denoted by $\{\mathbf{MG}_0, \mathbf{MG}_1, \mathbf{MG}_2, \mathbf{MG}_3\}$, as shown in Fig. 3(b-e). The differences in spectral feature curves between **MS** and **MG** indicated that combination of the two types of spectral features could improve the evaluation results of pork quality levels. Hence, PCA was applied to both **MS** and **MG** and principal components (PCs) generated from the two types of spectral features were used to classify the quality levels.

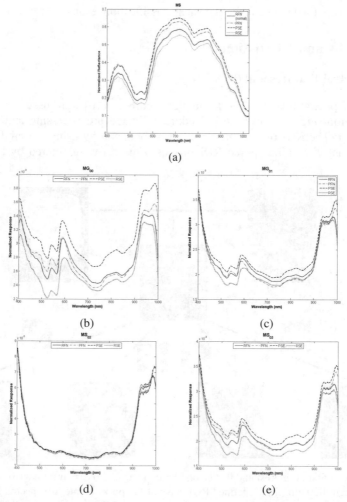

Fig. 3. (a) Spectral features of ROI and (b-e) spectral features/responses of the corresponding Gabor-filtered ROI

3.2 Linear Discriminant Analysis (LDA)

Table 1 listed the average accuracies and corresponding errors (standard deviation) over 210 partitions by all of 5 pure PCs as well as by 5 hybrid PCs which produced average accuracies above 80%. The highest average accuracy by 5 *hybrid* PCs reached 84% with a standard deviation of 10%, while the average accuracy by 5 **MS** PCs was 72% with a standard deviation of 12%. In [6], the corrected classification with Feed Forward Network was 69% using 5 PCs. Thus, a statistically significant improvement was achieved by using Gabor-filter based spectral features, i.e. image texture features. Furthermore, some perfect evaluation results, i.e. 100% classification for all 4 classes, were obtained for some partitions by 5 *hybrid* PCs, as shown in Table 1.

The corresponding selections of 5 *hybrid* PCs whose average accuracies were above 80% are also listed in Table 1. Comparing different Gabor filter-based spectral features, MG_1 was involved in all of six combinations, while MG_0, MG_2 and MG_3 only in one. This strongly suggested that the texture features along the direction of $\pi/4$ offered more useful information for pork quality level classification.

Table 1. Cross validation results over 210 partitions using LDA by 5 PCs

PCs	MS	MG_0	MG_1	MG_2	MG_3	Accuracy (%) Mean	STD	The Number of 100% Classification
	5	0	0	0	0	72	12	0
	0	5	0	0	0	46	11	0
Pure PC	0	0	5	0	0	55	12	0
	0	0	0	5	0	52	9	0
	0	0	0	0	5	48	10	0
	4	0	1	0	0	80	9	3
	1	0	4	0	0	81	9	5
Hybrid PC	2	0	3	0	0	81	9	7
	1	0	3	1	0	82	9	3
	1	0	3	0	1	84	9	9
	1	1	3	0	0	**84**	10	16

4 Conclusions

The study demonstrated that the hypserspectral imaging-based pork quality evaluation system greatly improved the accuracy of pork quality assessment by using texture features which were extracted by 2-D Gabor filters. In this study, the region of interest of each hypercube was filtered by a set of oriented Gabor filters to extract texture features. Spectral features created by Gabor-filtered ROI were used for pork quality evaluation with the spectral features of ROI. Dimensionality of spectral features was reduced by employing principal component analysis. Based on principal components, LDA was applied for pork quality evaluation and the cross-validation technique over a total of 210 partitions was employed for unbiased evaluation results. The average accuracy of pork quality evaluation reached 84% by 5 hybrid PCs and was 72% by 5 **MS** PCs, which suggested a statistically significant improvement using texture features. The evaluation results strongly indicated that the texture features along the direction of $\pi/4$ offered more useful information for pork quality level classification.

References

1. NPB (National pork board): Pork quality standards. Des Moines, IA, USA (1999)
2. Kauffman, R.G., Cassens, R.G., Scherer, A., Meeker, D.L.: Variations in pork quality. National Pork Producer Council, Des Moines (1992)
3. Nam, K.C., Du, M., Jo, C., Ahn, D.U.: Effect of ionizing radiation on quality characteristics of vacuum-packaged normal, pale-soft-exudative, and dark-firm-dry pork. Innovative Food Sci. & Emerging Tech. 3, 73–79 (2002)
4. Murray, A.C.: Reducing losses from farm gate to packer: A Canadian's perspective. In: Proceeding of 1st Int'l. virtual conference on pork quality, Concordia, Brazil, pp. 72–84 (2000)
5. Geesink, G.H., Schreutelkamp, F.H., Frankhuizen, R., Vedder, H.W., Faber, N.M., Kranen, R.W., Gerritzen, M.A.: Prediction of pork quality attributes from near infrared reflectance spectra. Meat Sci. 65, 661–668 (2003)
6. Qiao, J., Ngadi, M.O., Wang, N., Gariepy, C., Prasher, S.O.: Pork quality and marbling level assessment using a hyperspectral imaging system. J. Food Eng. 83, 10–16 (2007)
7. Lu, R.F., Chen, Y.R.: Hyperspectral imaging for safety inspection of food and agricultural products. In: SPIE Conference on Pathogen Detection and Remediation for Safe Eating, Boston (November 1998)
8. Daugman, J.G.: Uncertainty relation for resolution in space, spatial frequency, and orientation optimized by two-dimensional visual cortical filters. J. Opt. Soc. Am. A 2(7), 1160–1169 (1985)
9. Daugman, J.G.: High confidence visual recognition of persons by a test of statistical independence. IEEE Trans. Pattern Anal. Mach. Intel. 15(11), 1148–1161 (1993)
10. Clausi, D.A., Jernigan, M.: Designing Gabor filters for optimal texture separability. Pattern Recognit. 33, 1835–1849 (2000)
11. Ma, L., Wang, Y., Tan, T.: Iris recognition using circular symmetric filters. In: Proc. 16th Int'l. Conf. Pattern Recognit., vol. II, pp. 414–417 (2002)
12. Duda, R.O., Hart, P.E., Stork, D.: Pattern Classification. Wiley, Chichester (2000); Hartigan, J.A., Wong, M.A.: A k-means clustering algorithm. Applied Statistics 28(1), 100–108 (1979)

Abnormal Image Detection Using Texton Method in Wireless Capsule Endoscopy Videos

Ruwan Dharshana Nawarathna[1], JungHwan Oh[1], Xiaohui Yuan[1],
Jeongkyu Lee[2], and Shou Jiang Tang[3]

[1] Department of Computer Science and Engineering, University of North Texas, Denton,
TX 76203, U.S.A
rdn0025@unt.edu, {jhoh,xyuan}@cse.unt.edu
[2] Department of Computer Science and Engineering, University of Bridgeport, Bridgeport,
CT 06604, U.S.A
jelee@bridgeport.edu
[3] Shou Jiang Tang, Endoscopy Center, Trinity Mother Frances Hospitals and Clinics,
Tyler, TX 75702, U.S.A

Abstract. One of the main goals of Wireless Capsule Endoscopy (WCE) is to detect the mucosal abnormalities such as blood, ulcer, polyp, and so on in the gastrointestinal tract. Only less than 5% of total 55,000 frames of a WCE video typically have abnormalities, so it is critical to develop a technique to automatically discriminate abnormal findings from normal ones. We introduce "Texton" method which has been successfully used for image texture classification in non-medical domains. A histogram of Textons (exemplar responses occurring after convolving an image with a set of filters called "Filter bank") called a "Texton Histogram" is used to represent an abnormal or a normal region. Then, a classifier (i.e., SVM or K-NN, and etc.) is trained using the Texton Histograms to distinguish images with abnormal regions from ones without them. Experimental results on our current data set show that the proposed method achieves promising performances.

Keywords: Wireless Capsule Endoscopy, Abnormality, Texton, Filter bank, Texton histogram, Abnormality dictionary.

1 Introduction

Conventional endoscopies such as Gastroscopy, Push Enteroscopy, Colonoscopy, and etc have been playing a very important role as diagnostic tools for the digestive tract. However, they are limited in viewing the small intestine. To address the problem, Wireless Capsule Endoscopy (WCE) was first developed in 2000, which integrates wireless transmission with image and video technology to allow physicians to examine the entire small intestine non-invasively [1-3]. One of the main goals of WCE is to detect abnormal lesions. The important abnormal lesions (abnormalities) in WCE are fresh blood (bleeding), ulceration, erosion, angioectasia, polyps, and tumors. Details about these abnormalities or diseases can be found in [1-3]. A frame showing an abnormality is called an abnormal image. Fig. 1 shows four abnormal images from a real WCE video.

D. Zhang and M. Sonka (Eds.): ICMB 2010, LNCS 6165, pp. 153–162, 2010.
© Springer-Verlag Berlin Heidelberg 2010

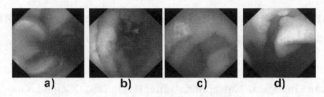

Fig. 1. Images showing abnormalities such as a) Blood, b) Erosion, c) Ulcer, and d) Polyp

A frame that does not show any abnormality is a normal image. In a typical WCE video, less than 5% of frames are abnormal images. Since there is a large number of images (i.e., 55,000) in a video, this examination is an extremely time consuming job for the physician. It limits its general application and incurs considerable amount of health-care costs. Therefore, it is highly desired to develop an automatic method to filter out the abnormal and normal images before GI investigates it. In an abnormal image, only a small region shows an abnormality. Thus, the automated procedure should be able to discern abnormal regions from normal ones. To address this requirement, we propose a new algorithm utilizing the Texton method [4-6]. Our method focuses on distinguishing regions showing abnormalities such as blood, erosion, polyp, and ulcer from WCE images.

First, we generate a set of Textons called "Texton Dictionary" from the abnormal and the normal regions of each image. Textons are the exemplar responses occurring after convolving an image with a set of filters called "Filter bank". These Textons can represent many different pixel relationships in a region, which is essential for image texture analysis. A filter in a filter bank is an $n \times n$ matrix of numbers that is convoluted with each pixel to find various different features of pixels. Mostly, n is 3, 5, 7, 25 or 49. Then, this Texton dictionary is used to generate a histogram called "Texton Histogram" to represent an abnormal or a normal region. Using the Texton Histograms, a classifier (i.e., SVN or K-NN, etc. [11]) is trained to distinguish images with abnormal regions from ones without them. Experimental results on our current data set show that the proposed method achieves promising performances, which can be used to detect abnormal images in practice.

The remainder of this paper is organized as follows. Background and related work are discussed in section 2. The basics of Texton method are discussed in Section 3. Our method for abnormality detection is proposed in Section 4. We discuss our experimental results in Section 5. Finally, Section 6 presents some concluding remarks.

2   Background and Related Work

The related works could be divided into three main categories: (1) preprocessing of WCE videos such as image enhancement, frame reduction, and so on, (2) detection of abnormalities such as bleeding, ulceration, and etc, and (3) video segmentation. Most of the research works done in WCE are aiming at developing automated algorithms for abnormality detection. In most of the WCE abnormality detection methods, either some have focused on detecting only one type of abnormality, or the overall accuracy are somewhat low if they tried to detect multiple abnormalities. According to the domain experts, if all the abnormal frames are filtered out from a WCE video, finding a particular abnormality is not a big task. A method to detect all abnormal frames is desired. Hence, our main aim is to develop a method that can discover any kind of abnormality. When developing such a method, image texture analysis is an ultimate choice because of the following. Generally, WCE images are distorted by various substances like stool, water and also by specular reflection. Therefore, color information in WCE images is unreliable. In WCE images, we have observed that each of the abnormalities has a unique texture. Texton method with its success proven in [4-6,9,10] is a good candidate for discriminating these various different texture patterns. We are proposing a technique utilizing the Texton method, which could detect abnormalities such as blood, polyp, ulcer, and erosion. This could be extended to detect any kind of abnormality in WCE.

3   Texton Method

In this section, we discuss the fundamentals of Texton method. Texton method is one of the most recent image texture analysis methods, which was proposed by Leung and Malik [4]. It has shown a tremendous success in detecting materials such as glass, wood, concrete and etc [4-6]. Basically, it tends to describe a small region of texture which is frequently occurring in an image, and uses it as a representative of the texture as a whole. In the simplest of terms, Textons are exemplar filter responses resulted in after convolving (with a collection of filters) a set of images belonging to one particular class having common texture. This collection of filters is called a filter bank. Each image in a particular class is convoluted by all the filters in the filter bank. Then, all the filter responses of a pixel are concatenated as a vector. If a filter bank has N filters, each pixel generates an N-dimensional vector. All the vectors are collected and clustered using a clustering algorithm such as K-means or DBScan [11]. The resulting cluster centers are called Textons.

As mentioned above, a filter bank is a set of filters. The goal of using filter banks in the Texton method is that, by using a set of filters, variety of different texture features such as directionality, orientation, line-likeness, repetitiveness, and etc could be extracted. This is the main advantage of using Texton over the other existing texture analysis methods such as concurrence matrix, discrete wavelet transform and etc. Most commonly used filter banks for Texton generation are Leung-Malik (LM), Schmid (S), Maximum Responses (MR) and Gabor [4-6, 9-10]. LM filter bank consists of 48 different filters. Six of first derivatives of Gaussian filters in three different scales and another six of second derivates of Gaussian filters in three different scales, give 36 filters in total. The remaining filters are eight Laplacian of Gaussian (LoG) filters and four Gaussian filters. A filter in the Schmid filter bank takes the form given in (1). In (1), τ is the number of cycles of the harmonic function within the Gaussian envelope of the filter and $F_0(\sigma, \tau)$ is used to obtain a zero DC component. The ranges of σ, and τ are between 2 and 10 and between 1 and 4, respectively.

$$F(x, y, \sigma, \tau) = F_0(\sigma, \tau) + \cos\left(\frac{\sqrt{x^2+y^2}\tau\pi}{\sigma}\right)e^{-\frac{x^2+y^2}{2\sigma^2}}. \qquad (1)$$

The Maximum Response (MR) filter is another successful filter bank consisting 38 filters. It has one Gaussian and one Laplacian of Gaussian filter with $\sigma=10$. Also, it has total 18 edge filters (three scales and six orientations), and 18 bar filters (three scales and six orientations). MR8 is a variation of MR filter bank which produces only 8 responses [6].

We tested LM, S, and MR8 filter banks, and found out that LM filter bank performs better than the other two for our domain. Therefore, our abnormal image detection is associated with the LM filter bank.

4   Our Method: Abnormal Image Detection Using Texton Method

Our proposed method has three major steps: (1) Texton generation, (2) Abnormality dictionary generation, and (3) Abnormality detection. The method is developed based on a set of images extracted from real WCE Videos. The set has images showing either some abnormality or no abnormality. Our method is designed to detect some major abnormalities such as blood, polyp, ulcer, and erosion. Therefore, all the images fall into one of four classes: *Blood*, *Polyp*, *Ulcer-Erosion*, and *Normal*. Since ulcer and erosion abnormalities show similar texture, those two are considered as one class named Ulcer-Erosion. In the following sections, each of the major steps is discussed in detail.

4.1   Texton Generation

The first step, Texton Generation, is a key step in our abnormality detection method. A main purpose of this step is to find the pixels (as Textons) that are most likely to occur in a particular class. First, we select a set of images from each of the four classes mentioned above so that each set has all possible texture patterns in that particular class. We extract the exact region showing an abnormality from each image in each abnormal class, and also extract some arbitrary image blocks (128 by 128 pixels) from each image in Normal class. Therefore, our image regions have arbitrary shapes. Fig. 2 a) and b) show how the abnormal regions (with blue-colored boundaries) are extracted from abnormal images.

Generally, WCE images are distorted by various illumination and lighting effects of the camera. Thus, before applying any filter bank to the extracted regions, two preprocessing steps are necessary on image regions. First, each region is converted into gray scale, and normalized to have zero mean and unit standard deviation. These can make sure that, we get exact textural information rather than color variations in the pixels. After preprocessing steps, each region is convoluted with a filter bank. We apply the Leung-Malik filter bank which produces 48 responses for each pixel in a region. We combine all the 48 responses into a vector. We call this vector as a response vector. In order to make sure that all filter responses are in the same range and vector quantization is trouble-free, each filter response is converted to L1-norm. Each vector is normalized using Webber's formula [6], which is defined in (2).

$$F(x) \leftarrow \frac{F(x)\left[\log\left(1+\frac{L(x)}{0.03}\right)\right]}{L(x)}. \tag{2}$$

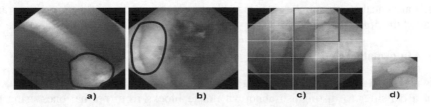

Fig. 2. a) Extraction of a polyp region, b) Extraction of an erosion region for Texton generation, c) Image divided into number of blocks, and d) Extracted block from c)

In (2), F(x) is the gray value and L(x) is L1-norm of the gray values. This procedure provides a set of 48-element response vectors. The total number of elements in this set is equal to the total number of pixels in all image regions in each class. From each class, K representative vectors are selected by applying K-Means clustering algorithm [11]. A representative vector is called a Texton. The collection of Textons from all classes is called the Texton dictionary.

4.2 Abnormality Dictionary Generation

This is the second step of our abnormal image detection method, which builds a model that distinguishes abnormal regions from normal ones. A main purpose of this step is to come up with a set of possible representations of abnormal and normal regions and use them to train a classifier. To generate the abnormality dictionary, we divide each image into fixed size image blocks, for instance 32 by 32, 64 by 64 or 128 by 128 pixels. We select blocks showing some kind of abnormality from each abnormal class and some random blocks from the normal class. Since our image has 320 x 320 pixels, it can be divided into 32 by 32 or 64 by 64 pixel blocks. But, it cannot be divided into 128 by 128 pixel blocks, so we allow some block overlapping of 64 by 64 pixel blocks. Total 16 (= 4 x 4) of 128 by 128 pixel blocks are available in an image. Fig. 2 c) and d) show 128 by 128 block division of an abnormal image and its extracted abnormal block. This process provides a set of image blocks for each class. The advantage of using fixed size blocks is that, we do not need to concern about the accuracy of region segmentation in either training or actual evaluation stages. Therefore, our proposed method distinguishes actually abnormal blocks from normal blocks. The preprocessing steps mentioned above are applied to each block to make filter responses compatible with the filter responses used to generate the Texton dictionary. Then, a histogram called Texton Histogram is generated using the Texton dictionary and the filter responses (i.e., applying LM filter bank) from each block. Textons in Texton dictionary are used as the bins in a Texton Histogram, and the number of similar response vectors is counted for each bin. The number of bins in the histogram is equal to the number of Textons in the Texton dictionary. The Euclidean distance is used as the similarity measure. In this way a group of histograms that represents a particular class could be derived. For instance, from a block of 128 by 128 we get 16,384 response vectors. Then, we map each response vector to the most similar Texton in the Texton dictionary, and increase the bin corresponding to the Texton. We select a Texton with minimum Euclidean distance. From this process we get a Texton Histogram for each image block, and in total we get Texton histograms which are equal to the number of image blocks. The collection of Texton histograms is called the *Abnormality dictionary*.

4.3 Abnormality Detection

This is the third step of our abnormal image detection method. In this step, first, we train a classifier to distinguish abnormal image blocks from normal ones using the abnormality dictionary. Using the trained classifier, previously unseen image blocks are evaluated. The abnormality dictionary contains a set of Texton histograms for image blocks. For each block, Texton histogram is considered as the feature vector which represents that particular block. Therefore, the dimension of the feature vectors is equivalent to the number of Textons in the Texton dictionary. We test our detection method on two of the most popular classifiers: Support vector machines (SVM) [7] and K-nearest neighbor (K-NN) [11]. Since SVMs are known to be showing good performance for high dimensional data samples, SVM classifier is a good candidate to our problem. Also, K-NN has shown excellent performance in recognizing material textures using Texton method [4-6]. Experimentally, we found out that K-NN classifier performs slightly better than SVM classifier for our abnormal image detection method.

5 Experimental Results

In this section, we assess the effectiveness of the proposed abnormal detection technique. For the assessment, we have performed the experiments with five wireless capsule endoscopy videos. The data set was prepared in the following way. First, our domain specialist (Tang, M.D.) examined the five videos, and annotated the frames for the abnormalities such as blood, erosion, ulcer, polyp, and etc. Second, we extracted all the frames annotated as showing abnormalities, and divided them into three classes: Blood, Polyp, and Ulcer-Erosion (UE). The UE class has images showing either ulcer or erosion. Also, we extracted a set of frames not showing any abnormalities as the Normal class. 50 images were selected for Blood and Polyp classes and 100 and 400 images were selected for UE and Normal classes. Then, we divided the set of images in each class into two categories: training and testing. From each class, 50% of the images were chosen for training and the other half images were chosen for testing.

From the training set of each class a set of image regions were obtained for Texton generation as mentioned in Section 4.1. 23 Blood regions, 22 Polyp regions, 49 UE regions and 50 Normal regions were extracted for Texton generation. Our method was tested on Texton dictionaries with three different numbers of Textons: 40, 80, and 160 Textons (i.e., 40-Texton Dictionary, 80-Texton Dictionary, and 160-Texton Dictionary). Three filter banks: Leung–Malik (LM), Maximum Response 8 (MR8), and Schmid (S) were used for generating Textons. Also, the K-means clustering algorithm with K=10, 20, and 40 were used for the Texton dictionary generation. That is, for instance, in the 80-Texton dictionary, there are 20 Textons from Blood, Polyp, UE and Normal classes. Then, fixed size image blocks were extracted from each class for the Texton histogram generation. The system was tested for three different image blocks: 32 by 32, 64 by 64 and 128 by 128 pixels. The distribution of extracted image blocks from each class is shown in Table 1. It is apparent that the performance of the

Table 1. Image Block Distribution (TR = Training, TE = Testing)

Class	32 by 32		64 by 64		128 by 128	
	TR	TE	TR	TE	TR	TE
Polyp	80	81	40	40	31	29
Blood	206	206	95	95	80	77
UE	280	281	129	128	130	93
Normal	400	400	400	400	400	400

classifier is depended on the distribution of blocks in the training and testing sets. After some testing we decided to use 50% of the blocks for training and 50% of the blocks for testing. Also, the proposed method was tested with two classifiers: K-Nearest Neighbor (K-NN) and Support Vector Machine (SVM) for abnormality detection. We have provided the results for only 160-Texton dictionary since there are little differences in terms of performances among the dictionaries. Finally, the results are summarized using a graph (see Fig. 3).

We present our results using commonly used performance matrices: recall, specificity, precision, and accuracy. Let, TP= instances which are actually positive and predicted as positive, FN= instances which are actually positive but predicted as negative, FP= instances which are actually negative but predicted as positives and TN= instances which are actually negative and predicted as negative. Then, Recall is the percentage of correctly classified positive instances and it is defined as TP/(TP+FN), precision which is equal to TP/(TP+FP), is the percentage of correctly classified positive instances from the predicted positives, specificity is the percentage of correctly classified negative instances and its formula is equal to TN/(TN+FN), and the accuracy which is equal to (TP+TN)/(TP+FP+FN+TN) is the number of correctly classified instances. Precision, recall, specificity and accuracy are reported for each test case.

Experiments with K-NN classifier were performed with different K values, however K=1 produced the best values for the four performance measures. Therefore, we report the performance measure only for K=1. Sequential Minimization Optimization (SMO) and C-SVM based implementation [7] were used for the SVM classifier. The Gaussian Radial Basis Function (RBF) kernel [7] was used as the kernel function. We obtained the best results when C=1. Our experiments were performed on an Intel Core 2 Duo 3.0 GHz CPU and 3GB memory computer. The programming environment was Matlab R2009b. Our experiments were performed demonstrating that:

- The proposed abnormal detection technique can distinguish abnormal and normal fixed size image blocks with average accuracy over 60% for all the image block sizes. When image blocks of 128 by 128 were used, the system performed with average accuracy over 85%.
- When considering the average performance measures, all filter banks showed almost identical performances, however, the LM filter bank showed slightly better values for all performance measures for image blocks of 128 by 128, which is the best test case in terms of performances.

- As the number of Textons for each class increases in the Texton dictionary, the performance of the system also increases. 160-Texton dictionary showed better values for all performance measures than the 80-Texton dictionary.
- When considering the average performance measures, K-NN classifier and SVM classifier showed almost similar performances for all the test cases. But, for image blocks of 128 by 128, K-NN classifier performed slightly better than SVM classifier.

The remainder of the experimental section is organized as follows. The results for image blocks of 128 by 128 that are classified using K-NN and SVM classifiers with a 160-Texton dictionary are discussed in Subsection 5.1, and the average accuracies of all block sizes for 80- and 160- Texton dictionaries are shown in Subsection 5.2.

5.1 Testing with Image Blocks of 128 by 128 and Different Filter Banks for 160-Texton Dictionary

In this experiment, three Texton dictionaries were generated by applying Leung-Malik, MR8, and Schmid filter banks using 40 Textons from each class. Then, for each image block size, the proposed system was tested by applying the same three filter banks for abnormality dictionary generation and K-NN and SVM classifiers for abnormal detection. In Table 2, the results are summarized for the image blocks of 128 by 128 using K-NN and SVM classifiers. In analyzing the results in Tables 2, it can be observed that for image blocks of 128 by 128, although values of performance measures for three filter banks are almost identical, the LM filter bank demonstrated slightly better values when all performance measures are considered. While the average accuracy values are almost identical for two classifiers, K-NN classifier has shown slightly better values than the SVM classifier when all performance measures are considered. The average recall and the average precision for the KNN classifier have reached 75% and the average accuracy has reached 90%.

Table 2. Image Blocks of 128 by 128 classified with 160-Texton Dictionary (P= Precision, R= Recall, S=Specificity, A= Accuracy)

Filter Bank	Class	KNN				SVM			
		P	R	S	A	P	R	S	A
LM	Polyp	80.28	74.03	96.80	93.40	86.15	72.73	97.71	93.62
	Blood	80.00	82.76	98.70	97.76	71.43	51.72	98.61	95.65
	UE	50.46	59.14	88.75	83.94	35.24	39.78	85.56	78.01
	Normal	88.69	86.25	75.56	82.93	81.37	83.00	58.70	75.34
	Average	74.86	75.55	89.95	89.51	68.55	61.81	85.14	85.66
MR8	Polyp	80.82	76.62	96.66	93.55	82.61	74.03	97.24	93.75
	Blood	55.26	72.41	96.30	94.89	72.22	44.83	98.94	95.81
	UE	50.00	48.39	90.30	83.30	61.54	34.41	95.73	85.56
	Normal	85.18	84.75	67.93	79.45	82.17	94.5	55.43	82.19
	Average	67.82	70.54	87.80	87.80	74.64	61.94	86.84	89.33
S	Polyp	80.77	81.82	96.33	94.03	87.32	80.52	97.91	95.27
	Blood	61.29	65.52	97.33	95.41	83.33	34.48	99.58	95.83
	UE	44.95	52.69	87.18	81.46	66.67	30.11	97.01	85.94
	Normal	85.56	81.50	70.43	77.99	80.80	95.75	52.36	81.73
	Average	68.14	70.38	87.82	87.22	79.53	60.22	86.72	89.69

5.2 Accuracy Comparison

Fig. 3 shows the average accuracy comparison of 80-Texton dictionary with 160-Texton dictionary for all image block sizes. In Fig. 3, in the x-axis, KNN-32 means, image blocks of 32 by 32 are classified using the K-NN classifier and the other labels are similarly named. In analyzing Fig 3, it can be clearly seen that as the size of the block is increased the accuracy is also increased. Also, 160-Texton dictionary performs slightly better than 80-Texton dictionary. Thus, it can be concluded that, as we increase the number of Textons the performances of our method also increase.

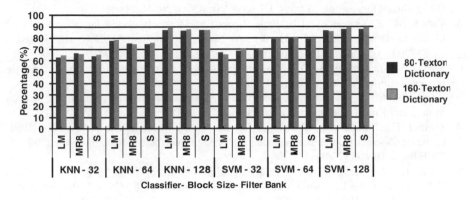

Fig. 3. Accuracy comparison of 80-Texton Dictionary with 160- Texton Dictionary

6 Concluding Remarks

Finding abnormalities in WCE videos such as Blood, Polyp, Ulcer, Erosion, and etc, is a major concern when a gastroenterologist reviews the videos. It is reported that the reviewing time takes up to one for just one video. This is very time consuming for the Endoscopist, which limits its general application, and incur considerable amount of health-care costs. In this paper, we propose a novel method for abnormal image detection in WCE videos based on "Texton" method. Our experimental results on real WCE videos indicate that the recall and the precision of the proposed abnormal detection algorithm achieve 75% and the accuracy reaches 90%. By achieving abnormal image detection for blood, polyp, ulcer, and erosion, we now have a basic foundation to provide optimal inspection to improve overall quality of WCE procedure, and reduce reviewing time. In the future, we are planning to extend our method to detect more minor abnormalities such as erythema, tumor, and etc.

Acknowledgments. This work was supported in part by the Texas Advanced Research Program 003594-0020-2007. Any opinions, findings, and conclusions or recommendations expressed in this paper are those of authors and do not reflect the views of the above funding source.

References

1. Adler, D.G., Gostout, C.J.: State of Art:Wireless Capsule Endoscopy. J. Hospital Physician 39(2), 14–17 (2003)
2. Coimbra, M., Mackiewicz, M., Fisher, M., Jamieson, C., Scares, J., Cunha, J.P.S.: Computer vision tools for capsule endoscopy exam analysis. EURASIP Newsletter 18(2), 1–19 (2007)
3. Eliakim, R.: Video Capsule Endoscopy of the Small Bowel. J. Current Opinion in Gastroenterology 24(2), 159–163 (2008)
4. Leung, T., Malik, J.: Representing and Recognizing the Visual Appearance of Materials Using Three-Dimensional Textons. J. Comp. Vis. 43(1), 29–44 (2001)
5. Varma, M., Zisserman, A.: Classifying Images of Materials: Achieving Viewpoint and Illumination Independence. In: Heyden, A., Sparr, G., Nielsen, M., Johansen, P. (eds.) ECCV 2002. LNCS, vol. 2352, pp. 255–271. Springer, Heidelberg (2002)
6. Varma, M., Zisserman, A.: A Statistical Approach to Texture Classification from Single Images. J. Comp. Vis. 62(1-2), 61–81 (2005)
7. Burges, C.J.C.: A Tutorial on Support Vector Machines for Pattern Recognition. J. Data Mining and Knowledge Discovery 2(2), 121–167 (1998)
8. Schmid, C.: Constructing Models for Content-Based Image Retrieval. In: 2001 IEEE Computer Society Conference on Computer Vision and Pattern Recognition, pp. 39–45. IEEE Press, Hawaii (2001)
9. Lei Z., Li S.Z., Chu R., Zhu X.: Face Recognition with Local Gabor Textons. In: In: Goos G.,Hartmanis J.,van Leeuwen J.(eds.) Advances in Biometrics. LNCS, vol. 4642,pp.49–57. Springer, Berlin / Heidelberg (2009).
10. Liu, G.H., Yanga, J.-Y.: Image Retrieval Based on the Texton Co-occurrence Matrix. J. Pat. Rec. 41(12), 3521–3527 (2008)
11. Duda, R.O., Hart, P.E., Stork, D.G.: Pattern Classication. John Wiley & Sons, Chichester (2001)

Active Contour Method Combining Local Fitting Energy and Global Fitting Energy Dynamically

Yang Yu[1], Caiming Zhang[1,2], Yu Wei[1], and Xuemei Li[1]

[1] School of Computer Science and Technology, Shandong University, Jinan, China
[2] Shandong economicUniversity, Jinan, China

Abstract. To get better segmentation results, local information and global information should be taken into consideration together. In this paper, we propose a new energy functional which combines a local intensity fitting term and an auxiliary global intensity fitting term, and we also give the method to adjust the weight of auxiliary global fitting term dynamically by using local contrast of the image. The combination of the two terms improves the accuracy of segmentation results obviously while reduces dependence on location of initial contour. The experiment results proved the effectiveness of our method.

1 Introduction

Image segmentation is the foundation of image analysis and computer vision. It aims to extract certain meaningful regions from an image. In the past several decades, a large number of segmentation methods have been proposed. Among variety of segmentation algorithms, the active contour method could give promising results.The existing active contour models can be mainly categorized into two classes: edge-based models[2]-[5] and region-based models[6]-[8].

In edge-based models, an edge detector based on image gradient is used to control the evolution speed of contours, attracting the contours evolve to the desired object boundaries. Snake[3] and GAC[2] are two typical models of this kind. For images with obvious edge, models of this kind can get promising result, but for images with weak boundaries or discrete boundaries, edge-based models are prone to suffer from edge leakage.

In region-based models, certain region descriptors are usually used to control the evolution of contours. Comparing to edge-based models, models of this kind have more advantages. First, they are less sensitive to noise and contour initialization. Second, they do not use the image gradient, so they can successfully segment images with weak boundaries or even without boundaries. C-V[6]model is a famous region-based active contour model. This model has the advantage of insensitivity to contour initialization and noise. However, for images with intensity inhomogeneity, C-V model does not work well.

In order to segment images with intensity inhomogeneity efficiently, Li[1] proposed an active contour model based on local binary fitting (LBF). It can handle

D. Zhang and M. Sonka (Eds.): ICMB 2010, LNCS 6165, pp. 163–172, 2010.

images with intensity inhomogeneity better than C-V model. However, because of using local information, this model introduces many local minima into the energy functional, so the minimal energy obtained by gradient descent flow method may be a local minima, and this consequently makes segmenting contour stuck in an incorrect location in some situation.

In this paper, we propose a new region-based active contour model in variational level set formulation. The new model combines global fitting energy used in C-V model and local fitting energy used in LBF model, and adjusts the weight of the global fitting term dynamically. The new model is capable of segmenting images with intensity inhomogeneity, while it reduces dependence on initial contour.

The rest of this paper is organized as follows. In Section 2, we review some classic models and indicate their limitations. Section 3 describes the new energy model and its level set formulation. Section 4 validates our method by extensive experiments on synthetic and real images. Section 5 concludes the paper.

2   Background

2.1   Mumford-Shah Energy Functional

In [11], D.Mumford and J.Shah proposed an object functional about image segmentation. For a given image, Mumford-Shah(MS) model describes image segmentation problem as: to find a curve C which segments image into several disjoint regions, and a fitting function u smooth within each sub-regions, which approximates the intensity of original image I. The energy functional defined by MS model is

$$E^{MS}(C, u) = \int_{\Omega} |u - I|^2 \, dx + \mu \int_{\Omega \backslash C} |\nabla u|^2 + \nu \cdot Length(C) \qquad (1)$$

Where Ω is the image domain, I is the intensity of original image, $Length(C)$ represents the length of curve C, μ and ν are fixed parameters. An optimal segmenting curve C and a fitting function u which approximates image I can be obtained by minimizing this energy functional. However, this energy functional is difficult to be minimized due to the unknown set C and the non-convexity of the this functional.

2.2   C-V Model

To overcome the complexity of MS model, Chan and Vese[6] proposed a simplified MS model. C-V model gives more simple and direct assumption, that is the intensity of image within each region is constant.The energy functional of the segmenting curve C defined by C-V model is

$$E(C) = \nu \cdot Length(C) + \nu \cdot Area(inside(C))$$
$$+ \lambda_1 \int_{inside(C)} |I(x) - c_1|^2 \, dx + \lambda_2 \int_{outside(C)} |I(x) - c_2|^2 \, dx \qquad (2)$$

Where $\mu \geq 0, \nu \geq 0, \lambda_1 \geq 0, \lambda_2 \geq 0$ are fixed parameters. $Inside(C)$ and $outside(C)$ are the regions of image inside and outside C respectively. $Area(inside(C))$ represents the area of the inside region. c_1 and c_2 are two constants which approximate the image intensities inside and outside the contour C, respectively.

For images with only two homogeneous regions, this model is able to get satisfactory result. However, for images with intensity inhomogeneity, the constant values c_1 and c_2 got by global fitting of this model may be not accurate, and some part of foreground/background may be identified as background/foreground incorrectly consequently.

2.3 LBF Model

In order to segment images with intensity inhomogeneity, Li[1] proposed an active contour model based on local binary fitting(LBF). For a pixel x of image, define its energy as

$$\varepsilon_x^{LBF}(C, f_1, f_2) = \lambda_1 \int_{inside(C)} K_\sigma(x - y)(I(y) - f_1(x))^2 dy$$

$$+ \lambda_2 \int_{outside(C)} K_\sigma(x - y)(I(y) - f_2(x))^2 dy \tag{3}$$

Where K_σ is a Gaussian kernel function with standard deviation σ , minimizing this energy functional can get contour C optimally segmenting image within the local region around pixel x. For the whole image, the energy functional is defined as the integral of each pixel's energy over the image domain. The energy functional defined by LBF model is

$$\varepsilon^{LBF}(C, f_1, f_2) = \int_\Omega \varepsilon_x^{LBF}(C, f_1, f_2)dx + \mu \cdot Length(C) \tag{4}$$

The energy functional of LBF model can be represented in level set formulation. Minimizing the energy functional $\varepsilon^{LBF}(C, f_1, f_2)$ with respect to ϕ, and incorporating a distance regularized term in [9], the following partial differential equation(PDE) can be got

$$\frac{\partial \phi}{\partial t} = \nu \cdot (\nabla^2 \phi - div(\frac{|\nabla \phi|}{\nabla \phi})) - \delta(\phi)(\lambda_1 e_1 - \lambda_2 e_2) \tag{5}$$

$$\begin{cases} e_1(x) = \int_\Omega K_\sigma(y - x) |I(x) - f_1(y)|^2 dy \\ e_2(x) = \int_\Omega K_\sigma(y - x) |I(x) - f_2(y)|^2 dy \end{cases} \tag{6}$$

$$\begin{cases} f_1(x) = \frac{K_\sigma * [H_\varepsilon(\phi(x))I(x)]}{K_\sigma * H_\varepsilon(\phi(x))} \\ f_2(x) = \frac{K_\sigma * [(1 - H_\varepsilon(\phi(x)))I(x)]}{K_\sigma * (1 - H_\varepsilon(\phi(x)))} \end{cases} \tag{7}$$

The data fitting term $\delta(\phi)(\lambda_1 e_1 - \lambda_2 e_2)$ plays the major role in the evolution process. In practice, this term draw upon local information within a window

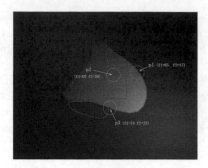

Fig. 1. Explanation of incorrect segmentation result of LBF model

to decide the direction to evolve. However, in some situations, such local information is not enough to give a correct decision. Fig.1 indicates this problem. In Fig.1, the blue rectangular is the initial contour, and the red contour is the segmenting curve got by LBF model. It can be seen that the contour is stuck in the middle of the object. This can be explained as follow, the pixel p1 is on the desired boundary, so the contour stops evolving in this pixel. On pixel p2 and p3, the fitting values f_1 and f_2 got by LBF are almost the same, so the data fitting term $\delta(\phi)(\lambda_1 e_1 - \lambda_2 e_2)$ is close to zero, and this consequently make curve evolution losing motivation on these points.

3   Our Method

3.1   The New Energy Functional

For a segmenting contour C of an image, we define its local fitting energy as

$$\varepsilon_{local}(C) = \lambda_1 \int_{inside(C)} K_\sigma(x - y)(I(y) - f_1(x))^2 dy dx$$
$$+ \lambda_2 \int_{outside(C)} K_\sigma(x - y)(I(y) - f_2(x))^2 dy dx \qquad (8)$$

And define its global fitting energy as

$$\varepsilon_{global}(C) = \lambda_1 \int_{inside(C)} |I(x) - c_1|^2 dx + \lambda_2 \int_{outside(C)} |I(x) - c_2|^2 dx \qquad (9)$$

They are the first two terms of LBF model and C-V model, respectively. By combining the local fitting energy and global fitting energy, we propose the new energy functional

$$\varepsilon(C) = \varepsilon_{local}(C) + \omega \cdot \varepsilon_{global}(C) + Length(C) \qquad (10)$$

It adds an auxiliary global fitting energy to the local fitting energy. ω represents the weight of the global fitting energy, and this weight value is a dynamic value that varies with location of image.

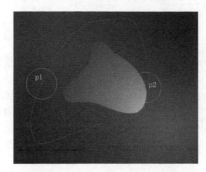

Fig. 2. Explanation of how to determine the weight of global term dynamically

Now we discuss how to combine $\varepsilon_{local}(C)$ and $\varepsilon_{global}(C)$ in (10) to get better result. The key point is how to determine the weight of the auxiliary global fitting energy dynamically. In regions far away from the desired boundary of objects, such as p1 in Fig.2, where the intensity varies slowly, the fitting values f_1 and f_2 got by local fitting are almost the same, and this means that f_1 and f_2 are not capable of reflecting the true background and foreground values correctly in such regions. The cause of this problem lies in that only local information is considered,while local information is not enough to describe the true background and foreground in such regions. So we increase the weight of the global fitting energy, so as to make the active contour evolving towards the right direction under the driving of global fitting energy. On the contrary, in the regions close to the desired boundary of objects, such as p2 in Fig.2, the foreground f_1 and background f_2 got by local fitting can reflect the foreground and background values correctly, and if the intensities of the image are inhomogeneous, c_1 and c_2 got by global fitting may be far away form the true foreground/background, and it will affect the accuracy of segmentation in such regions, so we need to reduce the weight of global fitting energy in such regions , so as to guarantee the accuracy of segmentation.

Based on the discussion above, we need to choose ω to make the weight of global fitting term bigger dynamically in the regions where intensity varies greatly, while make it smaller in the regions where intensity varies smoothly. In this paper, we make use of the local contrast ratio of an image , which is defined as

$$C_N(x) = \frac{M_{max} - M_{min}}{M_g} \tag{11}$$

Where N defines the size of the local window , M_{max} and M_{min} are the maximum and minimum of intensity within this local window respectively, M_g represents the intensity level of image, for grey level images, it is usually 255.

$C_N(x)$ varies between 0 and 1. It reflects how rapidly the intensity changes in a local region. It is smaller in the smooth regions and bigger in the regions close to boundary of objects. Now we define the weight function as

$$\omega = \gamma \cdot average(C_N) \cdot (1 - C_N) \tag{12}$$

$average(C_N)$ is the average value of C_N over the whole image, and it reflects the overall contrast information of the image. For an image with a strong overall contrast, we believe that the image has much more obvious background and foreground, so we increase the weight of global term on the whole. $(1 - C_N)$ adjusts the weight of global term dynamically in all regions, making it smaller in regions with high local contrast and larger in regions with low local contrast. γ is a fixed parameter.

3.2   Variational Level Set Formulation of the New Model

The new energy functional proposed in this paper can be represented by level set formulation as

$$\begin{cases} \varepsilon_{local}(\phi) = \lambda_1 \int \int K_\sigma(x-y) \left(I(y) - f_1(x)\right)^2 H\left(\phi(x)\right) dy dx \\ \qquad + \lambda_2 \int \int K_\sigma(x-y) \left(I(y) - f_2(x)\right)^2 H\left(1 - \phi(x)\right) dy dx \\ \varepsilon_{global}(\phi) = \lambda_1 \int |I(x) - c_1|^2 H\left(\phi(x)\right) dx \\ \qquad + \lambda_2 \int |I(x) - c_2|^2 \left(1 - H\left(\phi(x)\right)\right) dx \\ Length(\phi) = \int_\Omega |\nabla H\left(\phi(x)\right)| dx \end{cases} \tag{13}$$

In numerical calculation, regularized Heaviside function H_ε and Dirac function δ_ε are used, which are defined as

$$\begin{cases} H_\varepsilon(x) = \frac{1}{2}\left[1 + \frac{2}{\pi}arctan\left(\frac{x}{\varepsilon}\right)\right] \\ \delta_\varepsilon(x) = H'_\varepsilon(x) = \frac{1}{\pi}\frac{\varepsilon}{\varepsilon^2 + x^2} \end{cases} \tag{14}$$

Additionally, in the numerical calculation of level set function, it is necessary to keep the level set function close to signed distance function in order to ensure stable evolution of the level set function. We introduce the level set regularization term in [9] into our variational level set formulation, to make the level set function keep close to signed distance function automatically. This regularization term is defined as

$$P(\phi) = \int_\Omega \frac{1}{2} \left(|\nabla \phi(x)| - 1\right)^2 dx \tag{15}$$

Combining the terms introduced before, the level set formulation of the new energy functional is

$$\varepsilon(\phi) = \varepsilon_{local}(\phi) + \omega \cdot \varepsilon_{global}(\phi) + \mu \cdot Length(\phi) + \nu \cdot P(\phi) \tag{16}$$

3.3   Gradient Descent Flow

Standard gradient descent (or steepest descent) method is used to minimize the proposed energy functional (15), by calculation of variation, we obtain the following partial differential equation

$$\frac{\partial \phi}{\partial t} = \delta(\phi)\left(F_1 + \omega \cdot F_2 + \mu \cdot div\left(\frac{|\nabla \phi|}{\nabla \phi}\right)\right) \tag{17}$$

Here, F_1 and F_2 are local force and global force respectively.

$$\begin{cases} F_1 = -\lambda_1 \int K_\sigma(x-y)(I(y)-f_1(x))^2 dy \\ \qquad +\lambda_2 \int K_\sigma(x-y)(I(y)-f_2(x))^2 dy \\ F_2 = -\lambda_1 (I(y)-c_1)^2 dy + \lambda_2 (I(y)-c_2)^2 dy \end{cases} \tag{18}$$

c_1 and c_2 are global fitting background and foreground

$$\begin{cases} c_1 = \frac{H_\varepsilon(\phi(x))I(x)dx}{\int H_\varepsilon(\phi(x))dx} \\ c_2 = \frac{H_\varepsilon(1-\phi(x))I(x)dx}{\int H_\varepsilon(1-\phi(x))dx} \end{cases} \tag{19}$$

$f1$ and f_2 are local fitting background and foreground,and they are defined as (7). For an initial contour, its level set function evolutes under the control of the PDE (17) until convergence, and it will stops on the desired boundaries of the objects.

4 Algorithm Implementation and Experiment Results

Both synthetic and natural images have been used to test our method. For all the images in the experiment, we choose the following default parameters $\nu = 0.001 \times 255 \times 255$, $\gamma = 0.1$, $\lambda_1 = \lambda_2 = 1$, $\sigma = 3$, $\mu = 1$, time step $\tau = 0.1$. In Fig.3, (a) shows a synthetic image with intensity inhomogeneity. The global fitting method(C-V mode)fails to get the right segmentation result, as showed by(c). (d)shows the segmentation result got by LBF model, for the sake of incorrect initialization, the segmenting contour is trapped into a local minimal energy in the lower right part,(e) shows the result got by the LGIF model (proposed in[10]), because it uses a constant weight of global term, the segmenting curve was influenced by global fitting energy too great in the upper left part, consequently, it got an incorrect result.(f)shows the segmentation result got by our method, which is the best among the four methods.

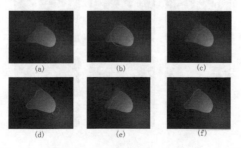

Fig. 3. Segmentation of a synthetic image: (a) original image (b) the initial contour (c) result of C-V model (d) result of LBF model (e) result of LGIF model (f) result of our model

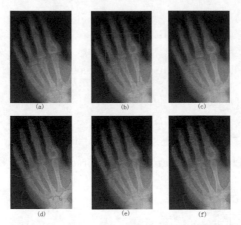

Fig. 4. Segmentation results of X ray photograph: (a) original image (b) the initial contour (c) result of C-V model (d) result of LBF model (e) result of LGIF model (f) result of our model

Fig.4 shows the segmentation results of an X-ray photograph got by various methods. There usually exists strong intensity inhomogeneity in X-ray images. So the global fitting method is usually not adequate to obtain accurate segmentation result. In addition, the topological structure of the bones is usually complex, so it is difficult to give an appropriate initial contour for local fitting method. (a) exhibits the original images, (b)shows the initial contour, (c)-(f) shows the segmentation result of C-V model, LBF model, LGIF model and our method, respectively. Obviously, our method gave the result superior than other models.

Fig.5 shows the comparison of the segmentation results got by various methods applied to a slice image of human brain. Exact segmentation of such images is crucial for three-dimensional reconstruction of human body. As can be seen from the original image (a), there exists obvious background and foreground in this

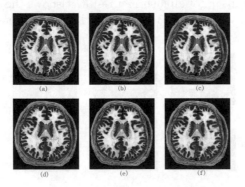

Fig. 5. Segmentation result of human brain slice photo (a) original image (b) the initial contour (c) result of C-V model (d) result of LBF model (e) result of LGIF model (f) result of our model

image, the intensity of ectocinerea is much higher than that of other parts, so C-V model can give a promising result as shown in (c), on the contrary, the result got by LBF model is not so good because it is trapped in local energy minima in many regions as shown in (d). our method give the result very close to that of C-V model, this proves that our method can process dynamically and can give a better result than C-V model and LBF model.

In Fig.6, we give some other segmentation results got by our method, the upper row shows the original images with initial contours. the under row shows the segmentation results. It can be seen that our method are able to give promising results in all the conditions.

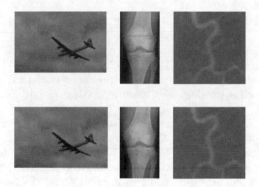

Fig. 6. Some other segmentation results of our model: upper row: the original image with initial contour; under row: the final segmentation results

5   Conclusion

A new region-based active contour method is proposed in this paper. The new model combines local information and global information of image efficiently, and adjust the weight of local fitting term with local contrast dynamically. The new model has the following advantages: first, by using local information, the new model can segment images with intensity inhomogeneity; second, the new model reduces the dependency to the location of initial contour, so the robustness of the model is improved; and finally, by incorporating the level set regularization term, it avoids expensive re-initialization procedures and thus increase the speed of contour evolution. Experimental results show the advantages of our approach.

Acknowlegements. This work is supported by National 863 High-Tech programme of China(2009AA01Z304), the National Natural Science Foundation of China(60933008), Shandong Province National Nature Science Foundation (No.Z2006G05), National Research Foundation for the Doctoral Program of Higher Education of China (20070422098).

References

1. Li, C., Kao, C., Gore, J., Ding, Z.: Implicit active contours driven by local binary fitting energy. In: CVPR 2007, pp. 1–7 (2007)
2. Caselles, V., Kimmel, R., Sapiro, G.: Geodesic active contour. IJCV 22(1), 61–79 (1997)
3. Xu, C., Prince, J.: Snakes, shapes, and gradient vector flow. IEEE T-IP 7(3), 359–369 (1998)
4. Paragios, N., Mellina-Gottardo, O., Ramesh, V.: Gradient vector flow geometric active contours. IEEE T-PAMI 26(3), 402–407 (2004)
5. Li, C., Liu, J., Fox, M.: Segmentation of edge preserving gradient vector flow: an approach toward automatically initializing and splitting of snakes. In: CVPR 2005, pp. 162–167 (2005)
6. Chan, T., Vese, L.: Active contours without edges. IEEE T-IP 10(2), 266–277 (2001)
7. Paragios, N., Deriche, R.: Geodesic active regions and level set methods for supervised texture segmentation. IJCV 46(3), 223–247 (2002)
8. Cremers, D., Rousson, M., Deriche, R.: A review of statistical approaches to level set segmentation: Integrating color, texture, motion and shape. IJCV 72(2), 195–215 (2007)
9. Li, C., Xu, C., Gui, C., Fox, M.D.: Level set evolution without re-initialization: a new variational formulation. In: CVPR, vol. 1, pp. 430–436 (2005)
10. Li, W., Li, C., Sun, Q., Xia, D., Kao, C.: Active contours driven by local and global intensity fitting energy with application to brain MR image segmentation. Computerized Medical Imaging and Graphics 33, 520–531 (2009)
11. Mumford, D., Shah, J.: Optimal approximation by piecewise smooth function and associated variational problems. Communication on Pure and Applied Mathematics 42, 577–685 (1989)
12. Osher, S., Sethian, J.A.: Fronts Propagating with Curvature dependent Speed: Algorithms Based on Hamilton-Jacobi Formulation. Journal Computational Physics 79, 12–49 (1988)
13. Vese, L., Chan, T.A.: Multiphase level set framework for image segmentation using the Mumford and Shah model. Int. J. Comput. Vision 50, 271–293 (2002)
14. Tsai, A., Yezzi, A., Willsky, A.S.: Curve evolution implementation of the Mumford-Shah functional for image segmentation, denoising, interpolation, and magnification. IEEE Transaction on Image Processing 10, 1169–1186 (2001)

Optic Disc Detection by Multi-scale Gaussian Filtering with Scale Production and a Vessels' Directional Matched Filter

Bob Zhang and Fakhri Karray

Department of Electrical and Computer Engineering, University of Waterloo,
Waterloo, ON, Canada N2L 3G1
{yibo,karray}@pami.uwaterloo.ca

Abstract. The optic disc (OD) is an important anatomical feature in retinal images, and its detection is vital for developing automated screening programs. In this paper we propose a method to automatically detect the OD in fundus images using two steps: OD vessel candidate detection and OD vessel candidate matching. The first step is achieved with multi-scale Gaussian filtering, scale production, and double thresholding to initially extract the vessels' directional map. The map is then thinned before another threshold is applied to remove pixels with low intensities. This result forms the OD vessel candidates. In the second step, a Vessels' Directional Matched Filter (VDMF) of various dimensions is applied to the candidates to be matched, and the pixel with the smallest difference designated the OD center. We tested the proposed method on a subset of a new database consisting of 139 images from a diabetic retinopathy (DR) screening programme. The OD center was successfully detected with an accuracy of 96.4% (134/139).

Keywords: optic disc, Matched Filter, Vessels' Directional Matched Filter, diabetic retinopathy.

1 Introduction

The OD is a vertical oval with average dimensions of 1.76mm (horizontally) × 1.92mm (vertically), and situated 3-4mm to the nasal side of the fovea [1]. In fundus imaging the OD is usually brighter than its surrounding area, and is the convergence of the retinal blood vessel network. Detection of the OD is useful in the diagnosis of glaucoma, optic neuropathies, optic neuritis, anterior ischemic optic neuropathy or papilledema, and optic disc drusen. It can also be used as a marker to help locate fovea/macula [2], [3], [4], as well as decide if the image is the left or right eye. For DR, detection of the OD assists physicians identifying neovascularization of the disc (NVD) in the advanced stage of DR, proliferative diabetic retinopathy (PDR). This makes the task of automatic OD detection both relevant and necessary. Automatic detection of the OD refers to the location of the disc center.

In literature, OD detection can be categorized into various groups. The first group uses properties of the OD [5], [6], [7], [8] such as high pixel intensity and its oval

D. Zhang and M. Sonka (Eds.): ICMB 2010, LNCS 6165, pp. 173–180, 2010.
© Springer-Verlag Berlin Heidelberg 2010

shape. Morphology [9] is also used where the OD center is the center of the brightest connected object found by thresholding an intensity image. [10], [11], [12] applied template matching to locate the center. [10]'s template was the average gray-level of 25 normalized images, [11] used the Hausdorff-based template, and [12] employed a VDMF. Two different kinds of transforms, Hough [13], [14], [15] and Watershed [16] have also been applied to locate the edges of the OD and subsequently its center. Supervised learning is another group consisting of feature extraction, and classification with a Bayesian [4] or k-NN [2], [3] classifier. A geometric model was built in [17] to represent the main retinal vessels which pass the OD center. Fuzzy convergence developed by [18] determined the originating vessel map convergence point near the OD center. Even though the OD features and characteristics are well defined in fundus images, the task of automatically identifying it is still challenging.

This paper proposes a method to detect the OD using both vessel and intensity information. In the first step, a vessels' directional map representing the OD vessel candidates is calculated using multi-scale Gaussian filtering with scale production, and double thresholding. A VDMF template is matched to each OD vessel candidate as part of the second step. The pixel candidate having the least difference with the template is assigned the OD center.

The remainder of this paper is organized as follows. Section 2 describes the material used, Section 3 presents the proposed method, and experimental results are relayed in Section 4. Finally, a conclusion is given in Section 5.

2 Material

We used a subset of the database obtained from a DR screening programme in Harbin China. This database will be referred to as the HIT database. The subset has 139 images, containing 46 normal and 93 pathological (all with DR). The 93 DR images were further divided into mild non-proliferative diabetic retinopathy (NPDR), moderate NPDR, severe NPDR and PDR based on [19]. The images were captured in digital form using a Canon CR-DGi Non-Mydriatic Retinal Camera at 45° field of view (FOV). The size of each image is 1936×1288 pixels with 24 bits and in compressed JPEG-format. Figure 1 shows examples of images from HIT, the one on the left is a healthy retinal while the other has PDR. The OD in each image was manually segmented by the first author, and its center subsequently calculated.

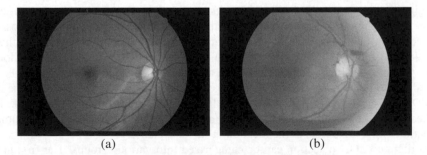

(a)   (b)

Fig. 1. (a) Normal fundus image from HIT. (b) Fundus diagnosed with PDR from HIT.

3   Proposed Method

As mentioned above, the proposed method consists of two steps, OD vessel candidate detection, and matching. In the following section each step is explained with more detail.

3.1   OD Vessel Candidate Detection

Multi-scale Gaussian filtering is based on Matched Filters first proposed in [20] to detect vessels. It makes use of the prior knowledge that the cross-section of vessels can be approximated by a Gaussian function. Therefore, a Gaussian-shaped filter can be used to "match" the vessels. The idea of multi-scale allows more than one scale to be used which can match vessels of various widths. The Multi-scale Gaussian filter is defined as:

$$f_i(x, y) = \frac{1}{\sqrt{2\pi}s_i}\exp\left(-\frac{x^2}{2s_i^2}\right) - m, \quad |x| \le t \cdot s_i, \quad |y| \le L_i/2 \ . \tag{1}$$

where s_i represents the scale of the filter; $m = \left(\int_{-ts}^{ts}\frac{1}{\sqrt{2\pi}s_i}\exp\left(-\frac{x^2}{2s_i^2}\right)dx\right)\Big/(2ts_i)$

is used to normalize the mean value of the filter to 0 so that the smooth background can be removed after filtering; and L_i is the length of the neighborhood along the y-axis to smooth noise; t is a constant and usually set as 3 because more than 99% of the area under the Gaussian curve lies within the range of $[-3 s_i, 3 s_i]$. The parameter L_i is also chosen based on s_i. When s_i is small, L_i is set relatively small, and vice versa. In the actual implementation $f_i(x, y)$ will be rotated to detect the vessels of different orientations.

The response of multi-scale Gaussian filtering can be expressed by:

$$R_i(x, y) = f_i(x, y) * im(x, y) \ . \tag{2}$$

where $im(x, y)$ is a normalized green channel image and $*$ denotes convolution. The scale production is defined as the product of filter responses at two scales i and j:

$$P(x, y) = R_i(x, y) * R_j(x, y) \ . \tag{3}$$

Double thresholding is then applied to $P(x, y)$ to generate a binary image where a one pixel wide center-line of the vessel is detected using morphological thinning. The vessels' directional map is calculated by finding the corresponding orientation that produced the maximum response with $f_i(x, y)$ (use of the vessel feature). This map is thinned by multiplying with the center-line vessel. Using the notation that the OD has brighter pixel intensity (use of the intensity feature), any pixels less than 0.9 in $im(x, y)$ are removed. A 51×51 neighborhood of each remaining pixel is extracted in order to better represent the OD vessels. This results in the OD vessel candidates. In some situations hard exudates may also be part of the OD vessel candidates since their pixel intensity is also high. However, these objects are not made of vessels and will be removed in the following step.

3.2 OD Vessel Candidate Matching

We define a 9×9 template as the VDMF shown in Table 1, where 8 orientations are used. In order to account for the various OD sizes, bilinear interpolation was employed to resize the template into 61×21, 121×41, 181×61, and 241×81. The values and sizes are specifically tuned for HIT. Each of the four templates is matched to the candidates with an absolute difference calculated. The candidate pixel with least accumulated difference is assigned the OD center. Figure 2 illustrates the steps of the proposed method using an example.

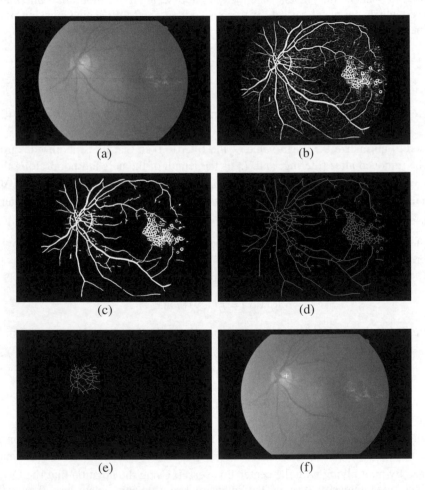

(a)

(b)

(c)

(d)

(e)

(f)

Fig. 2. Steps of the proposed method applied to a fundus image (a). (b) is the scale production of (a). The result of double thresholding on (b) is (c). The thinned vessels of (c) are (d). (e) is the OD vessel candidates after removing pixels with low intensities. The detected OD center is illustrated in (f) by a cross.

Table 1. Vessels' directional matched filter

7	6	6	6	1	4	4	4	3
8	7	6	6	1	4	4	3	2
8	8	7	6	1	4	3	2	2
8	8	8	7	1	3	2	2	2
5	5	5	5	1	5	5	5	5
2	2	2	3	1	7	8	8	8
2	2	3	4	1	7	8	8	8
2	3	4	4	1	6	7	8	8
2	3	4	4	1	6	6	7	8

4   Experimental Results

The key parameters in our experiments are set as follows: $s_1 = 1.5$, $s_2 = 1.8$, $s_3 = 2.0$, $s_4 = 2.4$, with corresponding $L_1 = 9$, $L_2 = 9$, $L_3 = 13$, $L_4 = 13$, and 8 orientations (refer to (1)). The scale production of s_1 and s_2 are combined along with the result of s_3 and s_4. It took 29secs to process each image using a 2.40GHz Intel Centrino Pro with 2GB RAM. In order to improve the computation time of the proposed method every image in HIT is resized by 0.5 to 968×644 pixels. In literature, the detected OD center is considered correct if it is positioned within 60 pixels of the manually identified center [12], [17], [18]. However, the images they used were 605×700 pixels. In order to compensate for the larger images in HIT, the distance from the manually identified center to the detected OD center is increased to 80 pixels.

Table 2 shows the result of the proposed method compared with [12]. There are 46 normal images, 36 with mild NPDR, 34 moderate NPDR, 11 severe NPDR, and 12 PDR. The proposed method detected all OD centers in normal, mild and moderate NPDR. For severe NPDR and PDR three and two were missed respectively. In [12]'s implementation which also used a VDMF, the vessels were extracted using the original Matched Filter [20]. This meant 12 orientations, a single scale, and thresholding by [21]. The dimensions of their templates were 241×81, 361×121, 481×161 and 601×201. [12]'s algorithm detected 25 OD centers for both normal and mild NPDR, 22 for moderate NPDR, 6 for severe NPDR and 9 for PDR. In total the proposed method detected 47 more OD centers accurately compared to [12].

Figure 3 illustrates examples of the OD center (marked with a cross) detected in HIT. In these examples you can see the proposed method is not swayed by other anatomical features such as microaneurysms (see Fig. 3(a)), hemorrhages (see Fig. 3(b)–(d)), or hard exudates (see Fig. 3(c)). Most of the missed OD centers shown in Fig. 4 are from

Table 2. OD detection result on HIT

	Number of images in each group				Total
	46	36	34	11	12	139
Method	Number correctly detected					
Proposed Method	46	36	34	8	10	134
[12]	25	25	22	6	9	87

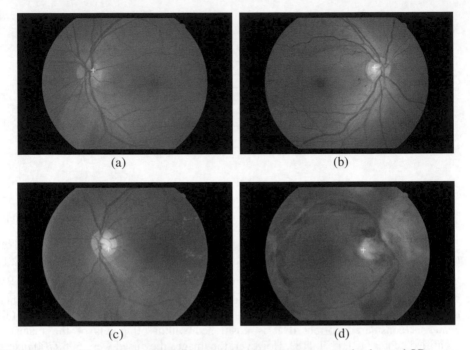

Fig. 3. Results of the proposed method where the white cross represents the detected OD center

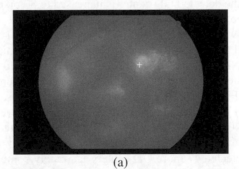

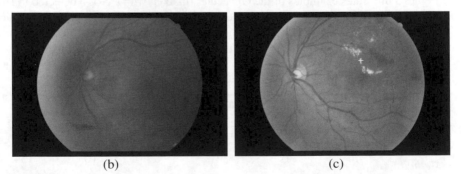

Fig. 4. Examples where the proposed method failed to correctly detect the OD center

severe NPDR and PDR. In a majority of these cases the OD did not have higher pixel intensity in the image as evident in Fig. 4(a)–(c), and as a result the OD vessel candidates were not detected. We admit in such cases our method may fail.

5 Conclusion

This paper presented a method for automatic OD detection using two steps, OD vessel candidate detection, and OD vessel candidate matching. It makes use of the OD's intensity information my removing low pixel values, and incorporates the vessel information in the form of a vessel's directional map. Compared with [12] using the same material, we demonstrated only 8 orientations are need rather than 12. Also, the use of multiple scales, scale production, and double thresholding produce a more accurate vessel directional map, as evident in Table 2.

Currently, there is no illumination correction in the proposed method. As part of the future work this aspect will be integrated to deal with cases where the OD is not the brighter object. After this development, further testing will be conducted on the full HIT database.

References

1. Duane's Ophthalmology. Lippincott Williams & Wilkins (2006)
2. Niemeijer, M., Abramoff, M.D., van Ginneken, B.: Automated localization of the optic disc and the fovea. In: Proceedings of International IEEE EMBS Conference, pp. 3538–3541 (2008)
3. Niemeijer., M., Abramoff, M.D., van Ginneken, B.: Segmentation of the optic disc, macula and vascular arch in fundus photographs. IEEE Transactions on Medical Imaging 26, 116–127 (2007)
4. Tobin, K.W., Chaum, E., Govindasamy, V.P., Karnowski, T.P.: Detection of anatomic structures in human retinal imagery. IEEE Transactions on Medical Imaging 26, 1729–1739 (2007)
5. Sinthanayothin, C., Boyce, J.F., Cook, H.L., Williamson, T.H.: Automated localisation of the optic disk, fovea, and retinal blood vessels from digital colour fundus images. Br. J. Ophthalmol. 83, 902–910 (1999)
6. Li, H., Chutatape, O.: Automatic location of optic disc in retinal images. In: Proceedings of IEEE International Conference Image Process. vol. 2, pp. 837–840 (2001)
7. Li, H., Chutatape, O.: A model-based approach for automated feature extraction in fundus images. In: Proceedings of IEEE International Conference Computer Vision, vol. 1, pp. 394–399 (2003)
8. Chrástek, R., Wolf, M., Donath, K., Michelson, G., Niemann, H.: Optic disc segmentation in retinal images., Bildverarbeitung für die Medizin. 263–266 (2002)
9. Walter, T., Walter, J.C.: Segmentation of color fundus images of the human retina: Detection of the optic disc and the vascular tree using morphological techniques. In: Crespo, J.L., Maojo, V., Martin, F. (eds.) ISMDA 2001. LNCS, vol. 2199, pp. 282–287. Springer, Heidelberg (2001)
10. Osareh, A., Mirmehdi, M., Thomas, B., Markham, R.: Comparison of colour spaces for optic disc localisation in retinal images. In: Proceedings of International Conference on Pattern Recognition, pp. 743–746 (2002)

11. Lalonde, M., Beaulieu, M., Gagnon, L.: Fast and robust optic disk detection using pyramidal decomposition and Hausdorff-based template matching. IEEE Transactions on Medical Imaging 20, 1193–1200 (2001)
12. Youssif, A., Ghalwash, A., Ghoneim, A.: Optic disc detection from normalized digital fundus images by means of a vessels' direction matched filter. IEEE Transactions on Medical Imaging 27, 11–18 (2008)
13. Abdel-Ghafar, R.A., Morris, T.: Progress towards automated detection and characterization of the optic disc in glaucoma and diabetic retinopathy Informatics for Health and Social Care, 32, 19–25 (2007)
14. Barrett, S.F., Naess, E., Molvik, T.: Employing the Hough transform to locate the optic disk. Biomed. Sci. Instrum. 37, 81–86 (2001)
15. Zhu, X., Rangayyan, R.M.: Detection of the optic disc in images of the retina using the hough transform. In: Proceedings of International IEEE EMBS Conference, pp. 3546–3549 (2008)
16. Hajer, J., Kamel, H., Noureddine, E.: Localization of the optic disk in retinal image using the "watersnake". In: Proceedings of International Conference on Computer and Communication Engineering, pp. 947–951 (2008)
17. Foracchia, M., Grisan, E., Ruggeri, A.: Detection of optic disc in retinal images by means of a geometrical model of vessel structure. IEEE Transactions on Medical Imaging 23, 1189–1195 (2004)
18. Hoover, A., Goldbaum, M.: Locating the optic nerve in a retinal image using the fuzzy convergence of the blood vessels. IEEE Transactions on Medical Imaging 22, 951–958 (2003)
19. Wilkinson, C.P., Ferris, F.L., Klein, R.E., Lee, P.P., Agardh, C.D., Davis, M., Dills, D., Kampik, A., Pararajasegaram, R., Verdaguer, J.T.: Proposed international clinical diabetic retinopathy and diabetic macular edema disease severity scales. Ophthalmology 110, 1677–1682 (2003)
20. Chaudhuri, S., Chatterjee, S., Katz, N., Nelson, M., Goldbaum, M.: Detection of blood vessels in retinal images using two-dimensional matched filters. IEEE Transactions on Medical Imaging, 263–269 (1989)
21. Otsu, N.: A threshold selection method from gray level histograms. IEEE Transactions on Systems, Man and Cybernetics 9, 62–66 (1979)

Retinal Vessel Centerline Extraction Using Multiscale Matched Filter and Sparse Representation-Based Classifier

Bob Zhang[1], Qin Li[2], Lei Zhang[2], Jane You[2], and Fakhri Karray[1]

[1] Department of Electrical and Computer Engineering, University of Waterloo,
Waterloo, ON, Canada N2L 3G1
{yibo,karray}@pami.uwaterloo.ca
[2] Department of Computing, The Hong Kong Polytechnic University,
Kowloon, Hong Kong
{csqinli,cslzhang,csyjia}@comp.polyu.edu.hk

Abstract. Retina located in the back of the eye contains useful information in the diagnosis of certain diseases. By locating a blood vessel's width, color, reflectivity, tortuosity and abnormal branching, one can deduce the existence of these diseases. In order for this to be achieved, blood vessels first need to be extracted from its background in fundus image. In this paper we propose a new method to extract vessels based on Multiscale Production of Matched Filter (MPMF) and Sparse Representation-based Classifier (SRC). First, we locate vessel centerline candidates using multi-scale Gaussian filtering, scale production, double thresholding and centerline detection. Then, the candidates which are centerline pixels are classified with SRC. Particularly, two dictionary elements of vessel and non-vessel are used in the SRC process. Experimental results on two public databases show that the proposed method is good at distinguishing vessel from non-vessel objects and extracting the centerlines of small vessels.

Keywords: retinal vessel, Matched Filter, Sparse Representation-based Classifier, multiscale Gaussian filtering.

1 Introduction

Careful examination of the blood vessels in the retinal has known to directly diagnosis diabetic retinopathy [1], [2], [3], [4], [5], [6], [7], [8], hypertension [9], glaucoma [10], obesity [11], arteriosclerosis and retinal artery occlusion. In each case if the patient is not correctly diagnosed or treated in a timely manner, the diseases will spread and cause adult blindness. That's why it is imperative vessels are extracted from its background as accurately as possible.

Vessel extraction is a form of line detection, and many methods have been proposed. A popular approach to vessel extraction is filtering-based methods [5], [12], [13], [14], [15], [16] where responses to vessel-like structures are maximized. Trace based methods [17] map out the global network of blood vessels after edge detection by tracing out the centerlines of vessels. Such methods rely heavily on the result of

D. Zhang and M. Sonka (Eds.): ICMB 2010, LNCS 6165, pp. 181–190, 2010.

edge detection. Machine Learning-based methods [1], [3], [17], [18] have also been proposed and can be divided into two categories: supervised methods [1], [3], [18] and unsupervised methods [17], [19]. Supervised methods utilize some prior labeling information to decide whether a pixel is a vessel or not, while unsupervised methods work without any prior labeling knowledge. Mathematical morphology [7], [20], [21] is another type of approach by applying morphological operators. However, the problem of separating vessel from non-vessel is still at large.

To this end, this paper proposes a novel method to extract vessel centerline using two steps. In the first step, the vessel centerline candidates are extracted using Multiscale Production of Matched Filter (MPMF), which consists of multiscale matched filtering, scale production, double thresholding, and centerline extraction. For the next step, the centerline candidates are classified as either vessel or non-vessel using Sparse Representation-based Classifier (SRC) with the help of two dictionary elements, vessel and non-vessel. Our method has two advantages: (1) The vessel and non-vessel objects are well separated due to SRC; (2) The small vessels, which are usually with low intensity contrast and are overwhelmed in background noises, are well extracted due to MPMF.

The rest of the paper is organized as follows. Sections 2 and 3 present the proposed method. Experimental results are presented in Section 4, and a conclusion is made in Section 5.

2 Vessel Centerline Candidates Extraction

MPMF is based on Matched Filters first proposed in [12] to detect vessels. It makes use of the prior knowledge that the cross-section of vessels can be approximated by a Gaussian function. Therefore, a Gaussian-shaped filter can be used to "match" the vessels. The idea of multiscale allows more than one scale to be used which can match vessels of various widths. The Multi-scale Gaussian filter is defined as:

$$f_i(x, y) = \frac{1}{\sqrt{2\pi}s_i} \exp\left(-\frac{x^2}{2s_i^2}\right) - m, \quad |x| \le t \cdot s_i, \quad |y| \le L_i/2 . \tag{1}$$

where s_i represents the scale of the filter; $m = \left(\int_{-ts}^{ts} \frac{1}{\sqrt{2\pi}s_i} \exp\left(-\frac{x^2}{2s_i^2}\right) dx\right) \Big/ (2ts_i)$

is used to normalize the mean value of the filter to 0 so that the smooth background can be removed after filtering; and L_i is the length of the neighborhood along the y-axis to smooth noise; t is a constant and usually set as 3 because more than 99% of the area under the Gaussian curve lies within the range of [-3 s_i, 3 s_i]. The parameter L_i is also chosen based on s_i. When s_i is small, L_i is set relatively small, and vice versa. In the actual implementation $f_i(x, y)$ will be rotated to detect the vessels of different orientations.

The response of multi-scale Gaussian filtering can be expressed by:

$$R_i(x, y) = f_i(x, y) * im(x, y) . \tag{2}$$

where $im(x, y)$ is a normalized green channel image and $*$ denotes convolution. The scale production is defined as the product of filter responses at two scales i and j :

$$P(x, y) = R_i(x, y) * R_j(x, y) .$$

(3)

Double thresholding is then applied to $P(x, y)$ using to generate a binary image. The one pixel wide centerline of the vessel is detected from this image using morphological thinning. These pixels represent the vessel centerline candidates.

3 Classification Using SRC

In this step we apply the concept of SRC to classify vessel candidates. Given a test sample and a set of training samples from each class, the idea of SRC is to represent the test sample as a linear combination of the training samples, while requiring the representation coefficients are as sparse as possible. If the test sample is from class k, then among its representation coefficients over all the training samples, only those from the samples in class k will be significant while others will be very small, and hence the class label of the test sample can be determined. In practice, the l_1-norm minimization is used to solve the sparest linear representation of the test sample over the dictionary of training samples. More information about SRC can be found in [22].

Vessel detection can be thought of as a two-class classification problem: vessel and non-vessel. Therefore, we need two sub-dictionary elements for SRC: the vessel elements and the non-vessel elements. The feature vectors for vessels are artificially generated using a series of Gaussians with various STDs. The feature vectors for non-vessels are artificially generated using a series of step-edges smoothed using Gaussians with various STDs. We define two dictionaries:

$$A = [a_1, a_2, ..., a_{30}, c_1, c_2, ..., c_{48}] .$$

(4)

where a_i and c_i are 11×1 column vectors containing artificially generated cross-sections of vessels and step-edges respectively (with small sigma values) and another dictionary:

$$B = [b_1, b_2, ..., b_{30}, d_1, d_2, ..., d_{85}] .$$

(5)

where b_i and d_i are 23×1 column vectors also consisting of artificially generated cross-sections of vessels and step-edges (using large sigma values). We considered two dictionaries of different dimensions to better match vessels and non-vessels of various shape and size. The columns of A and B are normalized to have unit l_2-norm.

For each vessel candidate we find its corresponding 11×1 and 23×1 column vectors $\perp$ to its orientation, denoted by y_A and y_B. We use then solve the l_1-norm minimization problem:

$$\min_{\alpha_A} \|y_A - A\alpha_A\|_2^2 + \lambda|\alpha_A|_1 \text{ and } \min_{\alpha_B} \|y_B - B\alpha_B\|_2^2 + \lambda|\alpha_B|_1 .$$

(6)

where α_A and α_B are the sparse representation coefficient vectors of y_A and y_B over A and B, and λ is a constant. α_A and α_B can be solved by convex optimization. If y_A and y_B are vessel objects the maximum coefficient in α_A and α_B will be at index ≤ 30, and if they are not, the maximum will be at index > 30. The final step combines the result of y_A and y_B using logical OR.

Figure 1 shows the result of each step in the proposed method using an example from STARE [5].

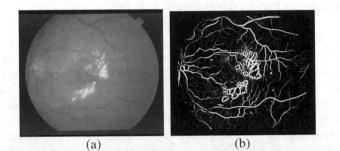

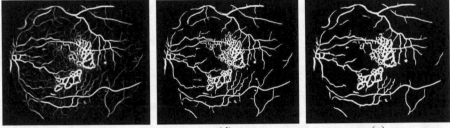

Fig. 1. Steps of the proposed method. (a) original image, (b) scale production using small scales, (c) scale production of large scales, (d) threshold result of (b), (e) threshold result of (c), (f) logical OR of (d) and (e), (g) detected centerline candidates, and (h) classified candidates.

4   Experimental Results

The key parameters in our experiments are set as follows: $s_i = \{1.5, 1.8, 2.0, 2.4\}$ and $L_i = \{9, 9, 13, 13\}$ (refer to (1)), and 8 directions were used in multiscale Gaussian filtering. These parameters were chosen based on our experimental experience.

We used two publicly available databases (STARE [5] and DRIVE [1]) to test the proposed method. The STARE database consists of retinal images captured by TopCon TRV-50 fundus camera at 35° field of view (FOV), which were digitized at 24-bits with a spatial resolution of 700×605 pixels. There are 20 images, 10 of which are healthy ocular fundus and the other 10 unhealthy. The database also provides hand-labeled images as the ground truth for vessel segmentation so that the algorithms can be evaluated for comparison. The DRIVE database consists of 40 images captured by Canon CR5 camera at 45° FOV, which were digitized at 24-bits with a spatial resolution of 565×584 pixels. The 40 images were divided into a training set and a test set by the authors of the database. The results of the manual segmentation are available for the two sets.

To compare different retinal vessel extraction algorithms, we selected (1) FPR (false positive rate), (2) TPR (true positive rate), and (3) PPV (positive predictive value) as performance measures. FPR is defined as the ratio of the number of non-vessel pixels inside FOV classified as vessel pixels to the number of non-vessel pixels inside FOV of the ground truth. TPR is the ratio of the number of correctly classified vessel pixels to the number of total vessel pixels in the ground truth. PPV (positive predictive value) represents the ratio of the number of correctly classified vessel pixels to the number of total predicted vessel pixels inside FOV. Please note that all comparisons are made with the centerline of each algorithm.

4.1   Visual Inspection

In this section, the comparisons of different retinal vessel extraction methods are visually demonstrated. The ability of our method to separate vessel from non-vessel is demonstrated in Figure 2. It can be seen that the proposed method detects less hard exudates as vessels compared to other algorithms, and is able to extract the small vessels around the lesion well. Figure 3 compares our method with the other stare-of-arts methods. We can see that the proposed method performs much better than the original Matched Filter and Hoover's method. And the proposed method is better than Soares' in extracting small vessels.

4.2   Quantitative Evaluation

In this section, our method is quantitatively evaluated. Table 1 presents the experimental results on the STARE database by different algorithms. The performance measures of Soares [2] and Hoover [5] were calculated using the segmented images from their websites. The FOV used for the STARE database was generated using code provided by Soares. All 20 images in STARE were used in this experiment. The hand-labeled images by the first human expert (labels-ah) were employed as ground truth. To facilitate the comparison of our method with Soares [2] and Hoover [5], we calculated the average TPR and PPV corresponding to an FPR of

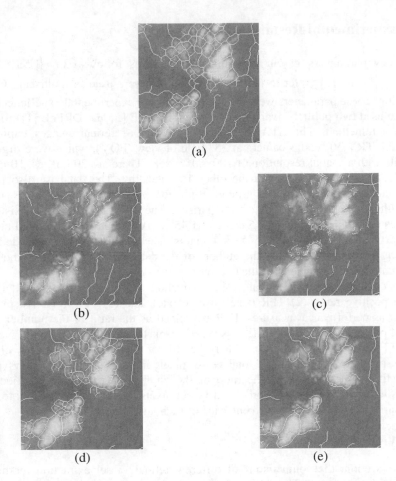

Fig. 2. Ability of the proposed method to distinguish between vessel and non-vessel objects using hard exudates and its surrounding vessels in Figure 1(a). (a) result of candidate detection, (b) candidate detection with SRC, (c) result of Soares [2], (d) result of Matched Filter [12], and (e) result of Hoover [5].

Table 1. Vessel extraction results on the STARE database

Method	FPR	TPR	PPV
Soares	0.0047	0.7494	0.7477
Matched Filter	0.0026	0.6177	0.8109
Hoover	0.0054	0.7192	0.7069
Proposed Method	**0.0041**	**0.7209**	**0.7596**

around 0.0045. The experimental results on the STARE database show that the proposed performs much better than the original Matched Filter, also outperforms Hoover's method and is slightly better than Soares (when comparing PPV).

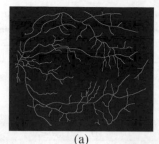

(a)

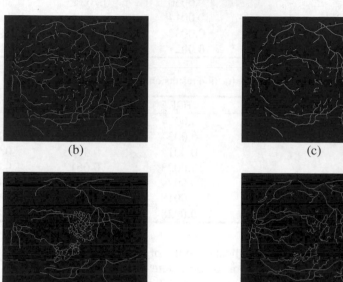

(b)   (c)

(d)   (e)

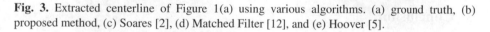

Fig. 3. Extracted centerline of Figure 1(a) using various algorithms. (a) ground truth, (b) proposed method, (c) Soares [2], (d) Matched Filter [12], and (e) Hoover [5].

In order to demonstrate the performance of our method in pathological cases, we compared the results by different algorithms on normal and abnormal images in STARE [5] (see Table 2). The experimental results show that for the abnormal cases the proposed method improves on Matched Filter and Hoover, and lags a bit behind Soares.

DRIVE [1] database results are presented in Table 3. The performance measures of Staal [1], Soares [2] and Niemeijer [3] were calculated using the segmented images from their websites. Jiang's [6] and Zana's [20] methods happened to be published a few years before DRIVE was established so their results in Table 3 were implemented by Staal [1] and Niemeijer [3] respectively. The DRIVE database came with its own FOV. All 20 images in the test set were used in the experiment with the hand-labeled images by the first human expert designated as ground truth. The experimental results

Table 2. Vessel extraction results on the STARE database (normal vs. abnormal cases)

Method	FPR	TPR	PPV
Abnormal cases			
Soares	0.0064	0.7355	0.6553
Matched Filter	0.0039	0.5614	0.7120
Hoover	0.0075	0.7133	0.5895
Proposed Method	**0.0063**	**0.7082**	**0.6315**
Normal cases			
Soares	0.0030	0.7634	0.8400
Matched Filter	0.0014	0.6740	0.9098
Hoover	0.0032	0.7251	0.8244
Proposed Method	**0.0021**	**0.7555**	**0.8860**

Table 3. Vessel extraction results on the DRIVE database

Method	FPR	TPR	PPV
Soares	0.0024	0.7043	0.8878
Matched Filter	0.0034	0.5161	0.8041
Jiang	0.0014	0.4888	0.9079
Niemejier	0.0023	0.6612	0.8883
Staal	0.0029	0.6991	0.8680
Zana	0.0018	0.6072	0.9045
Proposed Method	**0.0028**	**0.5766**	**0.8467**

on the DRIVE database again validate that the proposed method performs better than Matched Filter, while it is inferior to some state-of-the-art methods. This is because the proposed method has advantages in dealing with pathological retinal images but most of the images in the DRIVE test set are normal.

5 Conclusion

This paper proposed a novel retinal blood vessel extraction method using MPMF and SRC. The vessel candidates were first detected using multiscale matched filtering following by scale production, double thresholding and centerline calculation. These candidates were classified via SRC with two dictionary elements, vessel and non-vessel. The experimental results demonstrate the advantage of the proposed method at separating vessel from non-vessel, and at the same time detect small vessels due to the incorporation of multiscale matched filtering.

Currently, our SRC dictionaries are artificially generated using Gaussian functions. This may cause the low discriminative ability of SRC. In the future, we will further improve SRC by training optimal dictionaries on original color retinal images.

References

1. Staal, J.J., Abramoff, M.D., Niemeijer, M., Viergever, M.A., van Ginneken, B.: Ridge based vessel segmentation in color images of the retina. IEEE Transactions on Medical Imaging, 501–509 (2004)

2. Soares, J.V.B., Leandro, J.J.G., Cesar Jr., R.M., Jelinek, H.F., Cree, M.J.: Retinal vessel segmentation using the 2-d gabor wavelet and supervised classification. IEEE Trans. on Medical Imaging 25, 1214–1222 (2006)

3. Niemeijer, M., Staal, J.J., van Ginneken, B., Loog, M., Abramoff, M.D.: Comparative study of retinal vessel segmentation methods on a new publicly available database. In: SPIE Medical Imaging, vol. 5370, pp. 648–656 (2004)

4. Martínez-Pérez, M., Hughes, A., Stanton, A., Thom, S., Bharath, A., Parker, K.: Scale-space analysis for the characterisation of retinal blood vessels. Medical Image Computing and Computer-Assisted Intervention, 90–97 (1999)

5. Hoover, A., Kouznetsova, V., Goldbaum, M.: Locating blood vessels in retinal images by piecewise threshold probing of a matched filter response. IEEE Trans. Med. Imag. 19(3), 203–210 (2000)

6. Jiang, X., Mojon, D.: Adaptive local thresholding by verification based multithreshold probing with application to vessel detection in retinal images. IEEE Trans. Pattern Anal. Mach. Intell. 25(1), 131–137 (2003)

7. Mendonca, A.M., Campilho, A.: Segmentation of retinal blood vessels by combining the detection of centerlines and morphological reconstruction. IEEE Transactions on Medical Imaging 25(9), 1200–1213 (2006)

8. Martinez-Perez, M.E., Hughes, A.D., Thom, S.A., Bharath, A.A., Parker, K.H.: Segmentation of blood vessels from red-free and fluorescein retinal images. Medical Image Analysis 11(1), 47–61 (2007)

9. Leung, H., Wang, J.J., Rochtchina, E., Wong, T.Y., Klein, R., Mitchell, P.: Impact of current and past blood pressure on retinal arteriolar diameter in older population. J. Hypertens., 1543–1549 (2003)

10. Mitchell, P., Leung, H., Wang, J.J., Rochtchina, E., Lee, A.J., Wong, T.Y., Klein, R.: Retinal vessel diameter and open-angle glaucoma: the Blue Mountains eye study. Ophthalmology, 245–250 (2005)

11. Wang, J.J., Taylor, B., Wong, T.Y., Chua, B., Rochtchina, E., Klein, R., Mitchell, P.: Retinal vessel diameters and obesity: a population-based study in older persons. Obes. Res., 206–214 (2006)

12. Chaudhuri, S., Chatterjee, S., Katz, N., Nelson, M., Goldbaum, M.: Detection of blood vessels in retinal images using two-dimensional matched filters. IEEE Trans. Med. Imaging, 263–269 (1989)

13. Cinsdikici, M., Aydin, D.: Detection of blood vessels in ophthalmoscope images using MF/ant (matched filter/ant colony) algorithm. Computer Methods and Programs in Biomedicine 96, 85–95 (2009)

14. Al-Rawi, M., Qutaishat, M., Arrar, M.: An improvement matched filter for blood vessel detection of digital retinal images. Computers in Biology and Medicine 37, 262–267 (2007)

15. Chutatape, O., Zheng, L., Krishnan, S.: Retinal blood vessel detection and tracking by matched Gaussian and kalman filters. In: Proc. of Engineering in Medicine and Biology Society, pp. 3144–3149 (1998)

16. Rangayyan, R.M., Ayres, F.J., Oloumi, F., Oloumi, F., Eshghzadeh-Zanjani, P.: Detection of blood vessels in the retina with multiscale Gabor filters. J. Electron. Imaging 17, 023018 (2008)

17. Tolias, Y., Panas, S.: A fuzzy vessel tracking algorithm for retinal images based on fuzzy clustering. IEEE Trans. Med. Imaging, 263–273 (1998)

18. Sinthanayothin, C., Boyce, J., Williamson, C.T.: Automated Localisation of the Optic Disk, Fovea, and Retinal Blood Vessels from Digital Colour Fundus Images. British Journal of Ophthalmology, 902–910 (1999)
19. Garg, S., Sivaswamy, J., Chandra, S.: Unsupervised curvature-based retinal vessel segmentation. In: Proc. of IEEE International Symposium on Bio-Medical Imaging, pp. 344–347 (2007)
20. Zana, F., Klein, J.C.: Segmentation of vessel-like patterns using mathematical morphology and curvature evaluation. IEEE Trans. Image Process. 1010–1019 (2001)
21. Fraz, M.M., Javed, M.Y., Basit, A.: Evaluation of Retinal Vessel Segmentation Methodologies Based on Combination of Vessel Centerlines and Morphological Processing. In: IEEE International Conference on Emerging Technologies, pp. 18–19 (2008)
22. Wright, J., Yang, A., Ganesh, A., Sastry, S., Ma, Y.: Robust Face Recognition via Sparse Representation. IEEE Transactions on Pattern Analysis and Machine Intelligence 2, 210–227 (2009)

Pulse Waveform Classification Using ERP-Based Difference-Weighted KNN Classifier

Dongyu Zhang, Wangmeng Zuo, Yanlai Li, and Naimin Li

School of Computer Science and Technology
Harbin Institute of Technology, Harbin, 150001, China
cswmzuo@gmail.com

Abstract. Although the great progress in sensor and signal processing techniques have provided effective tools for quantitative research into traditional Chinese pulse diagnosis, the automatic classification of pulse waveform is remained a difficult problem. In order to address this issue, we propose a novel edit distance with real penalty-based k-nearest neighbor classifier by referring to recent progress in time series matching and KNN classifier. Taking advantage of the metric property of ERP, we develop an ERP-induced inner product operator and then embed it into difference-weighted KNN classifier. Experimental results show that the proposed classifier is more accurate than comparable pulse waveform classification approaches.

Keywords: pulse waveform; time series classification; edit distance with real penalty (ERP); k-nearest neighbor (KNN).

1 Introduction

Traditional Chinese pulse diagnosis (TCPD) is a convenient, non-invasive and effective diagnostic method used in traditional Chinese medicine (TCM) [1]. This diagnosis method requires practitioners to feel for the fluctuations in the radial pulse at the styloid processes of the wrist and classify them into distinct patterns which are related to different syndromes and diseases in TCM. Due to the limitation of experience and knowledge of different practitioners, the accuracy of pulse diagnosis could not be guaranteed. As a way to improve the reliability and consistency of diagnoses, in recent years techniques developed for measuring, processing, and analyzing the physiological signals [2, 3] have been considered in quantitative TCPD research [4, 5, 6, 7]. Since then, different pulse signal acquisition systems have been developed [8, 9, 10], and many methods are proposed in pulse signal preprocessing and analysis, including pulse signal denoising [11], baseline rectification etc. [14]. Moreover, a great progress have been made in pulse classification as a number of feature extraction and recognition methods have been studied for pulse signal classification and diagnosis [12, 13, 15, 16, 17, 18].

Although progress has been made, there are still problems in the automatic classification of pulse waveforms which involves classifying a pulse waveform as one of the traditional pulse patterns, e.g., moderate, smooth, taut, hollow, and unsmooth

D. Zhang and M. Sonka (Eds.): ICMB 2010, LNCS 6165, pp. 191–200, 2010.

pulse according to its shape, position, regularity, force, and rhythm [1]. Since the intra-class variation in pulse patterns is inevitable, each pulse pattern may have more than one typical waveform. Because of the adverse effect of local time shifting and noise, pulse waveform classification suffers from the low accuracy. The developed pulse waveform classification methods, such as neural networks and dynamical time warping (DTW) [19, 20, 27], achieve accuracies mostly below 90%, and usually are only tested on small data sets.

In this paper, by referring to the development in time series matching techniques, i.e., edit distance with real penalty (ERP), we investigate novel approach for pulse waveform classification. We first propose an ERP-induced inner product operator, and using the difference-weighted KNN (DFWKNN) framework we further present a novel ERP-based classifier, i.e., ERP-based difference-weighted KNN (ERP-DFWKNN), for pulse waveform classification. We evaluate the proposed method on a pulse waveform data set which includes 2470 pulse waveforms. Experimental results show that the proposed method achieves an average classification accuracy of 90.36%, which is higher than several other pulse waveform classification approaches.

The remainder of this paper is organized as follows. Section 2 introduces the main modules in pulse waveform classification. Section 3 first presents a brief survey on ERP and DFWKNN, and then proposes ERP-DFWKNN. Section 4 provides the experimental results on pulse waveform classification. Finally, a conclusion is drawn in Section 5.

2   The Pulse Waveform Classification Modules

A pulse waveform classification system usually includes three major modules: a pulse waveform acquisition module, a preprocessing module, and a classification module (Fig. 1). The acquisition module is essential, for a good quality of acquired pulse signal will allow effective preprocessing and accurate classification possible. The preprocessing module is used to remove the distortions of the pulse waveform caused by noise and baseline wander. In the third module, different feature extraction and classification methods are used to classify the pulse waveform into distinct patterns.

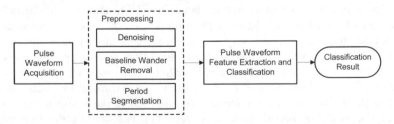

Fig. 1. Outline of pulse waveform classification modules

2.1   Pulse Waveform Acquisition

In this work, the pulse acquisition system was jointly developed by Hong Kong Polytechnic University and Harbin Institute of Technology. In pulse waveform

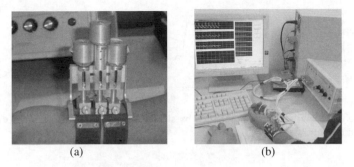

Fig. 2. Pulse waveform acquisition system: (a) three motors and pressure sensor embedded probes, and (b) the whole pulse waveform acquisition system

acquisition, the sensor (Fig. 2(a)) is attached to wrist and contact pressure is applied by the computer-controlled automatic rotation of motors and mechanical screws. The acquired pulse waveforms are transmitted to computer through the USB interface. Fig. 2(b) shows an image of the pulse waveform collection scene. The acquisition system is stable and can acquire satisfactory pulse signals for subsequent processing.

2.2 Pulse Waveform Preprocessing

Before the feature extraction, it is necessary to remove the random noise and baseline wander, which would greatly distort the pulse waveform, shown as in Fig. 3(a). Noise is suppressed by employing a *Daubechies* 4 wavelet transform according to the empirical comparison on several wavelet functions. Baseline wander is corrected using wavelet-based cascaded adaptive filter [11]. We then split each pulse waveform into several single periods based on the locations of onsets and select one of these periods as a sample for our pulse waveform data set. Fig. 3(b) shows the result of baseline wander correction and the locations of onsets.

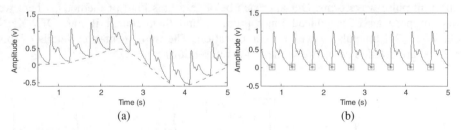

Fig. 3. Pulse waveform preprocessing: (a) pulse waveforms distorted by baseline wander, and (b) after baseline wander correction.

2.3 Pulse Waveform Feature Extraction and Classification

In TCPD, there are more than 20 kinds of pulse patterns defined according to the criteria such as shape, regularity, force and rhythm [1]. These are not settled issues in this field but there is general agreement that, according to the shape there are five pulse patterns, moderate, smooth, taut, hollow, and unsmooth pulse. Fig. 4 shows the typical

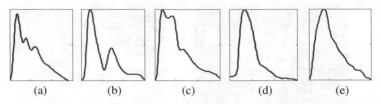

Fig. 4. Five typical pulse patterns classified by shape: (a) moderate, (b) smooth, (c) unsmooth, (d) taut, and (e) hollow pulse

waveforms of these five pulse patterns acquired by our pulse acquisition system. All of these pulses can be defined according to the presence, absence, or strength of three types of wave or peak, percussion, tidal, and dicrotic. A moderate pulse, for example, usually has all three of these peaks in one period. A smooth pulse has low dicrotic notch and unnoticeable tidal wave. A taut pulse frequently exhibits a high tidal peak. An unsmooth pulse exhibits no noticeable tidal or dicrotic wave. A hollow pulse has rapid descending part in percussion wave and unnoticeable dicrotic wave.

As shown as in Fig. 5, pulse waveform classification, however, may suffer from the problems of small inter-class and large intra-class variations. Moderate pulse with unnoticeable tidal wave is similar to smooth pulse, and for some taut pulse the tidal wave becomes very high and is even merged with the percussion wave.

By far, a number of pulse waveform classification approaches have been proposed, which can be grouped into two categories: representation-based methods and similarity measure-based methods. Representation-based classification methods first extract representative features using techniques such as FFT [28] and wavelet transform [29]. Then the classification is performed in the feature space using different classifiers.

For similarity measure-based methods, classification is usually performed in the original data space by using certain distance measures, e.g., Euclidean distance, DTW [21] and longest common subsequence (LCSS) [22], to measure the dissimilarity of different pulse waveforms. Our approach belongs to this kind of methods, where we first propose an ERP-induced inner product, and then embed it into DFWKNN classifier [24] for pulse waveform classification. The implementation details of this method are detailed described in the following section.

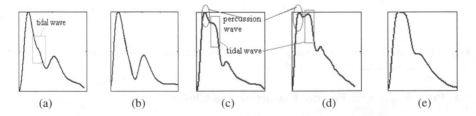

Fig. 5. Inter- and intra-class variations of pulse patterns: (a) a moderate pulse with unnoticeable tidal wave is similar to (b) a smooth pulse; Taut pulse patterns may exhibit different shapes, e.g., (c) typical taut pulse, (d) taut pulse with high tidal wave, and (e) taut pulse with tidal wave merged with percussion wave.

3   The ERP-DFWKNN Classifier

In this section, we first provide a brief introduction to ERP and DFWKNN, and then explain the operations and implementation of the proposed ERP-DFWKNN classifier.

3.1   ERP and DFWKNN

Edit distance with real penalty (ERP) is a recently developed elastic distance metric [23]. Given two time series $A = [a_1, a_2, \ldots, a_m]$ with m elements and $B = [b_1, b_2, \ldots, b_n]$ with n elements, the ERP distance $d_{erp}\left(A_1^m, B_1^n\right)$ of A and B is recursively defined as:

$$
d_{erp}\left(A_1^m, B_1^n\right) =
\begin{cases}
\displaystyle\sum_{i=1}^{m} |a_i - g|, & \text{if } n = 0 \\[2mm]
\displaystyle\sum_{i=1}^{n} |b_i - g|, & \text{if } m = 0 \\[2mm]
\min\begin{cases} d_{erp}\left(A_2^m, B_2^n\right) + |a_1 - b_1|, \\ d_{erp}\left(A_2^m, B_1^n\right) + |a_1 - g|, \\ d_{erp}\left(A_1^m, B_2^n\right) + |b_1 - g|, \end{cases}, & \text{otherwise}
\end{cases}
\tag{1}
$$

where $A_1^p = [a_1, \ldots, a_p]$ denotes the subsequence of A, $|\cdot|$ denotes the l_1-norm and g is a constant with the default value equal to 0 [23]. In [23], Chen and Ng also prove the metric property of ERP:

Theorem 1 [23]. The ERP distance satisfies the triangle inequality and is a metric.

DFWKNN is recently developed KNN classifier reported with classification performance comparable with or better than several state-of-the-art classification methods [24]. It treats the weight assignment as a constrained optimization problem of sample reconstruction from its neighborhood as

$$
\begin{aligned}
W &= \arg\min_{W} \frac{1}{2}\left\| x - X^{nn}W \right\|^2 \\
s.t. \quad &\sum_i w_i = 1
\end{aligned}
\tag{2}
$$

where $X^{nn} = \left[x_1^{nn}, x_2^{nn}, \ldots, x_k^{nn} \right]$ are k-nearest neighbors of x. Define Gram matrix G as

$$
G = \left[x - x_1^{nn}, \ldots, x - x_k^{nn} \right]^T \left[x - x_1^{nn}, \ldots, x - x_k^{nn} \right],
\tag{3}
$$

where the element in the ith row and the jth column of the Gram matrix G is

$$
G_{ij} = \left(x - x_i^{nn} \right)^T \left(x - x_j^{nn} \right) = \left\langle x - x_i^{nn}, x - x_j^{nn} \right\rangle,
\tag{4}
$$

where $\langle \cdot , \cdot \rangle$ denotes the inner product operation. Then the weights W of DFWKNN can be obtained by solving the following linear equation:

$$\left[G + \eta tr(G)/k \right] W = 1_k ,$$ (5)

where $\eta \in 10^{-2} \sim 10^{-3}$ is the regularization parameter, 1_k is $k \times 1$ vector with all entries equal to 1, and $tr(G)$ denotes the trace of matrix G.

After solving Eq. (5) and obtaining the weights W, the weighted KNN rule is employed to assign a class label to sample x.

3.2 ERP-Induced Inner Product and ERP-DFWKNN

In this section we first develop ERP-induced inner product and further propose an ERP based difference-weighted KNN (ERP-DFWKNN) for pulse waveform classification. Our motivations of developing the ERP-DFWKNN classifier can be stated from the following two aspects:

First, we want to utilize the metric property of ERP to improve the performance of pulse waveform classification. As an elastic measure, ERP is effective in handling the time axis distortion problem of time series. Furthermore, as we will show that, utilizing the metric property of ERP, we can easily get an ERP-induced inner product operator, which is essential for combining ERP with DFWKNN classifier. While for non-metric elastic measures, such as DTW and LCSS, we cannot get the similar result.

Second, we intend to design effective KNN classifier for pulse waveform classification. Current time series classification methods usually adopt the simple nearest neighbor or KNN classifiers. Recently, several weighted KNN methods have been proposed and improved classification performance has been reported [24, 25]. Thus, we will investigate the approach to incorporate ERP into these KNN classifiers to improve the pulse waveform classification accuracy.

Let $\langle \cdot , \cdot \rangle_{erp}$ denote the ERP-induced inner product. Since ERP is a metric, we can get the definition of $\langle x_i , x_j \rangle_{erp}$ by the following heuristic deduction:

$$
\begin{aligned}
d_{erp}^2 \left(x_i , x_j \right) &= \left\langle x_i - x_j , x_i - x_j \right\rangle_{erp} \\
&= d_{erp}^2 \left(x_i , 0 \right) + d_{erp}^2 \left(x_j , 0 \right) - 2 \left\langle x_i , x_j \right\rangle_{erp} \\
\Rightarrow \quad \left\langle x_i , x_j \right\rangle_{erp} &= \left(d_{erp}^2 \left(x_i , 0 \right) + d_{erp}^2 \left(x_j , 0 \right) - d_{erp}^2 \left(x_i , x_j \right) \right) \big/ 2
\end{aligned}
$$ (6)

where $d_{erp} \left(x_i , x_j \right)$ is the ERP distance of x_i and x_j and the 0 represents a zero-length time series.

In ERP-DFWKNN, we replace the regular inner product in Eq. (4) with ERP-induced inner product to calculate G_{ij},

$$
\begin{aligned}
G_{ij} &= \left\langle x - x_i^{nn} , x - x_j^{nn} \right\rangle_{erp} \\
&= k_{ij} + k_0 - k_c(i) - k_c(j)
\end{aligned}
$$ (7)

where $k_{ij} = \left\langle x_i^{nn}, x_j^{nn} \right\rangle_{erp}$, $k_0 = \left\langle x, x \right\rangle_{erp}$, and $k_c(i) = \left\langle x, x_i^{nn} \right\rangle_{erp}$. Then the Gram matrix in ERP-DFWKNN can be written as

$$G_{erp} = K_{erp} + k_0 1_{kk} - 1_k k_c^T - k_c 1_k^T , \qquad (8)$$

where K_{erp} is a $k \times k$ matrix with the element k_{ij}, 1_{kk} is a $k \times k$ matrix with each element equals 1, and 1_k is a $k \times 1$ vectors where all elements are 1.

Once we have the Gram matrix G_{erp}, we can directly use DFWKNN for pulse waveform classification by solving the linear system of equation defined in Eq. (5). The detailed algorithm of ERP-DFWKNN is summarized as follows.

Input: Unclassified sample x, training samples $X = \{x_1, \cdots, x_n\}$ with the corresponding class labels $Y = \{y_1, \cdots, y_n\}$, the regularization parameter η, and the number of nearest neighbors, k.

Output: Predicted class label $\omega_{j_{max}}$ of the sample x.

Step 1. Use the ERP distance to obtain the k nearest neighbors of sample x, $\{x_1^{nn}, \cdots, x_k^{nn}\}$ and their classes labels $\{y_1^{nn}, \cdots, y_k^{nn}\}$.

Step 2. Calculate the ERP-induced inner product of the samples x and each of its nearest neighbors, $k_c(i) = \left\langle x, x_i^{nn} \right\rangle_{erp} = \left(d_{erp}^2(x,0) + d_{erp}^2(x_i^{nn},0) - d_{erp}^2(x, x_i^{nn}) \right) / 2$.

Step 3. Calculate the ERP-induced inner product of the nearest neighbors of the sample x by using $k_{ij} = \left\langle x_i^{nn}, x_j^{nn} \right\rangle_{erp}$.

Step 4. Calculate the self-inner product of the sample x, $k_0 = \left\langle x, x \right\rangle_{erp}$.

Step 5. Calculate $G_{erp} = K_{erp} + k_0 1_{kk} - 1_k k_c^T - k_c 1_k^T$.

Step 6. Calculate W by solving $\left[G_{erp} + \eta \ tr(G_{erp}) / k \right] W = 1_k$.

Step 7. Assign class label $\omega_{j_{max}}$ to the sample x using the following rule,

$$\omega_{j_{max}} = \arg \max_{\omega_j} \left(\sum_{y_i^{nn} = \omega_j} w_i \right).$$

---

4 Experimental Results

The data set we used includes 2470 pulse waveforms of five pulse patterns including moderate, smooth, taut, unsmooth, and hollow pulse. Table 1 lists the number of pulse waveforms of each pulse pattern. We make use of only one period from each pulse signal and normalize them to the same length of 150 points. To reduce the bias in evaluating the performance of pulse waveform classification, we adopt the average classification accuracy of the 10 runs of the 3-fold cross validation. Using the stepwise selection strategy [24], for ERP-DFWKNN, we choose the following values of hyper-parameters k, and η: $k = 4$, $\eta = 0.01$. The classification rate of ERP-DFWKNN is 90.36%. Table 2 shows the confusion matrix of ERP-DFWKNN.

Table 1. Data set used in our experiments

Pulse	Moderate	Smooth	Taut	Unsmooth	Hollow	Total
Numbers	800	550	800	160	160	2470

Table 2. The confusion matrix of ERP-DFWKNN

		Predicted				
		M	S	T	H	U
Actual	M	720	59	19	2	0
	S	68	473	3	6	0
	T	22	5	764	3	6
	H	7	9	4	139	1
	U	1	1	20	2	136

To provide an objective comparison, we independently implemented two other pulse waveform classification methods, i.e., Improved DTW (IDTW) [20] and wavelet network [26], and evaluate their performance on our data set. The average classification rates of these two methods are listed in Table 3. Besides, we also compare the proposed method with several related classification methods, i.e., 1NN-Euclidean, 1NN-DTW, and 1NN-ERP. These results are also listed in Table 3. From Table 3, one can see that, in term of overall average classification accuracy, our method outperforms all the other methods.

Table 3. Average classification rates (%) of different methods

Pulse Patterns	1NN-DTW	1NN-ERP	1NN-Euclidean	Wavelet Network [26]	IDTW [20]	ERP-DFWKNN
Moderate	82.44	88.31	86.11	87.23	87.31	89.94
Smooth	81.16	86.31	85.02	85.36	80.38	86.00
Taut	87.95	95.10	95.76	89.63	93.15	95.50
Hollow	82.44	87.56	86.75	85.63	80.44	86.88
Unsmooth	70.81	84.75	84.06	80.63	89.50	85.00
Average	83.19	89.79	87.36	87.08	88.90	**90.36**

5   Conclusion

By incorporating the state-of-the-art time series matching method with advanced KNN classifier, we develop an accurate pulse waveform classification method, ERP-DFWKNN, to address the intra-class variation and the local time shifting problems in pulse patterns. To evaluate the classification performance, we construct a data set of 2470 pulse waveforms, which may be the largest data set yet used in pulse waveform classification. The experimental results show that the proposed ERP-DFWKNN achieves an average classification accuracy of 90.36%, which is higher than other state-of-the-art pulse waveform classification methods.

Acknowledgment. This work is partially supported by the CERG fund from the HKSAR Government, the central fund from Hong Kong Polytechnic University, and the NSFC/SZHK-innovation funds of China under Contract Nos. 60620160097, 60902099, 60872099, 60871033, and SG200810100003A.

References

1. Li, S.Z.: Pulse Diagnosis. Paradigm Press (1985) (in Chinese)
2. Dickhaus, H., Heinrich, H.: Classifying biosignals with wavelet networks. IEEE Engineering in Medicine and Biology 15, 103–111 (1996)
3. Adeli, H., Dastidar, S.G., Dadmehr, N.: A wavelet-chaos methodology for analysis of EEGs and EEG subbands to detect seizure and epilepsy. IEEE Transaction on Biomedical Engineering 54, 205–211 (2007)
4. Fei, Z.F.: Contemporary Sphygmology in Traditional Chinese Medicine. People's Medical Publishing House, Beijing (2003) (in Chinese)
5. Wang, H., Cheng, Y.: A quantitative system for pulse diagnosis in traditional Chinese medicine. In: Proceedings of the 27th Annual International Conference of the IEEE Engineering in Medicine and Biology Society, pp. 5676–5679 (2005)
6. Fu, S.E., Lai, S.P.: A system for pulse measurement and analysis of Chinese medicine. In: Proceedings of the 11th Annual International Conference of the IEEE Engineering in Medicine and Biology Society, pp. 1695–1696 (1989)
7. Lee, J., Kim, J., Lee, M.: Design of digital hardware system for pulse signals. Journal of Medical Systems 25, 385–394 (2001)
8. Hertzman, A.: The blood supply of various skin areas as estimated by the photo-electric plethysmograph. American Journal Physiology 124, 328–340 (1938)
9. Tyan, C.C., Liu, S.H., Chen, J.Y., Chen, J.J., Liang, W.M.: A novel noninvasive measurement technique for analyzing the pressure pulse waveform of the radial artery. IEEE Transactions on Biomedical Engineering 55, 288–297 (2008)
10. Wu, J.H., Chang, R.S., Jiang, J.A.: A novel pulse measurement system by using laser triangulation and a CMOS image sensor. Sensors 7, 3366–3385 (2007)
11. Xu, L.S., Zhang, D., Wang, K.Q.: Wavelet-based cascaded adaptive filter for removing baseline drift in pulse waveforms. IEEE Transactions on Biomedical Engineering 52, 1973–1975 (2005)
12. Xu, L., Zhang, D., Wang, K., Wang, L.: Arrhythmic pulses detection using Lempel-Ziv complexity analysis. EURASIP Journal on Applied Signal Processing, Article ID 18268, 12 pages (2006)
13. Wang, K.Q., Xu, L.S., Li, Z.G., Zhang, D., Li, N.M., Wang, S.Y.: Approximate entropy based pulse variability analysis. In: Proceedings of 16th IEEE Symposium on Computer-Based Medical Systems, pp. 236–241 (2003)
14. Xu, L.S., Meng, M.Q.-H., Liu, R., Wang, K.Q.: Robust peak detection of pulse waveform using height ratio. In: Proceedings of 30th Annual International IEEE EMBS Conference, pp. 3856–3859 (2008)
15. Zhang, P.Y., Wang, H.Y.: A framework for automatic time-domain characteristic parameters extraction of human pulse signals. EURASIP Journal on Advances in Signal Processing, Article ID 468390, 9 Pages. (2008)
16. Allen, J., Murray, A.: Comparison of three arterial pulse waveform classification techniques. Journal of Medical Engineering and Technology 20, 109–114 (1996)

17. Wang, B.H., Xiang, J.L.: ANN recognition of TCM pulse state. Journal of Northwestern Polytechnic University 20, 454–457 (2002)
18. Xu, L.S., Meng, M.Q.-H., Wang, K.Q., Wang, L., Li, N.M.: Pulse images recognition using fuzzy neural network. Expert Systems with Applications 36, 3805–3811 (2009)
19. Folland, R., Das, A., Dutta, R., Hines, E.L., Stocks, N.G., Morgan, D.: Pulse waveform classification using neural networks with cross-validation techniques: some signal processing considerations. In: Proceedings of Biomedical Engineering, 4 Pages. (2004)
20. Wang, L., Wang, K.Q., Xu, L.S.: Recognizing wrist pulse waveforms with improved dynamic time warping algorithm. In: Proceedings of the Third International Conference on Machine Learning and Cybernetics, pp. 3644–3649 (2004)
21. Yi, Z.B.-K., Jagadish, H., Faloutsos, C.: Efficient retrieval of similar time sequences under time warping. In: Proceedings of 14th International Conference on Data Engineering, pp. 23–27 (1998)
22. Vlachos, M., Kollios, G., Gunopulos, D.: Discovering similar multidimensional trajectories. In: Proceedings of 18th Internal Conference on Data Engineering, pp. 673–684 (2002)
23. Chen, L., Ng, R.: On the marriage of Lp-norms and edit distance. In: Proceedings of 30th Very Large Data Bases Conference, pp. 792–801 (2004)
24. Zuo, W.M., Zhang, D., Wang, K.Q.: On kernel difference-weighted k-nearest neighbor classification. Pattern Analysis Application 11, 247–257 (2008)
25. Gupta, M.R., Gray, R.M., Olshen, R.A.: Nonparametric supervised learning by linear interpolation and maximum entropy. IEEE Transactions on Pattern Analysis and Machine Intelligence 28, 766–781 (2006)
26. Xu, L.S., Wang, K.Q., Wang, L.: Pulse waveforms classification based on wavelet network. In: Proceedings of the 27th Annual Conference on IEEE Engineering in Medicine and Biology, Shanghai, China, pp. 4596–4599 (2005)
27. Trajcevski, H.D.G., Scheuermann, P., Wang, X.Y., Keogh, E.: Querying and mining of time series data: experimental comparison of representations and distance measures. In: Proceedings of the VLDB Endowment, vol. 2, pp. 1542–1552 (2008)
28. Yang, H., Zhou, Q., Xiao, J.: Relationship between vascular elasticity and human pulse waveform based on FFT analysis of pulse waveform with different age. In: Proceedings of International Conference on Bioinformatics and Biomedical Engineering, 4 Pages (2009)
29. Guo, Q., Wang, K., Zhang, D., Li, N.: A wavelet packet based pulse waveform analysis for Cholecystitis and nephrotic syndrome diagnosis. In: Proceedings of the 2008 International Conference on Wavelet Analysis and Pattern Recognition, Hong Kong, pp. 513–517 (2008)

Development of a Ubiquitous Emergency Medical Service System Based on Zigbee and 3.5G Wireless Communication Technologies

Ching-Su Chang[1], Tan-Hsu Tan[1,*], Yung-Fu Chen[2], Yung-Fa Huang[3],
Ming-Huei Lee[4], Jin-Chyr Hsu[4], and Hou-Chaung Chen[4]

[1] Department of Electrical Engineering, National Taipei University of Technology,
No.1, Sec. 3, Chung-hsiao E. Rd., Taipei, 10608, Taiwan, R.O.C.
Tel.: +886-2-27712171#2113; Fax: +886-2-27317187
{chingsu,thtan}@ntut.edu.tw
[2] Department of Health Services Administration, China Medical University,
No. 91, Hsuch-shih Rd., Taichung, 40402, Taiwan, R.O.C.
yungfu@mail.cmu.edu.tw
[3] Department of Information and Communication Engineering,
Chaoyang University of Technology, No.168, Jifeng E. Rd., Wufeng Township,
Taichung County, 41349, Taiwan, R.O.C.
yfahuang@mail.cyut.edu.tw
[4] Department of Health, Executive Yuan, Taichung Hospital, Taiwan, R.O.C.
taic66006@mail.taic.doh.gov.tw, jinchyr.hsu@msa.hinet.net,
action4health@yahoo.com

Abstract. This study builds a ubiquitous emergency medical service (EMS) system with sensor devices, a smartphone, a webcam, Zigbee, and a 3.5G wireless network. To avoid excessive wiring, a Zigbee-based wireless sensor network within the ambulance records the patient's biosignals, which includes an electrocardiogram (ECG), pulse, blood oxygen content, and temperature. Those biosignals are then transmitted to the Server located in the hospital emergency room via 3.5G wireless network for immediate first-aid preparation. This process significantly enhances EMS quality. Our experiment demonstrates biosignal transmission in real time for EMS application. In the future, we will promote the proposed system to WiMAX network when it becomes more pervasive in Taiwan to offer a much higher data throughput.

Keywords: Zigbee, Emergency Medical Service (EMS), 3.5G Wireless Networks, ECG.

1 Introduction

Accidental injury ranks third on the list of death causes in Taiwan [1]. According to a report released by the Taiwan Department of Health (DOH) [1], 57% of accidental injuries occurred in traffic accidents, 11% in falls, 9% in drowning or submersion, 3% in fire accidents, and 20% in other situations. The DOH also reported inadequate

---

* Corresponding author.

D. Zhang and M. Sonka (Eds.): ICMB 2010, LNCS 6165, pp. 201–208, 2010.
© Springer-Verlag Berlin Heidelberg 2010

emergency medical care, resulting in a survival rate of only 1.4%. Other studies indicate that providing proper treatments at the accident sites or en route to the hospital can prevent 20-35% of deaths. The need for proper emergency medical care is evident. Despite the increasing number of medical institutions, most medical resources are concentrated in urban areas. Fewer medical resources, especially emergency care, are located in suburban or remote areas. Increasing medical resources between urban and remote areas and accelerating emergency procedures are important issues in medical information system development.

Medical advances have greatly reduced the threat of disease for human beings. However, terrorist attacks, natural calamities, and incidental accidents are not easy to prevent. These unnecessary deaths are not preventable without a proper emergent rescuing mechanism. Consequently, researchers have developed several telemedicine systems [2–4] to improve emergency care quality. A wireless telemedicine system based on IEEE 802.16/WiMAX has been presented in [2]. It transmits various biosignals from health care centers to hospitals via WiMAX for preparing emergency operations. In [3], an emergency care system integrating biotelemetry devices and 3.5G HSDPA (High Speed Downlink Packet Access) has been proposed for use in ambulance. In [4], the author points out that American healthcare industry faces several challenges, including increase of medical costs, mistakes, and insufficient facilities, personnel, and budgets. A solution to this crisis is a pervasive healthcare policy, which relies on an integrated use of mobile devices, wireless LANs (WLANs), and middleware systems. Healthcare information systems based on mobile phones and handheld computers, namely m-health, efficiently provides quality of services (QoS) in various healthcare settings, especially in areas with insufficient medical resources [5]. For areas lacking healthcare providers, m-health systems enable general healthcare workers to perform clinician tasks [6]. Several m-health systems monitor patients during transport [7], chronic disease care [8], and unattended patients at a disaster site or overcrowded emergency department [9].

The aforementioned systems provide extensive services, including mobile telemedicine, patient monitoring, emergency management, and instant access to medical information. However, several difficulties remain unresolved:

(1) Managing cables in the ambulance: While delivering patients to hospitals, sensor wires may cause inconvenience to both the patients and healthcare providers.
(2) Ubiquitous accessibility of patient biosignals in an emergency: Healthcare providers cannot readily obtain patient information, such as in a remote area where ambulance cannot directly reach. This may cause a delay in the rescue.

To address these issues, this study integrates sensor devices, an ECG smartphone, Zigbee, a webcam, and a 3.5G network to develop a ubiquitous healthcare system for emergency healthcare support. First, an ECG smartphone acquires ECG signals from the patient. The ECG smartphone directly communicates with the remote Server. In the ambulance, all biosignals detected by the sensor devices transmit to the in-car application Server via the Zigbee wireless sensor network. Then, the biomedical information, including ECG report, blood pressure, breath, pulse, blood oxygen concentration, body temperature, and real-time video, are instantly delivered via 3.5G mobile network to the e-Health Server located in the emergency room.

Section 2 of this paper describes implementation of the proposed emergency medical service system. Section 3 demonstrates experimental results. Section 4 includes conclusions and future possibilities.

2   System Implementation

2.1   System Framework

Figure 1 shows the framework of the proposed system. This system is built on a client-server architecture, where the client consists of wireless biosignal sensors, webcam, ECG smartphone, and application Server installed inside the ambulance. The server consists of a hospital-side Server, ECG smartphone, and SQL Server. The main function of each component is explained as follows:

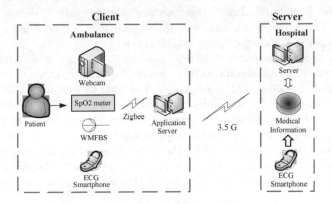

Fig. 1. System framework

1. **Wireless sensors:** There are three types of wireless sensors, including EEG smartphone, SpO2 meter, and Wireless multi-functional biomedical sensors, as detailed below:
   A. ECG smartphone:
   This mobile phone measures and records ECG signals, and uploads them to the hospital-side Server via the 3.5G mobile network.
   B. SpO2 meter:
   This mobile phone measures and records ECG signals, and uploads them to the hospital-side Server via the 3.5G mobile network.
   C. Wireless multi-functional biomedical sensors (WMFBS):
   This device measures ECG signals and body temperature and delivers results to the application Server for storage via the Zigbee module.
2. **Human-machine interface (HMI) application Server:** The HMI application Server displays biosignals measured by WMFBS and videos captured by a webcam showing real-time physiological and symptomatic conditions of patients. A monitor displays various types of information for diagnostic and caring references. HMI delivers the data to the hospital-side Server via the 3.5G wireless network.

3. **Server-side HMI:** This interface establishes connections with client-side application Servers and displays real-time biosignals and videos of patients.
4. **Medical Information:** This module stores ECG signals uploaded by the ECG smartphone. Healthcare providers can view the ECG using the web browsers.

2.2 Implementation of Client-Side

Emergency personnel arriving at an incident in a remote area can use the ECG smartphone to instantly measure the victim's ECG. The analog signals detected by ECG electrodes are converted to digital signals and displayed on the built-in HMI. This smartphone also stores and uploads ECG graphs.

After loading the patient into the ambulance, healthcare providers can use WMFBS to measure ECG and body temperature, and a wireless SpO2 meter to measure oxygen concentration and pulse. The digitized measurements are delivered to the application Server via Zigbee. Visual Basic 6.0 was used to develop the HMI of the application Server installed in the ambulance. This interface integrates the real-time videos and biosignals of the patient, and connects to the hospital-side Server. Figure 2 illustrates the workflow of the proposed client-side HMI service.

The HMI program determines if the real-time videos are applicable. It automatically displays the real-time videos if the webcam has been correctly set up. Users can adjust the video-encoding method for video quality. The SpO2 meter and WMFBS deliver biosignals in packets, distinguished by packet header and encoding method. The program receives and identifies biosignal packets through a COM port. The program extracts ECG signals, body temperature, oxygen saturation, and pulse, from received packets and displays them on the HMI. The webcam is set to the server mode. It delivers real-time videos of the patient upon connection requests. As Winsock delivers biosignals, the HMI displays the hospital Server IP after establishing a connection.

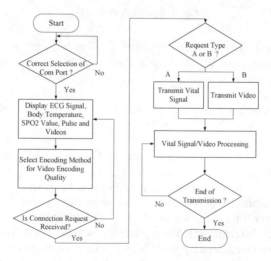

Fig. 2. Workflow of the client-side HMI service

2.3 Implementation of Server-Side

After uploading ECG signals, healthcare providers can use web browsers to open a designated page and access the results with their accounts and passwords. This study also develops HMI using Visual Basic 6.0 for hospitals. This hospital-side HMI allows healthcare providers to connect with the ambulance and view patient biosignals and real-time videos. Figure 3 illustrates the workflow of the server-side HMI service. Hospital healthcare providers input the IP address of the ambulance application Server to send a request for connection. Upon connection acceptance, the system receives packets of biosignals and real-time videos of patients via Winsock. Users can record real-time videos at a rate of 30 frames per second and save them in Audio Video Interleave (AVI) format for further diagnosing reference.

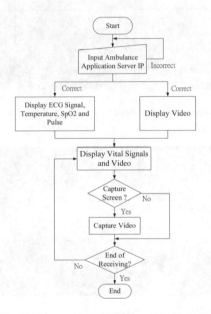

Fig. 3. Workflow of the Server-side HMI service

3   Experimental Results

We conducted a series of experiments in Taichung Hospital and a cooperated health care center. Figure 4 shows the process of measuring a patient's biometric information in the ambulance. Before loading the patient to the ambulance, the ECG smartphone was used to record one-lead ECG signals, as shown in Fig. 5. Emergency healthcare providers can select the recording interval (30 seconds in Fig. 5) and execute the "Transmit" function on the interactive touch panel. On request for transmission, the ECG smartphone uploads the measured ECG signals to the hospital-side Server via the 3.5G wireless network. Figure 6 shows that authorized healthcare providers can use web browsers to observe ECG signals stored in the SQL Server. Figures 7 and 8 give examples of Client-side HMI and Hospital-side HMI displayed biosignals. Since the wireless network card adopts a dynamic IP mode, i.e., each reconnection to the Internet at the aiding side produces a different IP address, the hospital needs to relocate the current IP address at the aiding side to successfully receive the biosignals and videos.

Fig. 4. Scenario of measurement of patient's biometric information in ambulance

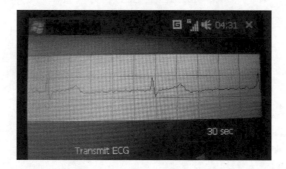

Fig. 5. Measured one-lead ECG signals on ECG smartphone

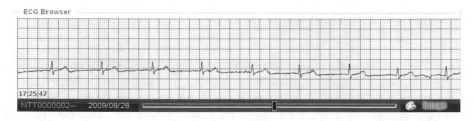

Fig. 6. ECG signals stored in the SQL Server

Fig. 7. Measured results shown in Client-side HMI

Fig. 8. Measured results shown in Hospital-side HMI

4   Discussions and Conclusions

In this study, we constructed a ubiquitous EMS by using sensor devices, a smartphone, a webcam, Zigbee, and a 3.5G network. This system can provide real-time displays of physiological and symptomatic conditions in two stages. In the first stage, the advance emergency rescuer uses the ECG smartphone to upload ECG signals of the patient to the hospital-side Server before the patient is loaded into the ambulance. In the second stage, the emergency rescuer uses wireless biosignal sensors to measure ECG, body temperature, oxygen saturation, and pulse after the patient is in the ambulance. All biosignals can be wirelessly transmitted to the application Server located inside the ambulance. The real-time videos are captured by a webcam and are transmitted to the hospital-side Server via the 3.5G mobile network. A series of experiments was conducted in Taichung Hospital and a cooperated health care center. The experimental results demonstrate that the proposed system effectively transmits real-time biosignals and videos of the patient for accelerated emergency medical services.

The experimental results show that under normal circumstances, the biosignals and videos of the subjects could be transmitted successfully, but the electrocardiographic (ECG) signals experienced some distortion, such as the two cases described below:

(1)   ECG signal noise was observed when the subjects moved their bodies in a fierce manner. This could be because the biomedical sensors brushed against the subjects' skin during the process.

(2)   ECG signal noise was observed when the subjects were very skinny. This may be because the biomedical sensors were unable to completely contact the subject's waist during the process.

Apart from the above cases, the remaining subjects did not present any problems. With respect to network transmission quality, the data transmission rate of the 3.5 G wireless network card used was sufficient for simultaneous biosignal and video transmission.

In the future, we plan to extend the proposed system to WiMAX network when it becomes more pervasive in Taiwan to offer a much higher data throughput.

Acknowledgments

The authors would like to thank the National Science Council of Taiwan, R.O.C. for financially supporting this research under Contract Nos. NSC 97-2221-E-027-095, NSC 98-2221-E-027-062, and NSC 98-2410-H-039-003-MY2.

References

1. Department of Health, Executive Yuan, R.O.C. (TAIWAN),
   http://www.doh.gov.tw/cht2006/index_populace.aspx
2. Niyato, D., Hossain, E., Diamond, J.: IEEE 802.16/WiMAX-based Broadband Wireless Access and Its Application for Telemedicine/E-health Services. IEEE Wireless Communications 14(1), 72–83 (2007)
3. Kang, J., Shin, H., Koo, Y., Jung, M.Y., Suh, G.J., Kim, H.C.: HSDPA (3.5G)-Based Ubiquitous Integrated Biotelemetry System for Emergency Care. In: Proc. of IEEE International Conference on EMBS, pp. 23–26 (2007)
4. Varshney, U.: Pervasive Healthcare. IEEE Computer Mag. 36(12), 138–140 (2003)
5. Blaya, J.A., Fraser, H.S.F., Holt, B.: E-health Technologies Show Promise in Developing Countries. Health Affairs 29(2), 244–251 (2010)
6. Rosen, S., Fox, M.P., Gill, C.J.: Patient Retention in Antiretroviral Therapy Programs in sub-Saharan Africa: A Systematic Review. PLoS Med. 4(10), e298 (2007)
7. Lin, Y.H., Jan, I.C., Ko, P.C.I., Chen, Y.Y., Wong, J.M., Jan, G.J.: A Wireless PDA-based Physiological Monitoring System for Patient Transport. IEEE Trans. Inf. Tech. Biomedicine 8(4), 439–447 (2004)
8. Lee, R.G., Chen, K.C., Hsiao, C.C., Tseng, C.L.: A Mobile Care System with Alert Mechanism. IEEE Trans. Inf. Tech. Biomedicine 11(5), 507–517 (2007)
9. Curtis, D.W., et al.: SMART—An Integrated Wireless System for Monitoring Unattended Patients. Journal of the American Medical Informatics Association 15(1), 44–53 (2008)

Digital Skating Board with RFID Technique for Upper Extremity Rehabilitation

C.-C. Chen[1], J.-C. Lin[2], C.-H. Chou[3], Y.-Y. Shih[4], and Y.-L. Chen[2,3,*]

[1] Department of Information Management, Hwa-Hsia Institute of Technology
[2] Department of Digital Technology Design, National Taipei University of Education
[3] Department of Computer Science, National Taipei University of Education
[4] Department of Rehabilitation Medicine, Chang Gung Memorial Hospital

Abstract. This research involves developing a hand-skate training system for the upper extremity rehabilitation. Included in this system are a hand-skating board, radio frequency identification (RFID) reader, and a computer. The hand-skating board is a platform with multiple RFID-tags and a RFID reader under the hand-skate for the patients to operate "∞" figure. The selected items and figures on the computer activate specific designs linking with each tags and are displayed on the computer screen. The patients then draw the same figures on the platform by the arm-skate with their hands. The corrections and time elapsed can be recorded and sent back to computer unit through the blue tooth interface. The flexion, extension, and coordination of hand muscle will be achieved after repeatedly performing this training. Meanwhile, in addition to multiple training items of corresponding figures, this system also provides functions for building a patient database to offer clinicians scientific data for long-term monitoring the rehabilitation effects.

Keywords: upper extremity rehabilitation, RFID-tags, hand-skate, blue tooth, "∞" figure.

1 Introduction

This research developed an aid training system for upper extremity in coordination rehabilitation. The system has manual and virtual specific designs whose hands are paralyzed caused by hemiplegics. Through a specific processing system, corrections can be determined as the system is trained on the upper extremity coordinating.

Most of the patients whose IQ is normal only have language expression and muscle coordination obstacles. Therefore, rehabilitation training for muscle coordination in the upper extremity is essential and urgent. However, the present training system for upper extremity is limited to traditional remedies or some simple training tools that cannot efficiently guide patients into reconstruction procedures and targets. The current systems are also unable to offer clinicians with scientific data to evaluate the rehabilitation process [1-4].

---

* Corresponding author.

D. Zhang and M. Sonka (Eds.): ICMB 2010, LNCS 6165, pp. 209–214, 2010.
© Springer-Verlag Berlin Heidelberg 2010

The main purpose of this research is to provide a specific design and determine its effectiveness through a specific processing system for repeatedly practicing, further enhancing a rehabilitation training system for upper extremity in coordination.

2 Methodology

The configuration of the hand-skate system for upper extremity rehabilitation is shown in Fig.1 which includes: (1) hand-skating board and a hand-skate, (2) a radio frequency identification (RFID) reader under hand-skate, (3) computer unit linking with hand-skating unit. The hand-skating unit as shown in Fig.2 includes a platform and a hand-skate, the platform has a operating areas suitable for manual actions. Its surface is smooth and flat, beneficial for hand-skate to move on the surface easily.

As shown in Fig.1 and Fig.2, there are multiple sets of RFID-tags set in the bottom of the platform. The RFID-tags are constructed in the form of a 2-D of array. In the examples, 120 sets of tags are set in bottom area of the platform, 20(sets)* 6(sets) as a 2-D of array used for checking hand-skate position for patients' hands to move. Each tags is linked with a computer through RFID reader and blue tooth interface.

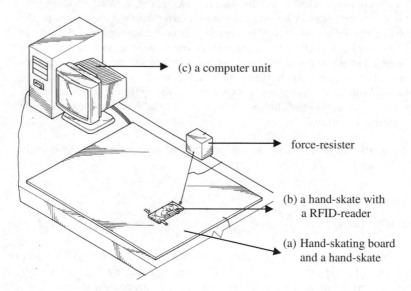

(c) a computer unit

force-resister

(b) a hand-skate with
a RFID-reader

(a) Hand-skating board
and a hand-skate

Fig. 1. The configuration of the aid training system for upper extremity rehabilitation

For the hand-skate moved by patients' hands (as shown in Fig.2), 360° slipping wheels provide free revolving set at the four corners of the hand-skate. The slipping wheels which resemble a universal joint, making the hand-skate can move in all directions. There is a RFID reader at the central position under the hand-skate. The

RFID reader and platform retains a regular and suitable distance (about 1mm). When the hand-skate moving on the platform, the RFID reader under its bottom will sense tags at corresponding position in the bottom of the platform.

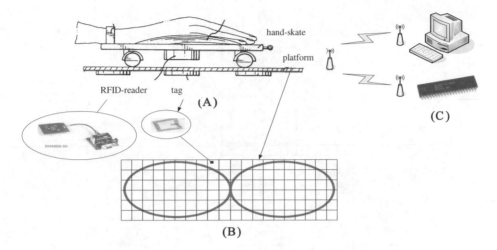

Fig. 2. The diagram of the upper extremity rehabilitation system

A holding part slightly protrudes on the hand-skate surface in a specific shape, suitable for patient hand to support on it, and move on the platform arbitrarily. A sticky buckle firmly set on the wrist to prevent the hand-skate interfering with the patients' actions.

After repeatedly training, actions and muscle coordination in the upper extremity will gradually be improved and able to handle the hand-skate. For promoting the training effects, difficult hand-skate levels can be increased step by step in order to enhance the physical strength of the patient's hand and provide increased flexibility. The hand-skating unit related could also include a force-resister (as shown in Fig.1), which is connected to the hand-skate by a traction lead. This force-resister can modulate the resistance of the hand-skate. In other words, when operating a hand-skate, in addition to drawing designs according to specific routes, resistance can also be applied. The resistance levels can be modulated according to the real demands to comply with the patients' physical conditions and the rate of training progress.

After the operator has set up the selected items and figures on the computer unit, specific designs linking with to each RFID-tags will be displayed on the computer screen. The patients can then draw the same designs on the platform using the hand-skate with their hand. The correction and time elapsed can be recorded and sent back to the computer through blue tooth interface. The training results will be saved in the patient database. The software control flow-chart as shown in Fig.3.

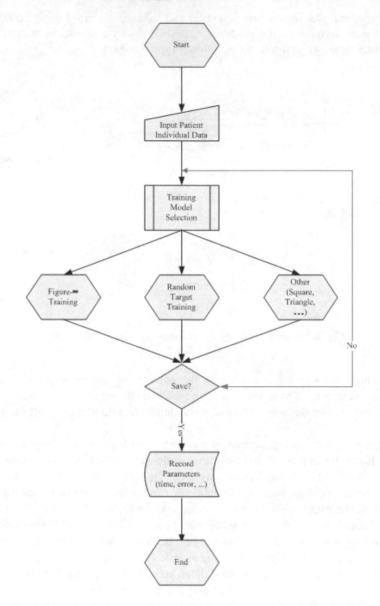

Fig. 3. The software flow-chart of the upper extremity rehabilitation system

The circuit include a skate with a RFID reader, multiple RFID-tags which are arranged as a 2-D array, a set of decoding and encoding circuits, and a microprocessor (89C51). Among then, RFID reader is placed at central position under the hand-skate operated by patients to react to the RFID-tags under the operating platform. The tags are arranged in a 2-D array that resembles to the keyboard circuit. The array scans if tags are closed through the scanning circuit. The microprocessor generates a set of scanning signals to the tags by its Port1 through the decoding circuit. If the tags are sensed, the Port2 receives another set of signals through encoding circuit. Therefore, the microprocessor can instantly read where the tags operated by patients and respond to a closed circuit. Finally, the microprocessor is connected to the computer's communication port (COM1 through RS-232) through its serial transmitting port (Port3), transmitting data of tags response to computer for analysis and processing. The computer unit includes a processing unit, a indicating unit (screen), a input unit (keyboard, mouse) and a alarm unit. Its construction is the same as common computer, but can execute hand coordination training by using window software in connection the apparatus described above.

The concrete methods to take effect for the hand coordination training related above are mentioned in detail as follows. When performing hand coordination training, the patient seated in front of the platform. The patient supports one hand with the hand-skate, and then fixes the wrist with the sticky buckle on hand-skate. According to the system operation flow, when system starts, operator selects item "hand rehabilitation", and begins coordination training. On the computer screen there is a picture of a 2-D array displayed, composed by 20*6= 120 plain pictures. Meanwhile, specific designs on arranged pictures, as related above, serve as orbit demonstrators of a rectangle, triangle, circle or symbol "∞". Patients look at the designs on the screen, move the arm-skate on the platform and draw routes that are the same as the appointed designs. While the arm-skate is moving, the magnet under it bottom will close the access to the reed-relays located at the corresponding position under the platform. The sensing signal access for tags will be sent back to the computer unit through blue tooth interface.

The computer processes the route orbit contrast with specific designs generated using position signals gained from detection by the tags. The differences in accuracy and time elapsed are calculated. If saving the information is not needed, return back to original state, and restart for training following the steps related above. If the information is saved, the rehabilitation training is saved as the contrast for the next training.

The hand coordinating training part can be proceeded repeatedly with patients according to the steps, using hand-training unit in connection with computer unit. Due to contrast correction function of this research, on the computer screen synchronically displays differences between the orbit and original appointed designs drawn on the platform. In other words, patients can use the synchronically display function on the computer screen and orbit correction sound effects of multimedia as patients' visual and audio feedback, instantly correcting the drawing design route. In the course of repeated training, patients' ability can be largely enhanced, further strength action the coordination of hand muscle.

3 Conclusions

The hand-skate system for upper extremity rehabilitation developed by this research has several features as follows:

1. By mean of drawing simple figures and lines, patients can easily practice rehabilitation training, and extend and bend the upper extremity, improving strength and coordination.

2. Using the instant correction function displayed synchronically by the computer unit, patients can be trained to improve their response of muscles, to reduce dullness during rehabilitation in connection with real clinical evaluation and friendly software interface.
3. Through connection with software and hardware, the course of instruction is be simple. Using a RFID-tag as corresponding elements, other signals cannot interfere easily. The price is much lower than other measuring sensors (infrared or laser receiver), comparatively reducing the production cost.
4. The computer unit can calculate and record the training results for continuity, integrity and proportion. The time elapsed can be recorded for clinicians as a reference in rehabilitation medical care.
5. Easy to operate, training can be achieved without long and dull practice steps.
6. The system clinical trail now is processing in this research, and actually put it into practice with the hemplegics.

References

1. Lum, P.S., Burgar, C.G., Shor, P.C., et al.: Robot-assisted Movement Training Compared with Conventional Therapy Techniques for the Rehabilitation of Upper-limb Motor Function After Stroke. Arch. Phys. Med. Rehabil. 83, 952–959 (2002)
2. Krebs, H.I., Volpe, B.T., Aisen, M.L., Hening, W., Adamovich, S., Poizner, H., Subrahmanyan, K., Hogan, N.: Robotic Applications in Neuromotor Rehabilitation. Robotica 21(1,3) (2003)
3. Fasoli, S.E., Krebs, H.I., Stein, J., Frontera, W.R., Hogan, N.: Effects of Robotic Therapy on Motor Impairment and Recovery in Chronic Stroke. Arch. Phys. Med. Rehabil. 84, 477–482 (2003)
4. Coote, S., Stokes, E.K.: Robot mediated therapy: Attitudes of patients and therapists towards the first prototype of the GENTLE/s system. Technology and Disability 15, 27–34 (2003)

Knowledge Acquisition and Reasoning in Web Based Epidemic Analysis

V.S. Harikrishnan, K. Pal Amutha, and S. Sridevi

Centre for Development of Advanced Computing (C-DAC),
Chennai, India
{harikrishnans,palamuthak,sridevis}@cdac.in

Abstract. Epidemic Analyzer: an intelligent query processor and health support model accepts Layman health information as input and tells the Layman about the percentage of chance for him/her to have an epidemic. This paper is an approach towards integrating knowledge acquisition and reasoning in Web based Epidemic Analysis. The information about allergic medicines for the patient, the environment which the patient resides, list of recently affected diseases is always useful in proper diagnosis of an epidemic. Immediate notification of possible emergency and health artifacts to nearest and possible communication with the personal medical assistant via medical language are always helpful in disease diagnosis at right time. We take the advantage of distributed space architecture and semantic web to achieve these goals.

Keywords: Tuple space, Semantic Web, Expert System.

1 Introduction

Epidemic Analyzer is a tool, which clarifies the curiosity of a person that whether he/she is suffering from a disease. This helps to remove panic from the minds of people, that he/she has an epidemic. By using a global Knowledge Base, the model tries to integrate the knowledge available from different parts of the world regarding an epidemic. A layman inputs the name of the disease that he suspects he has, due to the presence of certain symptoms. These symptoms are input as well. Epidemic Analyzer will analyze the input and derive inferences based on the knowledge stored in the Knowledge Base. The inference is let known to the user immediately.

In this model we are using tuple space to act as a Knowledge Base for the symptoms and other information regarding the particular disease. The data aggregation and integration capabilities of semantic web are utilized to do effective reasoning and query processing.

Maintaining the Layman user profile helps the epidemic analyzer to know the information such us the early diseases met by the patient, the places were the patient recently traveled and the epidemic possibilities of the environment in which user resides and the environment user earlier traveled, the climatic conditions of the region, the medicines which person is allergic to, the possibilities of occupational diseases. Epidemic analyzer considers these factor as input in its rule based reasoning

D. Zhang and M. Sonka (Eds.): ICMB 2010, LNCS 6165, pp. 215–223, 2010.
© Springer-Verlag Berlin Heidelberg 2010

strategies. In handling emergency situations, the epidemic analyzer can SMS (Short Message Service) to the personal consultant associated with the layman and also alert the nearest by SMS and E-mail. Communication of medical terms and medical diagnosis always helps the personal consultant to take quick decisions-which possibly saves a life. The user profile information is stored in form of tuples in tuple space.

2 Why Tuple Space?

It is quite natural that the following questions arise: 'why should we use a semantic web for such an application? Why can't we implement it in a system as any other normal software?' The following paragraph answers these.

Restricting the Epidemic Analyzer as system software forbids it from advantages of global knowledge acquisition and global knowledge availability. A global space model to analyze on an epidemic often helps the Knowledge Base to be filled with variety of information from all parts of the world. This can even help the underdeveloped countries to use the information from experts regarding the spread of epidemic and query the same. In case the result is positive the patient can be advised medical attention immediately. Models like Tripcom's European Patient Summary are evolving using tuple space. Epidemic Analyzer can be integrated with such semantic web services for health care.

3 An Introduction to Tuple Space and Semantic Web

Tuple space computing was introduced in Linda, a parallel programming language to implement communication between parallel processes. A tuple Space is an implementation of the associative memory paradigm for parallel/ distributed computing.

It is a globally shared, associatively addressed memory space that is organized as a bag of tuples. Instead of sending messages backward and forward a simple means of communication is provided. The basic element of tuple space system is the tuple. Processes can write, delete, and read tuples from a global persistent space. Patient records are stored in tuple space and are accessed using read, write and delete operations.

Tuple or space-based computing has one very strong advantage: It de-couples three orthogonal dimensions involved in information exchange: reference, time, and space.

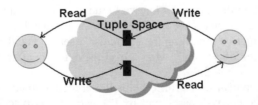

Fig. 1. Tuple Space Computing

- Processes communicating with each other do not need to know explicitly from each other. They exchange information by writing and reading tuples from the tuple space, however, they do not need to set up an explicit connection channel, i.e., reference-wise the processes are completely de-coupled.
- Communication can be completely asynchronous since the tuple space guarantees persistent storage of data, i.e., time-wise the processes are completely de-coupled.
- The processes can run in completely different computational environments as long as both can make access to the same tuple space, i.e., space-wise the processes are completely de-coupled.

Thus we see that it provides asynchronous communication, persistent storage space and reference autonomy best suited for distributed discovery of web services.

Web's content today is designed for humans to read, not for computer programs to manipulate meaningfully. The Semantic Web is an extension of the current Web that enables navigation and meaningful use of digital resources by automatic processes. It is based on common formats that support aggregation and integration of data drawn from diverse sources.

A program that wants to compare or combine information across the two databases has to know that two fields of the two different databases are being used to mean the same thing. Ideally, the program must have a way to discover such common meanings for whatever databases it encounters. Semantic Web helps at this point with ontology. Ontology is a document or file that formally defines the relations among terms. It uses rules to make inferences and helps to choose courses of action and answer questions. The patient health record stored by different hospitals may be in different formats. We use ontology to inter-relate the different health records.

4 Working of Epidemic Analyzer

This analyzer is aimed for laymen who have a doubt in mind that whether he/she is infected by an epidemic or not. After an epidemic outbreak in a certain region, the local doctors acquire substantial knowledge about the symptoms and the cure for the disease. This model allows experts in that region to put their valuable ideas into tuple space. The data in tuple space will be stored in the form of RDF (Resource Description Framework) structures. This system involves following important modules:

4.1 User Interfaces

User Interface helps the Layman and Expert to interact with the system. The proposed system has architecture of two User Interfaces:

Doctor/Expert User Interface. By using this interface, the Doctor/Expert can input the medical data into the tuple space. The data can be in the format of rules. The Doctor/Expert can provide comments on these rules so that any other expert can read and understand the set of rules. Doctors can query the knowledge base to know the path of reasoning i.e. the rules used to diagnose a disease.

Layman User Interface. Layman can use this user interface to check for the disease he himself doubt to have. Also he can specify the symptoms he wants to say to the expert. If the semantic matchmaker senses that there is insufficient data or clarifications are needed, it posts further queries to the user, to which the user can reply.

Layman can create a user profile for storing his health record data. Health record data plays a major supporting role in semantic reasoning. The profile input interface generally asks for age, sex, places recently visited, the diseases recently met with, specify any epidemic break out happened in his region, the climatic conditions in his region and his occupation. He can also specify the phone number of a person to be contacted in case of emergency and his personal consultants contact number.

4.2 Knowledge Base

In our system, tuple space acts as a knowledge base for medical rules and facts. Storing data into distributed space helps to bring a space based communication between expert/doctor and the layman.

4.3 Controversy and Mapping Analyzer

When an expert inserts new information about a disease, Controversy and Mapping Analyzer will compare with the existing information in the space about that particular disease. If the new information is a redundant data, the data is accepted and the Expert/Doctor is given response that his contribution already exists in the tuple space. If the new information is not already present in the space, the analyzer can add the information and respond that his contribution is accepted. The Controversy and Mapping Analyzer also checks whether the information is controversial. For example, Doctor A says X & Y symptoms lead to disease M.

$$X \& Y \longrightarrow M$$

Doctor B says X & Y symptoms lead to disease N.

$$X \& Y \longrightarrow N$$

Then there is a controversy. In such cases, doctor A and doctor B are notified by the tuple space that a controversy has occurred. In such cases the web service provides provisions to discuss on the topic and arrive at a conclusion and finally deciding which rule should be updated. The model also provides mechanism for the doctor to query on the details about a particular disease stored in space and add ontology mappings to the existing terms, if different representation is used for same data at different places.

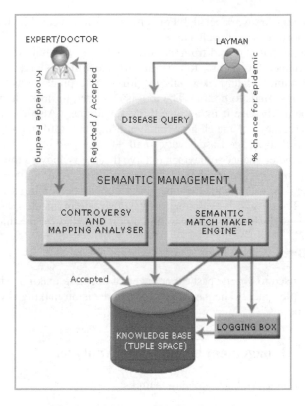

Fig. 2. Architecture of Epidemic Analyzer

4.4 Semantic Matchmaker Engine

The Semantic Matchmaker Engine uses the properties of semantic web to do effective reasoning and query processing. The user, when in doubt about the presence of a disease or diseases in his/her body, inputs the symptoms. At the same time, he can also enter laboratory examination results, if asked by the semantic matchmaker. The disease query goes both to tuple space and semantic matchmaker module. The layman can input the list of diseases he is suspecting to have. Based on this, semantic matchmaker engine retrieves the information regarding all these diseases from the Knowledge Base and do reasoning on it. Semantic Matchmaker on receiving the symptoms from layman and Knowledge Base of diseases from tuple space performs the inference.

The inference mechanism used in epidemic analyzer also uses the profile information to perform intelligent reasoning. Semantic Matchmaker Engine uses Layman's profile information to provide additional weight to certain set of rules based on the environment factors such us climatic condition, recent epidemic break outs in layman's current area, recent travels made by him and the epidemic out breaks in that region, allergic factors, any occupational disease related to his work environment and so on. For example consider the scenario where, over a million people died of cancer in the European Union in 2006. Many of these deaths were the direct result of workers being exposed to carcinogens at work. The scientific consensus is that on average,

8% of cancer deaths are work-related. For some types of cancer, like bladder and lung cancer, the proportion is even well above 10%.

Based on the output obtained from the inference, the Semantic Matchmaker calculates the % of chance for the user to have the disease which is given back as a report to the user. Semantic Matchmaker tells the immediate precautions the user should take- incase the test finds to be true. The tool is designed to insist the patient to 'take advice from doctor' in case it finds some thing suspicious. Also it can give negative results if the percentage of match between symptoms and information about the particular disease in Knowledge Base is very small.

If the percentage of chance for layman to have disease is above a threshold value the personal consultant associated with the layman is notified. And in the case of emergencies both personal consultant and the emergency contact person in user profile is notified via SMS. The diagnosis reasoning path is also communicated with the personal consultant via E-mail which helps him to find quick remedies for the diagnosis.

4.5 Logging Box

Logging Box is used to log the past experiences of decision making. The log output is also stored in tuple space. The past experience in decision making is often useful for handling new queries.

5 Semantic Management in Epidemic Analyzer

The proposed architecture has following features:

5.1 Semantic Discovery

Customers of Epidemic Analyzer (Doctor/Expert or Layman) need to perform semantic discovery to find out the matching set of rules from the tuple space. Doctor/Expert may need to check for redundancy of rules and also to take intelligent decisions regarding controversy. When a layman queries about a particular disease, semantic discovery retrieves the disease related rules. In order to perform successful semantic discovery we need to perform tasks like discovery, selection, composition, mediation, execution and monitoring.

5.2 Representing Semantic Data in Tuple Space

The data fetched by the Doctor/Expert interface is stored into Tuple Space. The representation of semantic data within tuple space requires new types of tuples which are tailored to standard Semantic Web languages RDF(S) and OWL (Web Ontology Language). RDF is one of the fundamental building blocks of the Semantic Web, and gives a formal specification for the syntax and semantics of statements. The Semantic Web schema languages, RDFS (RDF Schema) and OWL, offer the potentialto simplify the management and comprehension of a complicated and rapidly evolving set of relationships that need to be recorded among the data describing the products of the life and medical sciences. RDF triples can be represented in a four field tuple of form (subject, predicate, object, id). As foreseen by the RDF abstract model, the first three

fields contain URIs (Uniform Resource Identifier) (or, in the case of the object, literals). These URIs identify RDF resources previously available on the Web. Fields are
typed by RDFS/OWL classes (URIs) and XML-based data types (literals). The standard Linda matching approach is extended in order to efficiently manage the newly
defined tuple types. The former includes procedures to deal with the types associated
with the subject, predicate and object fields of each RDF tuple.

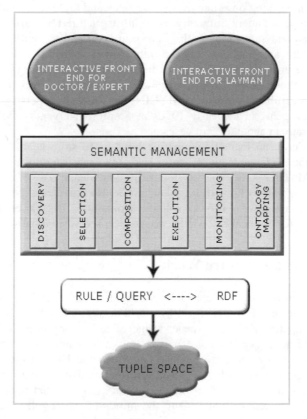

Fig. 3. Role of Semantic Web in Epidemic Analyzer

5.3 Mediation Tasks

In this architecture, we publish rule descriptions into the space and later for service
discovery; we perform reading them from the shared space and discovering the required data. Semantic matching techniques may then make use of distributed knowledge base with the help of reasoning services in order to refine the retrieval capabilities
of the tuple space.

Orthogonal to these dimensions Semantic Web Spaces contain a tuple space ontology, which is used as a formal description of the tuplespace, its components and
properties. Doctor/Expert can add ontology mappings to establish the relationship
between related terms used in ontology. Using ontologies in this context allows a

more flexible management of the tuple space content and of the interaction between tuplespace and information providers and consumers.

6   Related Work

A number of research works are going on in semantic web. Tripcom is a project which had been working on semantic web services and tuple space based computing. Tripcom's European Patient Summary, is an infrastructure for sharing and coordinating access to patient summaries at the European level and to facilitate the pervasive delivery of healthcare, thus ensuring the right to patient mobility. Easy Diagnosis is an instant online analysis of important medical symptoms in a user friendly format. It asks questions about symptoms of a disease to the user and gives the percentage of chance for them to have the disease MYCIN was an early expert system developed over five or six years in the early 1970s at Stanford University. MYCIN does an acceptable therapy in about 69% of cases.

Tripcom's European Patient Summary allows us to store health records into the space, where the epidemic analyzer stores rules obtained from doctors to process laymen queries. MYCIN is an expert system using knowledge bases, but the knowledge base storage capability is limited. Easy Diagnosis also provides expert advice through web, but there is no provision for knowledge acquisition. But our epidemic Analyzer provides facility for Medical Knowledge acquisition and vast global storage space.

7   Conclusion and Future Work

This model combines the concept of Expert System and Semantic Web to make it useful for the attempts towards a healthy world scenario. For today's expert systems to perform competently over a broader range of tasks, they need to be given more knowledge. The next generation of expert systems will require large knowledge bases. This model just acts as an expert system in which the Knowledge Base is a tuple space. Tuple Space model helps to collect information from global resources whereas in a local model, it is restricted to the locality or company that builds the software. The model helps to remove panic from the minds of the people, and confirm their current health condition and be aware of the precautions that need to be taken in case of an epidemic spreading outburst scenario.

Epidemic Analyzer prototype is under the initial stages of its development as part of the Ubiquitous Computing Project. We are trying to create disease symptom ontology based on the inputs from the doctor/expert. We use OWL format for the storage of disease and symptoms. The usage of tuple space is not mandatory. But for efficient retrieval and storage, the storage can be done in a regionalized manner i.e., if the region X is highly affected with disease A, its related symptoms and rules be stored in a tuple space data servers in that region X.

The notification mechanism provided by epidemic analyzer to alert the personal consultant of the layman with diagnosis reasoning path helps the consultant to arrive at quick conclusions.

We can enhance the existing model in which user doesn't have to say in which disease, the user want to perform a diagnosis on. Here we say that 'the layman have to specify for which disease user has to perform query' because it will be too tedious to perform an inference using all the diseases on the Knowledge Base (tuple space). In future as semantic web becomes more reliable and faster, we can try to integrate the patient health records and their diagnosis reports to this knowledge Base for testing epidemics.

Acknowledgement

This project is funded under National Ubiquitous Computing Research Initiative by Department of Information Technology, Ministry of Communications and Information Technology, Government of India.

References

1. Sapkota, B., Roman, D., Kruk, S.R., Fensel, D.: Distributed Web Service Discovery Architecture. In: Proceedings of the Advanced Int'l. Conference on Telecommunications and Int'l. Conference on Internet and Web Applications and Services, p. 136 (2006), ISBN:0-7695-2522-9
2. Cerizza, D., Valle, E.D., Foxvog, D., Krummenache, R., Murth, M.: Towards European Patient Summaries based on Triple Space Computing, http://www.tripcom.org
3. Fensel, D.: Triple-space computing: Semantic Web Services based on persistent publication of information, http://www.ai.sri.com
4. Gelernter, D.: Generative communication in Linda. ACM Transaction on Prolog Langauge and Systems 7(1), 80–112 (1995)
5. Easy Diagnosis Project, http://www.easydiagnosis.com/
6. http://hesa.etui-rehs.org/uk/dossiers/dossier.asp?dos_pk=20, Appendix: Springer-Author Discount
7. Nixon, L., Paslaru, E., Simperl, B., Antonechko, O., Tolksdorf, R.: Towards Semantic Tuplespace Computing: The Semantic Web Spaces System. In: Proceedings of the ACM symposium on Applied computing, Seoul, Korea. SESSION: Coordination models, languages and applications (2007)
8. Harrison, R., Chan, C.W.: Distributed Ontology Management System. In: IEEE Canadian Conference on Electrical and Computer Engineering, 2005, May 1–4, pp. 661–664 (2005), doi:10.1109/CCECE.2005.1557017
9. Sudha, R., Thamarai Selvi, S., Rajagopalan, M.R., Selvanayaki, M.: Ubiquitous Semantic Space: A context-aware and coordination middleware for Ubiquitous Computing Communication Systems Software and Middleware. In: 2nd International Conference on COMSWARE 2007, January 7-12, pp. 1–7 (2007), doi: 10.1109/COMSWA.2007.382562
10. Berners-Lee, T., Hendler, J., Lassila, O.: The Semantic Web. Scientific American, http://www.scientificamerican.com

Assistive Communication Robot for Pre-operative Health Care

R. Khosla[1], M.-T. Chu[1], K. Denecke[1,2], K. Yamada[3], and T. Yamaguchi[3]

[1] La Trobe University Melbourne, Melbourne, Australia
[2] L3S Research Centre, Hannover, Germany
[3] NEC Corporation, Nara, Japan
r.khosla@latrobe.edu.au

Abstract. Health care costs and ageing population are two factors which are of major concern to western governments in the 21st century. Existing work in affective health care is primary focused on developing avatars in the tele-health space. This paper reports on the modeling of emotions (anxiety level) of patients in pre-operative stage using communication robots to assist nurses in providing personalized care and determining the sedation level of patients. The research is based on the need for better management of IT resources for social innovation as well as need to improve operational efficiency and quality of care.

Keywords: Healthcare, Emotion, Communication Robot, Personalized Care.

1 Introduction

Information and Communication Technologies (ICT) are today integral part of our lives. However, organizations are unable to fully harness their potential because of the underlying design constraints. Traditionally, design of ICT has been underpinned in the laws of physics. They have been good at churning out information and to an extent provide some convenience. Our research aims at shifting the design focus of ICTs from purely as information providers to social innovation. In this context, we have designed affective communication robots with human sensing and tracking attributes like emotion, speech, face tracking, touch and motion sensors to work with their human partners in addressing social problems and issues in 21st century.

Historically health care systems have been designed around technology-push models and have not included human context or non-verbal/emotional data [1, 7]. More recently, the focus has shifted towards design of more context-centered and patient-centered health care systems [1, 2, 3]. Patient-centered systems require a social, semantic and pragmatic context based design approach. Emotions form an important part human to human communication and decision making [4, 11]. From a social and pragmatic context, people's interpretation of meaning and application of knowledge is not entirely based on their cold cognitive scripts or rules but are mediated by their emotional attitudes [7]. This is also consistent with an emerging paradigm in intelligent systems, namely, sense-making [6].

D. Zhang and M. Sonka (Eds.): ICMB 2010, LNCS 6165, pp. 224–230, 2010.

In the context of health care not only emotional state assessment is a key indicator of the patient's mental or physical health status, but the power of emotions themselves over the recovery process has also been well documented [3]. Therefore, it is important that the caregiver and care recipient communicate along the socio-emotional channel to provide better assessment and responsiveness.

From an organizational perspective, in order to bring down the spiraling health care costs and improve operational efficiency ICT need to play a more integral part in supporting health care professionals (e.g., nurses) in delivering patient-centered personalized care [5, 12]. In contrast to existing work which is largely focused on tele-health [3, 14], in this paper, a context-aware architecture of affective communication robot for health care is outlined. Situation-affect-action profiles of a patient in pre-operative stage are developed by modeling transient changes in their emotional states. The work is part of a larger exercise to demonstrate how ICT and affective communication robots can play an effective role in improving operational efficiency, alleviating stress levels of nurses and establishing a socio-emotional communication channel between the care giver (nurse) and care receiver in pre-operative health care. In due course such affective communication robots with effective knowledge bases can also be used for training new nurses in emotionally intelligent patient-care.

The paper is structured as follows: Section 2 outlines an architecture for a seven layered Affective Communication Robot System (AHRS). Section 3 illustrates the application of this architecture in health care. The paper finishes with conclusions in Section 4.

2   Affective Communication Robot Architecture

The affective communication robot (which is an extension of NEC's PaPeRo series http://en.wikipedia.org/wiki/PaPeRo) includes emotion, speech/voice, touch, motion and face tracking sensors. The emotion, motion and face tracking sensors allow it move its head and body to track the face of the patient and emotional feedback of a patient. The speech/voice sensors allow the communication robot to engage in pre-operative dialog with patient for different types of procedures. Figure 1 provides a high level context-aware view of the AHRS architecture.

The AHRS architecture in affective communication robot is defined in terms of three behavior levels [8]. These are perception-action, procedural and construction levels respectively. At the perception-action level, inference is direct. This level represents hard-coded inference mechanisms that provide immediate responses to appropriate sensed information. They correspond to the need for urgency of action. In terms of the AHRS architecture, this level is represented by the reactive layer. The reactive layer consists of agents which represent stimulus-response phenomenon in a user defined environment which may not need learning. This includes for example, the ability of communication robot to respond to emergency situations where a patient is showing high levels of distress involving primary emotions.

At the procedural level, a set of rules or learnt patterns are used for inference. This set may be large, but is always closed, i.e., it corresponds to pre-formed, pre-determined insights and learnt patterns of behavior. In the AHRS architecture, this level is represented by the intelligent technology layer. The intelligent technology layer contains agents which involve learning to find patterns from data involving soft computing techniques. This layer deals with well defined procedures related to minor or major surgery for assessing the patients at pre-operative stage.

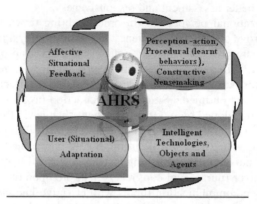

Fig. 1. Affective Communication Robot System (AHRS) Architecture

At the constructive level, inference is a creative and constructive process based on an open world of actions, constraints and resources. In the AHRS architecture, this level is represented by three layers, namely, sense-making (cognitive) layer, sense-making (affective) layer and the situation-adaptation layer. The agents in the sense-making (cognitive) layer are defined in terms of problem solving or information processing phase. An application of the cognitive layer and its five phases are based on earlier work [2]. Applications in clinical health care applications have been reported in [13]. The sense-making (affective) layer includes agents for capturing non-verbal communication of patients through facial expressions, gestures and eye-gaze. An example in the next section will illustrate this aspect. This layer is also connected with voice recognition system in the communication robot and built in corpus of dialogs with the patient for different types of procedures.

The situation-adaptation layer consists of situation monitoring and situation adaptation agents. They monitor or measure the result of the action of the communication robot on the patient /environment in a given health care situation (e.g., acceptance/ rejection of a recommendation in a scenario involving change of dietary habits) and incorporate this feedback to adapt the actions of the situation-action phase agents. In order to model this phenomenon three types of knowledge components (solution knowledge, dynamic knowledge and selection knowledge).

These three knowledge components are modelled in the affective communication robot to enable it to learn, adapt and refine its solution knowledge related to pre-operative care on an on going basis. The solution knowledge includes one or more prediction models of situation-action-affect profile or pattern of a patient. The situation can include 'pre-operative stage hernia patient 1.' The action includes nurse response in terms of sedative used (or care rendered) and affect (as illustrated in nest section) which includes the emotional or anxiety coefficient of the patient derived from their emotional profile. In other words, the communication robot may predict a certain level of care or sedative based on the anxiety coefficient of the patient which may be accepted or rejected by an expert human nurse. This feedback forms the dynamic knowledge component and is monitored by the performance measurement agent. In case the feedback is rejected, the selection knowledge component or performance optimization agent is activated. This agent then refines or alters the parameters of the solution knowledge component.

It may be noted that human context is thus modelled in four layers or levels, reactive, cognitive, affective and situation adaptation layers which involves human feedback.

The AHRS architecture also includes two more layers. The distribution and coordination layer consists of agents who process data on behalf of agents in other layers in a parallel and distributed manner to meet the real-time needs of the health care system. The coordination layer agents are used to coordinate communication between the user and the agents in sense-making (cognitive), optimization and sense-making (affective\emotion) layers of the architecture. The coordination layer agents are also used to coordinate communication between agents in the seven layers and maintaining a blackboard type global system state representation.

Finally, the object layer is used to represent ontology of the domain objects which are manipulated by the different layers (e.g. emotion sensing objects). Given the space limitations, in the rest of paper we focus on illustrating the role of sense-making (affective) layer and situation adaptation layer.

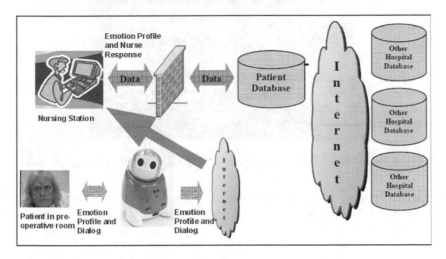

Fig. 2. Patient-centered pre-operative health care data flow

3 Application in Health Care

The purpose of this section is to illustrate how meaningful interpretation of patient's pre-operative condition is mediated by the emotional or affective data using the affective layer of ACRS. This process also helps to construct a corpus of situation-action affect profiles of a patient for solution knowledge or predictive modeling component of the situation adaptation layer in the affective communication robot.

Figure 2 shows data flow between patient, affective communication robot, nursing station, patient and other hospital databases. As shown in Figure 2 the emotion/anxiety profile and dialog of a patient are transmitted to the nursing station to enable them to adapt or customize care and /or sedation level of the patient. This data can then also be stored in other hospital data bases for post-operative care of the patient.

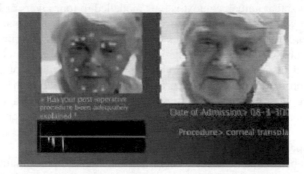

Fig. 3. Few negative emotional spikes in the graph indicating patient is generally calm

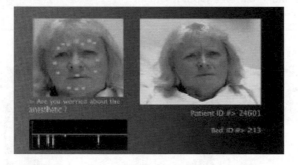

Fig. 4. Patient with higher anxiety level requiring more nursing or therapeutic care

Facial expressions are an important non-invasive physiological indicator of human emotions [9]. An affect space model used by psychologists with three dimensions, namely, Valance (measured on a scale of pleasure (+) to displeasure (-)), Arousal (measured on scale of excited/aroused (+) to sleepy (-)) and Stance (measured on a scale of high confidence (+) and low confidence (-)) has been used [10]. The psychologists point out that facial expression alone may not be an accurate indicator of the emotional state of a person but changes in facial expression may indicate a change in emotional state. In our case, affective communication robot monitors these transient positive and negative emotional state changes (as against typical emotional states – happy, sad) of a patient in pre-operative stage in response to a dialog which includes questions related to preparedness of the patient for an upcoming operation. The questions include "do you have any allergies?" has the anaesthetic come to see you?," "Has you post-operative procedure been adequately explained (shown in Figure 3)?," etc.

The methodology for concurrent validation of emotional state changes involved comparing positive and negative emotional state changes with corresponding similar changes in facial features of two or more human subjects. A correlation exceeding 0.9 (i.e. 80 percent) was established over 75 human subjects.

Figures 3 and 4 illustrate two patients with varying anxiety levels. The spikes in the graphs below the line represent negative emotional state and above the line represent positive emotional state. On the line represents neutral or no change in emotional state. The nursing station can thus monitor the anxiety levels of the different patients on an on going basis, adapt care, make more informed decisions on sedation levels in pre and post-operative stages, and optimize deployment of nursing resources and improve operational efficiency. The emotional profiles have been developed using optical flow method [9], Facial Action Coding System [11] and Affect Space model [10].

4 Conclusion

Health care in light of ageing population in western world is a worrying phenomenon for western governments. The research reported in this paper makes novel contributions through i) fusion of human senses in an affective communication robot, ii) socially innovative ICT design with application in health care to assist health care professionals in hospitals in improving quality of care, and iii) a multi-layer architecture with human context modelling and feedback at reactive, cognitive, affective and situation-adaptation levels. This approach towards design of ICTs also facilitates a better management of IT as well as reduction of barriers between technology and people.

References

1. Harkness, J.: Patient involvement: a vital principle for patient-centered health care. World Hosp. Health Serv. 41(2), 12-6, 40-3 (2005)
2. Khosla, R., Sethi, I., Damiani, E.: Intelligent Multimedia Multi-Agent Systems- A Human-Centered Approach, 333 pages. Kluwer Academic Publishers, Massachusetts (2000)
3. Lisetti, C., Nasoza, F., LeRougeb, C., Ozyera, O., Alvarezc, K.: Developing multimodal intelligent affective interfaces for tele-home health care. International Journal of Human-Computer Studies 59(1-2), 245–255 (2003)
4. Picard, R.: Affective Computing. MIT Press, Cambridge (1997)
5. Shapiro, S.L., Astin, J.A., Bishop, S.R., Cordova, M.: Mindfulness-Based Stress Reduction for Health Care Professionals: Results from a Randomized Trial. International Journal of Stress Management 12(2), 164–176 (2005)
6. Klein, G., Moon, B., Hoffman, R.R.: Making Sense of Sensemaking 1: Alternative Perspectives. In: IEEE Intelligent Systems, pp. 70–73 (2006), doi:10.1109/MIS.2006.75
7. Helfenstein, S.: Product Meaning, Affective Use Evaluation, and Transfer: A Preliminary Study. Human Technology 1(1), 76–100 (2005)
8. Rasmussen, J.: Information Processing and Human-Machine Interaction: An Approach to Cognitive Engineering. North-Holland, Amsterdam (1986)
9. Irfan, A., et al.: Coding, Analysis, Interpretation, and Recognition of Facial Expressions. IEEE Trans. on Pattern analysis and Machine Intelligence 19(7) (July 1997)

10. Lang, P.J.: The emotion probe: Studies of motivation and attention. American Psychologist 50(5), 372–385 (1995)
11. Facial Action Coding System,
    http://www.cs.cmu.edu/afs/cs/project/face/www/facs.htm
12. Swindale, J.E.: The nurse's role in giving pre-operative information to reduce anxiety in patients admitted to hospital for elective minor surgery. Journal of Advanced Nursing 14, 899–905 (1989)
13. Khosla, R., Chowdhury, B.: Real-Time RFID-Based Intelligent Healthcare Diagnosis System. In: Zhang, D. (ed.) ICMB 2008. LNCS, vol. 4901, pp. 184–191. Springer, Heidelberg (2007)
14. Lisetti, C., LeRouge, C.: Affective Computing in Tele-home Health. In: 37th Hawaii International Conference on System Sciences – 2004, pp. 1–8 (2004)

Design and Implementation
of a Digital Auscultation System

Yuan-Hsiang Lin[1,*], Chih-Fong Lin[1], Chien-Chih Chan[1], and He-Zhong You[2]

[1] Department of Electronic Engineering,
National Taiwan University of Science and Technology, Taipei, Taiwan
[2] Graduate Institute of Biomedical Engineering,
National Taiwan University of Science and Technology, Taipei, Taiwan
{linyh,M9802132,M9802131,M9823013}@mail.ntust.edu.tw

Abstract. In this paper, we developed a digital auscultation system which provides a useful tool for users to measure the heart sounds and to see the waveform of sounds such as phonocardiogram. This system includes an ARM-based digital stethoscope with a USB port for data transmission and a PC-based graphic user interface for waveform and heart rate display and data recording. The functions of this digital stethoscope include play back, filtering, volume adjustment of sounds, and data transmission. Besides, the digital stethoscope can measure the sounds and then forward the data to the PC via the USB interface. On the PC, user can easy to see the waveform of sounds on the graphic user interface. For heart sounds measurement, we also implemented a heart rate detection algorithm so that heart rate can also be shown on the graphic user interface. This system would be useful for aiding the auscultation diagnosis and telemedicine applications.

Keywords: auscultation, stethoscope, ARM-based, USB, telemedicine.

1 Introduction

Telemedicine is a rapidly developing application of clinical medicine where medical information is transferred through the Internet or other networks for the purpose of consulting, and sometimes remote medical procedures or examinations. Remote monitoring enables medical professionals to monitor a patient remotely using various technological devices. This method is primarily used for managing chronic diseases or specific conditions, such as heart disease, diabetes mellitus, or asthma. Stethoscope is one of important devices for examinations, which is used as a fundamental device for auscultation [1]. However, traditional stethoscope cannot satisfy telemedicine applications due to the sounds cannot be transmitted. Therefore, digital stethoscopes are used for the telemedicine applications [2].

---

* Corresponding author. He is also an IEEE member.

D. Zhang and M. Sonka (Eds.): ICMB 2010, LNCS 6165, pp. 231–240, 2010.

Digital stethoscopes have some advantages, which include data recording, signal processing, and data transmission. Lukkarinen S. and others [3] presented a system for recording, processing, and analyzing the heart sounds. This equipment consists of a separate handheld electronic stethoscope, headphones connected to the stethoscope, personal computer with sounds handling capabilities, and special software for recording and analyzing the heart sounds. Brusco M. and Nazeran H. [4] developed a PDA-based biomedical instrument which is capable of acquisition, processing, and analysis of heart sounds. Johnson J. and others [5] also presented a low power, portable electronic stethoscope system, which can allow more convenient health care at different time and place.

In the past, there are many researchers working on analysis of heart and lung sounds using different digital signal processing (DSP) methods. Forkheim K. E. and others [6] presented the analysis of the use of neural networks to process lung sounds and identify wheezes. Hung K. and others [7] developed a multifunction stethoscope for telemedicine. Providing functions such as video-conferencing, heart sounds and lung sounds separation for tele-consultation. Bai Y. W. and Lu C. L. [8] used a 10-order band-pass IIR filter, in order to obtain a clearer heart and lung sounds. Bai Y. W. and Yeh C. H. [9] designed an embedded DSP stethoscope with the functions of network transmission and the judging of heart murmur by using a diagram which includes both the time domain and the frequency domain.

The aims of this research are providing a useful auscultation system for telemedicine applications and aiding diagnosis of heart diseases. Therefore, we designed a digital stethoscope, which can amplify, filter and convert the weak heart sounds to digital signal. To provide a distinct sound signals and heart rate information for users. Medical staffs can use this digital stethoscope to hear heart sounds clearer. This digital stethoscope also has a USB transmission function, let patient's sound signals can be sent to the other devices such as a PC for data recording and analysis. The frequency range of heart and lung sound signals is generally located from 20 Hz to 1K Hz. The first and second heard sounds are located between 20 Hz to 120 Hz. Therefore, we provide four kinds of 11-tap FIR band-pass filters with different cut-off frequency ranges for users to select for different auscultation of sounds. Besides, we also implemented a graphic user interface (GUI) on the PC for displaying the waveform of sounds and heart rate, and perform data recording.

2   Methods

2.1   System Architecture

The system architecture of this auscultation system is shown in Fig. 1. The system is consisted of the ARM-based digital stethoscope and a PC-based GUI. The ARM-based digital stethoscope integrates with the microphone sensor, analog circuits including amplifiers, filters, level shift circuits and headphone driver, and the USB interface. Measuring results can be transmitted to the PC via USB. On the GUI, waveform of sounds and heart rate can be displayed and recorded. The data can also be transmitted to a remote server such as a server on the hospital.

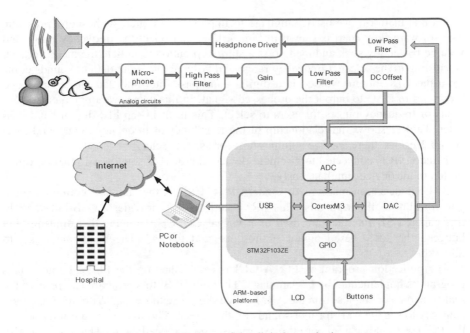

Fig. 1. System architecture of the digital auscultation system

1) Analog circuits: Microphone sensor receives the sound signals from the probe. A buffer is used to take this signal and then the output signal of buffer goes through a 2-order high-pass filter, gain amplifier and a 2-order low-pass filter. Finally, the sound signals are sent to ARM platform and transformed into digital signal by the built-in ADC of STM32F103ZE, sampling rate is 8K Hz. The cut-off frequency of the high-pass filter is 20 Hz, and the cut-off frequency of the low-pass filter is 2K Hz. Total gain is 30. After digital signal processing, the sound signals will be outputted by the built-in DAC of STM32F103ZE. The DAC output signal is sent to a 2-order low-pass filter with 2K Hz cut-off frequency and then through a driver to drive a headphone or a speaker.

2) ARM-based platform: We used the STM3210E-LK ARM-based platform, which is made by STMicroelectronics Company. It integrates a microcontroller STM32F103ZET6, one 128 64 dot-matrix LCD, one joy stick, buttons, UARTs, and USB peripherals. The STM32F103ZET6 is a high density ARM 32-bit microcontroller based on the Cortex-M3 core, with 512 Kbytes of embedded Flash memory, 64 Kbytes of embedded SRAM and a rich set of on-chip peripherals, including 12-bit ADCs and DACs, GPIOs, UARTs and USB, et al..

Fig. 2 shows the user's operating interfaces on the ARM platform, including LCD display, joy stick, start/stop button and four different color LEDs which are used for display the status of frequency range selection. From left to right, the LED colors are blue, red, yellow and green separately. The initial screen is shown on this display, including current frequency range and volume.

On this platform, we use C language to implement a digital stethoscope program. The software flow chart is shown in Fig. 3. User can select the frequency range and volume by the joy stick and then press a start/stop button to initiate the measurement. According to the selected frequency range and volume, the input data of sounds come from the ADC output would via a digital band-pass filter and data processing, and then send to DAC to output the processed signal. In this paper, we implemented four kinds of frequency ranges for users to select. This is an 11-tap FIR digital band-pass filter. Table 1 shows the relationship of the four kinds of frequency ranges and LED colors. Besides, there are five volume steps for users to select.

If the USB is connected to a remote device, the data of sounds will also be transmitted to the device simultaneously.

3) USB data transmission: In modern PCs, USB is the standard communication port for almost all peripherals. This ARM platform also provides a USB interface. In this paper, USB interface are used as a virtual COM port for the data transmission between the ARM-based digital stethoscope and the PC. The baud rate is set to 230.4kb/s.

4) Application program on PC: A GUI is established on the PC or Laptop. This program is implemented by C# language. Through USB, this program can receive the real-time data of sounds from the ARM-based digital stethoscope. With this GUI, user can easy to see the waveform of sounds on the screen. Through the audio interface on the PC, the sounds can play to the external speaker or headphone. This program also lets user to store the data of sounds into a file which is the .wav format. Besides, we developed a heart rate detect methods to measure the heart rate form the sound signals. Users and doctors would be easily to know the heart rate information.

Current
Frequency range
and volume

Start/Stop

Joy Stick
Up/Down: Volume adjustment
Left/Right: Frequency range selection

Fig. 2. User interfaces on the ARM platform

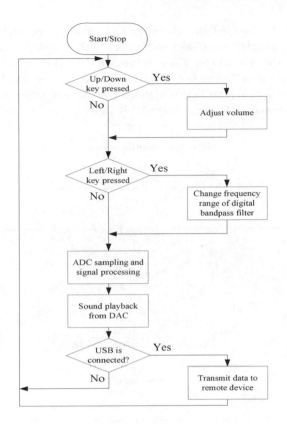

Fig. 3. Flow chart of software for the ARM platform

Table 1. Frequency ranges of filter and LED colors

Frequency range selection	Low cut-off frequency (Hz)	High cut-off frequency (Hz)	LED colors
1	20	1K	Green
2	20	120	Yellow
3	30	250	Red
4	100	2K	Blue

2.2 Heart Rate Detection Methods

Heart rate detection is finished on the program of PC. Fig. 4 shows this detection flow. The data of sounds received from the USB will first put into a data buffer. The data on the buffer are then sent to the audio interface of PC for sounds playing and the data also used for waveform display and heart rate detection. For heart rate detection, the data are first filtering by a 101-tap digital band-pass filter with cut-off frequency is 20Hz and 120Hz. Second, the filtered data via an envelope detection function and then each data value is comparing to a threshold *Vth*. Envelope detection function is used to convert the data into the absolute value of data. If the absolute value is larger than the *Vth* and the delay counter is also larger than a value which is equal to *Tth* × sample rate, then a new heart beat is detected. The delay counter will be cleared. The period between two detected heart beat (*ΔT*) is used to calculate the heart rate.

$$\text{Heart rate (bpm)} = \frac{60}{\Delta T} \tag{1}$$

where $\Delta T = t_i - t_{i-1}$, t_i is the time of i_{th} heart beat that is detected.

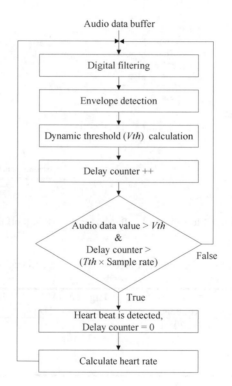

Fig. 4. The flow of heart rate detection

Vth is dynamic adjusted according to the equation (2). In equation (2), the *C* is set to 2.5. Delay counter is used to avoid the heart beat be detected more than one time in a normal heart beat period. *Tth* is a mask time which is smaller than a normal heart beat period. There is no new heart beat can be detected during this mask period. We set *Tth* = 0.5 sec in this study. Fig. 5 demonstrates the heart detection process and results.

$$Vth = \frac{\sum_{n=1}^{N} x[n]}{N} \times C \qquad (2)$$

where $x[n]$ is the sample of input data, N is the total number of samples from t_{i-1} to t_i, and C is a constant.

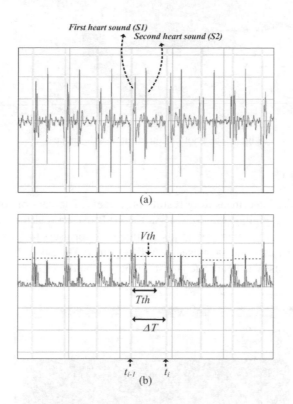

(a)

(b)

Fig. 5. Heart rate detection process and results. (a) the waveform of sounds after filtering, (b) the waveform of sounds after envelope detection.

3 Results

3.1 ARM-Based Digital Stethoscope

This digital stethoscope is consisted of an ARM platform, a sound probe and a self-designed analog circuit's module. Fig. 6 shows the photo of this digital stethoscope. The size of the analog module is 2 by 4.5 cm. The all functions were tested, including frequency range selection, volume selection, sounds playing and data transmission.

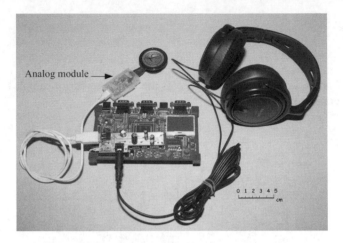

Fig. 6. The photo of the digital stethoscope

3.2   GUI on the PC

Fig. 7 shows the GUI and measuring results on the PC. The data of sounds received from the ARM-based digital stethoscope via the USB can be real-time displayed as a waveform on the GUI. There are six kinds of time base for users to select. Data is also recording simultaneously and save as a .wav file format. The saved file can be played back on off-line status. The heart rate is also displayed on the GUI.

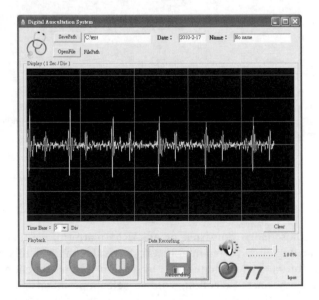

Fig. 7. Graphic user interface and measuring results on the PC

3.3 Heart Rate Detection Results

We compared our heart rate detection results with a commercial pulse oximeter. There are 4 people were tested, each person records 3 times and each time records 1 min. We calculated the average of the 1 min heart rates for our method and the commercial pulse oximeter. The range of all averaged heart rates is from 65.9 to 81.4 bpm. The results show the average of the differences of the two methods is 0.3 bpm. The standard deviation (SD) of the difference of the two methods is 0.7 bpm. Table 2 shows the results.

Table 2. Heart rate detection results

Personal ID	Commercial pulse oximeter (bpm)	Our method (bpm)	Difference (bpm)
1	65.9	66.5	0.6
	72.0	71.6	-0.4
	67.4	68.3	1.0
2	79.2	80.2	1.0
	77.0	77.3	0.3
	78.8	78.2	-0.6
3	67.7	68.0	0.3
	69.0	69.1	0.1
	67.7	68.5	0.8
4	81.4	80.3	-1.2
	73.8	74.4	0.6
	75.1	75.8	0.7
Average	72.9	73.2	0.3
SD	5.4	5.1	0.7

4 Discussion and Conclusion

A prototype of a digital auscultation system is designed and implemented. This system includes an ARM-based digital stethoscope with a USB port for data transmission and a PC-based GUI for waveform and heart rate display and data recording. Through digital signal processing, the stethoscope can select different frequency ranges for different kinds of sounds to obtain better effect for auscultation. Via the USB interface, the data of sounds can be transmitted to other devices such as PC for broadcast to a remote server. Through the GUI on the PC, users can easy to see the waveform of sounds. For heart sounds measurement, we also implemented a heart rate detection algorithm so that heart rate can also be shown on the interface. This system would be useful for aiding the auscultation diagnosis and telemedicine applications.

The used heart detection methods are easy to implement, so that they can easy to migrate to the ARM platform. However, there is several problems still need to be improved. Such as the heart rate detection is limited to 120 bpm now due to the Tth is a fixed value. We still need to improve this problem in the future to increase the range of heart rate detection. Besides, the motion artifacts will also cause false heart rate detections. Therefore, a robust heart rate detection method for the stethoscope is under studying.

Acknowledgments. This work was supported in part by a research grant from the National Science Council under Grant NSC 98-2218-E-011 -007 -.

References

1. Kumpituck, S., Kongprawechnon, W., Kondo, T., Nilkhamhang, I., Tungpimolrut, K., Nishihara, A.: Stereo Based Vision System for Cardiac Auscultation Tele-Diagnostic System. In: ICCAS-SICE, pp. 4015–4019. IEEE Press, Fukuoka (2009)
2. Nakamura, M., Yang, Y., Miura, Y., Takizawa, M.: Telemedicine for mountain climbers with high quality video and stethoscope sound transmission. In: IEEE EMBS Asian-Pacific Conference on Biomedical Engineering, pp. 78–79. IEEE Press, Kyoto (2003)
3. Lukkarinen, S., Noponen, A.-L., Sikio, K., Angerla, A.: A new phonocardiographic recording system. In: Computers in Cardiology, pp. 117–120. IEEE Press, Lund (1997)
4. Brusco, M., Nazeran, H.: Digital Phonocardiography: A PDA-based Approach. In: 26th Annual International Conference on Engineering in Medicine and Biology Society, vol. 1, pp. 2299–2302. IEEE Press, San Francisco (2004)
5. Johnson, J., Hermann, D., Witter, M., Cornu, E., Brennan, R., Dufaux, A.: An Ultra-Low Power Subband-Based Electronic Stethoscope. In: 2006 IEEE International Conference on Acoustics, Speech and Signal Processing, vol. 3, pp. 1156–1159. IEEE Press, Toulouse (2006)
6. Forkheim, K.E., Scuse, D., Pasterkamp, H.: A comparison of neural network models for wheeze detection. In: IEEE WESCANEX 1995, Communications, Power, and Computing, vol. 1, pp. 214–249. IEEE Press, Winnipeg (1995)
7. Hung, K., Luk, B.L., Choy, W.H., Tai, B., Tso, S.K.: Multifunction stethoscope for telemedicine. In: IEEE International Workshop on Computer Architectures for Machine Perception, pp. 87–89. IEEE Press, Hong Kong (2004)
8. Bai, Y.W., Lu, C.L.: The Embedded Digital Stethoscope Uses the Adaptive Noise Cancellation Filter and the Type I Chebyshev IIR Bandpass Filter to Reduce the Noise of the Heart Sound. In: Proceedings of 7th International Workshop on Enterprise networking and Computing in Healthcare Industry, pp. 278–281. IEEE Press, Kuala Lumpur (2005)
9. Bai, Y.W., Yeh, C.H.: Design and implementation of a remote embedded DSP stethoscope with a method for judging heart murmur. In: IEEE International Instrumentation and Measurement Technology Conference, pp. 1580–1585. IEEE Press, Singapore (2009)

The Application of Non-invasive Oxygen Saturation Sensor to Evaluate Blood Flow in Human

Yi-Te Lu[1], Kuen-Chang Hsieh[2], Ming-Feng Kao[3], Yu-Ywan Chen[3], Chiung-Chu Chung[4], and Hsueh-Kuan Lu[5,*]

[1] Department of General, National Tam-Shui Vocational High School, Taipei 251, Taiwan
[2] Office of Physical Education and Sport, National Chung Hsing University,
Taichung 402, Taiwan
[3] Department of Physical Education, National Taiwan College of Physical Education,
Taichung 404, Taiwan
[4] Athletics Department & Graduate School, National Taiwan College of Physical Education,
Taichung 404, Taiwan
[5] Sport Science Research Center, National Taiwan College of Physical Education,
Taichung 404, Taiwan
hklu@ntcpe.edu.tw

Abstract. Blood flow rate are one of the most important vital sign in clinical application. The one dimension laser Doppler flowmetry is currently used to non-invasive measuring blood flow *in vivo* for human. However, some limitations exist in usage of Doppler technique. To develop more precise method to measure blood flow rate, the two dimension blood flow technique and the blood flow index, L (cm/sec) were conducted. The L value is equivalent to participant height/$T_{lung\text{-}finger}$, where as $T_{lung\text{-}finger}$ is the time of blood flow from alveolar capillary to middle finger-tip capillary measured by oxy-meter with synchronized ventilator. Our results showed that the L values were examined with highly reliability.

Keywords: blood flow, oxy-meter.

1 Introduction

Some operative methods or technique are used to evaluate the blood flow on human including laser Doppler flowmetery [1, 2], fluorescent microspheres [3], radioactive microspheres [4], magnetic resonance imaging (MRI) [5], electro-magnetic flowmeter [6] and videomicroscopy [7]. Currently, only the laser Doppler flowmetry is the non-invasive for measurement. However, some limitations exist in usage of Doppler technique such as signal processing problem, choice of processing bandwidth, problems with artificial motion, and the effect of probe pressure on the skin [8]. It seems arbitrary to obtain the data of blood flow by measurement of one-point

---

* Corresponding author.

D. Zhang and M. Sonka (Eds.): ICMB 2010, LNCS 6165, pp. 241–248, 2010.
© Springer-Verlag Berlin Heidelberg 2010

flowmetry. Therefore, the issues about changes in blood flow relative to physiological implication have been retard to further profound study.

The blood flow will be affected by the intervention of drinking, smoking, exercise, medicine, anesthesia, disease …etc. The method for measuring blood flow rate in most published papers were invasive method such as, blood flow in glass capillaries [9], developer track [3-5], and invasive protocol in animal type [10], rather than non-invasive protocol *in vivo* for human. To develop the blood flow technique with both the non-invasive and more precise method to measure blood flow rate, the measurement of O_2 saturation in red blood cell at the middle finger tip by oxy-meter synchronized ventilator were conducted. The two point measurement by non-invasive and *in vivo* way will probably be successfully applied to be a reliable instrument, rather than one point Doppler method, for wild application of blood flow in nutrition, pathology, pharmacology, medicine, anesthesiology, rehabilitation, and physical activity.

2   Detector Design and Methods

2.1   The Idea of Design Was from Testing the Reliability of Oxy-Meter

Nattokinase is a potent fibrinolytic enzyme. In my research project about the effect of supplementation of Nattokinase on the blood circulation five years ago, I tried to measure the blood flow rate. The well known Doppler methods, which detect blood flow by one point assessment, have limited precision in data. When I stop breathing, the change of blood oxygen saturation in finger tip, which was measured by oxy-meter, were detected with delay for about 25 seconds. The phenomenon about detection with delay for 25 seconds after stopping breath may caused by the travelling time of de-oxygen red blood cells from lung vessel to finger vessel. The combination of delay time and the distance from lung to finger will figure out the blood flow rate.

2.2   Design Equipped with an Oxy-Meter Sensor and Served by a Synchronized Ventilator

The design equipped with an oxy-meter and served by a synchronized ventilator was shown with driftage control elements in Fig 1. The steps of the method includes keeping the person measured by an oxy-meter and served by a ventilator, setting an initial time point when the ventilator provides a first air to the person and starting to record blood-oxy values per a predict time interval, providing the person a second air by the ventilator at a first time point, providing the person a third air by the ventilator at a second time point, setting a reference time point according to a blood-oxy value which has a variation, and obtaining the person's blood circulation velocity is proportioned to the inverse of the difference of the reference time point and the first time point. The procedure of evaluating blood flow rate can be separated into six basic steps.

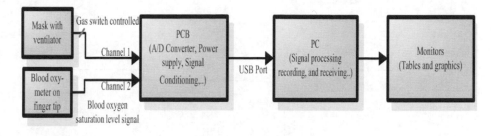

Fig. 1. Design equipped with an Oxy-meter and a ventilator synchronized by computer. The procedure of evaluating blood flow velocity was described on the text.

Step 1- Participant lies on the floor and is informed the basic procedure. After reaching the homeostatic in heart rate and breath frequency, the test begin.

Step 2- Turn on the system power, including OXY100C (Biopac systems, Inc., USA) with AcqKnowledge 3.7.3 software and a synchronized ventilator. The OXY100C sensor is attached on the middle finger-tip of subject's left hand. Subject is masked with a synchronized ventilator.

Step 3- Star testing and recording. The ventilator supply fresh air at the starting point.

Step 4- The CO_2 is supplied by ventilator for the next 20 seconds, and then the fresh air is supplied again.

Step 5- All the above steps are completed in 90 seconds.

Step 6- The blood flow rate, termed as L (cm/sec), as defined in equation 1.

$$L = Height/T_{lung\text{-}finger} \qquad (1)$$

Height: the body height of subject

$T_{lung\text{-}finger}$: the time of blood flow from alveolar capillary to middle finger-tip capillary measured by Oxy-meter.

Valid testing or invalid testing was judged by the oxygen saturation level whether the curve was stability or not before CO_2 was supplied (Fig. 2). The $T_{lung\text{-}finger}$ was calculated by the time of first down on oxygen saturation level (labled with * in Fig. 2) minus 20 seconds. The valid example has been shown in Fig. 3. The L value could be calculated while valid testing occurred.

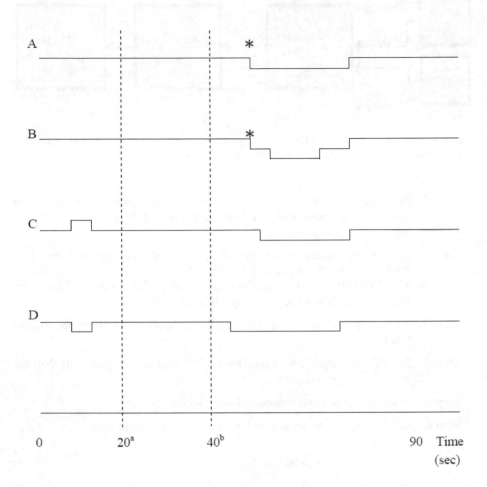

Fig. 2. The changes of O_2 saturation curve tested during 90 sec. The transverse axle is testing time, and the vertical axle is the level of O_2 saturation. The operative testing A and B results were collected as valid and testing C and D results were defined as invalid to be discarded. Testing A and B show with stable O_2 saturation while testing C and D show with unstable O_2 saturation before staring breath controller. The time of first down, marked with *, on the O_2 saturation curve minus 20 sec of before breath controlling time represents the time of blood flow between the capillaries of lung and middle finger of left hand ($T_{lung-finger}$). a: representing the time of staring breath gas controller supplying with CO_2; b: representing the time of rehabilitating normal gas supplied.

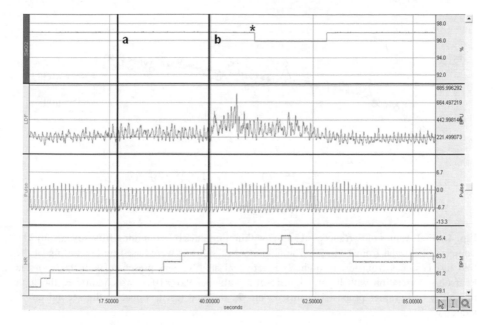

Fig. 3. Valid testing was shown on the block 1. Block from 1 to 4 showed oxygen saturation (SaO$_2$), blood perfusion unit (BPU), pulse and Heart rates (HR) by OXY100C and TSD143 (Biopac systems, Inc., USA), respectively. The time of first down, marked with *, on the O$_2$ saturation curve minus 20 sec of before breath controlling time represents the time of blood flow between the capillaries of lung and middle finger tip of left hand (T$_{lung-finger}$). a: representing the time of staring breath gas controller with CO$_2$; b: representing the time of rehabilitating normal gas supplied.

3   Results

The reliability of L values was examined by the repeatability of two testing results within 4 minutes. According to the instruction, participants are firstly asked to keep silent for 30 minutes, and then lie on the floor for testing. Basically, blood flow should be nearly kept stabilization under the testing conditions, at which the two successive tests within 4 minutes ideally, have the stable L value. The Pearson's correlation of L values by two successive tests on fifty-seven participants was 0.922 with highly significant P value of 9.67×10^{-25} (Fig. 4).

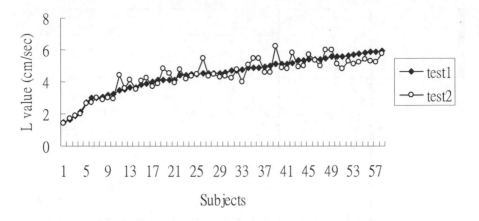

Fig. 4. The L values of two successive tests within 4 minutes by examining fifty-seven subjects. Subjects were aged from 18 to 23. The alignment of subjects were based on the sorting of the first test L values from low to high. The statistical analysis was presented with 0.922 of the Pearson's correlation with highly significant P value of 9.67×10^{-25}, which showed the high reliability of L values proposed in this paper. ◆, the first test L value; ○, the second test L value.

4   Conclusion and Discussion

The meaning of traditional one-point blood flow rate is quite different with that of estimated by two-points in this study. For example, the traffic velocity at a freeway location could not represent the condition of entire freeway. On the other word, if you monitor traffic velocity by each time one point, you will never comprehend the true traffic condition.

The blood flow rate can not be directly evaluated and must be converted from blood perfusion by one-dimension Doppler method. Microvascular blood perfusion unit (BPU) is the product of mean red blood cell velocity (Vb) and mean red blood cell concentration in the volume of tissue under illumination from the probe (Nb) [11]. However, the diameter of blood vessel is not always the same. The detected area may be changed while vasodilation or vasoconstriction occurs. Moreover, the blood flow in center area of vessel was faster than the margin area of blood vessel. The pre-set-down measuring depth and area in Doppler-metry will cause deviation in various measurements in different people [11]. In addition, the native shifting movement of normal finger will disturb the signals during measurement in Doppler-metry. The various presses by the operator on the detecting skin point will affect the detection by 1D Doppler-metry. The more loaded press in skin surface will render the blood flow faster. It was shown that the signal or detected curve of blood perfusion was very unstable for the moveable and unstable detected position or for the Doppler instrument internal factors during detecting procedure (Fig. 5). In addition, the blood perfusion curve from another commercial machine, HDI 5000 (Philips, USA) scanner with Doppler mode, showed in Fig. 6. The signal showed discontinuous and the pulses exist in naive heart beats. In conclusion, researcher could hesitate to convert

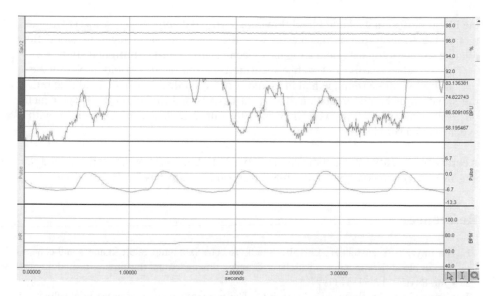

Fig. 5. Block from 1 to 4 showed oxygen saturation (SaO_2), blood perfusion unit (BPU), pulse and Heart rates (HR) by OXY100C and TSD143 (Biopac systems, Inc., USA), respectively. It was shown that the signals of blood perfusion were very unstable on the block 2 during detecting procedure. The blood flow velocity could be conversed from blood perfusion as mentioned in the text.

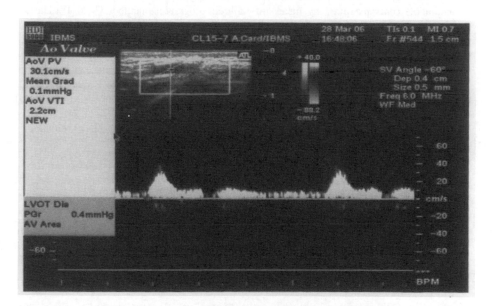

Fig. 6. The data from HDI 5000 scanner with Doppler mode (Philips, USA) showed with pulse of naive heart beats and with discontinuous signals. The blood flow rate could be conversed from blood perfusion as mentioned in the text.

the blood perfusion data into blood flow rate during data processing with 1D Doppler-metry. Up to now, this is the reason why the issues about changes in blood flow relative to physiological implication have been retard to further profound study.

The L values easily and stably measured in this study would probably be a prevailing two-dimension method of an index of blood flow rate and could true reflect the situation of blood circulation. The most defects and limitation in Doppler-metry on measuring blood flow were improved by our 2-D design of Oxy-meter with a synchronized ventilator.

Acknowledgments. We appreciate Dr. Jyh-Yih Leu on English writing and Dr. Yu-Ywan Chen to contribute the same corresponding effort as Dr. Hsueh-Kuan Lu.

References

1. Bornmyr, S., Svensson, H., Lilja, B., Sundkvist, G.: Skin temperature changes and changes in skin blood flow monitored with laser Doppler flowmetry and imaging: a methodological. Clin. Physiol. 17, 71–81 (1997)
2. Conroy, D.A., Spielman, A.J., Scott, R.Q.: Daily rhythm of cerebral blood flow velocity. J. Circadian Rhythms 3, 3–13 (2005)
3. Van Oosterhout, M.F., Prinzen, F.W., Sakurada, S., Glenny, R.W., Hales, J.R.: Fluorescent microspheres are superior to radioactive microspheres in chronic blood flow measurements. Am. J. Physiol. 275, H110–H115 (1998)
4. Chaloupecky, V., Hruba, B., Samanek, M.: Pulmonary circulation time in healthy children and adolescents measured by means of radiocirculographic method. Cesk. Pediatr. 30, 478–481 (1975)
5. Strouse, J.J., Cox, C.S., Melhem, E.R., Lu, H., Kraut, M.A., Razumovsky, A., Yohay, K., van Zijl, P.C., Casella, J.F.: Inverse correlation between cerebral blood flow measured by continuous arterial spin-labeling (CASL) MRI and neurocognitive function in children with sickle cell anemia (SCA). Blood 108, 379–381 (2006)
6. Cordell, A.R., Spencer, M.P.: Electromagnetic blood flow measurement in extracorporeal circuits: its application to cardiopulmonary bypass. Ann. Surg. 151, 71–74 (1960)
7. Hitt, D.L., Lowe, M.L., Tincher, J.R., Watters, J.M.: A new method for blood velocimetry in the microcirculation. Microcirculation 3, 259–262 (1996)
8. Obeid, A.N., Barnett, N.J., Dougherty, G., Ward, G.: A critical review of laser Doppler flowmetry. J. Med. Eng. Technol. 14, 178–181 (1990)
9. Sheppard, C.W.: Flow of Labeled Blood in Small Cylindrical Tubes and Glass Bead Systems. Circ. Res. 11, 442–449 (1962)
10. Hakimé, A., Peddi, H., Hines-Peralta, A.U., Wilcox, C.J., Kruskal, J., Lin, S., de Baere, T., Raptopoulos, V.D., Goldberg, S.N.: CT perfusion for determination of pharmacologically mediated blood flow changes in an animal tumor model. Radiology 243(3), 712–719 (2007)
11. Discovery Technology International. Principles of Laser Doppler Flowmetry,
   http://www.discovtech.com/
   physiology-principles_of_laser_doppler_flowmetry.htm
   (accessed January 15, 2010)

Context Aware Health Monitoring System

S. Sridevi, Bhattacharya Sayantani, K. Pal Amutha,
C. Madan Mohan, and R. Pitchiah

Centre for Development of Advanced Computing
Chennai, India
{sridevis,sayantanib,palamuthak,madanmohanc,rpitchiah}@cdac.in

Abstract. On one side of India, we have multi-specialty hospitals which satisfy the healthcare needs of people with specialized and speedy treatments. On the other side, a large part of the population in India resides in rural areas where basic medical facility is sometimes unavailable. People in the rural areas do not get proper treatment due to the non availability of required number of registered medical practitioners. This paper presents the architecture of Context Aware Health Monitoring System developed for connecting Primary Healthcare Centres in the rural areas with the sophisticated hospitals in the urban areas of India through mobile communication and IT infrastructure. The system aims to provide affordable, efficient and sustainable healthcare service by leveraging mobile communication and information technology. The system monitors and delivers patient's physiological readings to the hospitals and provides an alert mechanism triggered by the patient's vital signs which is linked to a medical practitioner's mobile device.

Keywords: Healthcare, Context awareness, Mobile healthcare, E-healthcare, Wireless medical sensors.

1 Introduction

Health care systems will integrate new computing paradigms in the coming years. Context aware computing is a research field which often refers to healthcare as an interesting and rich area of application [1]. Context Aware Systems can help health care professionals perform their tasks in a better way based on the context information and increase the quality of patient care. Today mobile phones have become ubiquitous device. Rapid development in the mobile technologies, falling cost of the mobile devices and ease of use are the factors that influenced the choice of mobile as communication device in our system for transmitting health data from Primary Healthcare Centres (PHC) to a centralized health database. Recent advancements in wireless technology has lead to the development of wireless medical devices for Electrocardiogram (ECG), BP Monitoring, Pulse Monitoring, etc. The proliferation of wireless devices and recent advances in miniature sensors prove the technical feasibility of a ubiquitous heath monitoring system [2]. We suggest using such kind of medical devices in the Primary healthcare centres for promoting preventive healthcare.

D. Zhang and M. Sonka (Eds.): ICMB 2010, LNCS 6165, pp. 249–257, 2010.
© Springer-Verlag Berlin Heidelberg 2010

Indian healthcare sector is expected to grow at 13-14 % per annum and currently represents 5% of the GDP [3]. This, compared with an average of 8-10% in developed countries, is clearly inadequate and fails to meet the healthcare needs of the country [3]. Further unlike developed nations, most (around 80%) of it consists of private expenditure. There is a large unmet demand which is compounded by the lack of quality healthcare infrastructure. There is a huge shortage of doctors, nurses and trained paramedical staff today and the situation will only worsen with the commissioning of new and expansion of current hospitals.

In order to overcome the current scenario, we have developed a Context Aware Health Monitoring system for Primary Healthcare Centres connecting them with the well equipped multi specialty hospitals and medical practitioners in the urban areas through mobile communications and IT infrastructure. The rest of the paper is organized in the following manner; Section 2 describes about current scenario of primary health care centre in India, Section 3 gives an insight on Context Aware Computing, Section 4 elaborates the architecture and the sub systems of Context Aware Health Monitoring System, Section 5 explains about medical sensors used in our experiments Section 6 talks about related works and finally Section 7 describes our future work plan.

2   Primary Health Care Centres in India

Primary Healthcare Centre (PHC) is the first contact point between village community and the Medical Officer. The PHCs were envisaged to provide an integrated curative and preventive health care to the rural population with emphasis on preventive aspects of health care. The activities of PHC involve curative, preventive and Family Welfare Services. There are 23236 PHCs functioning as on September, 2005 in the India [4]. The PHCs are hubs for 5-6 sub-centres that cover 3-4 villages and are operated by an Auxiliary Nurse Midwife (ANM).

3   Context Aware Computing

The term context-awareness in ubiquitous computing was introduced by Schilit (1994). Context Aware Computing is a key aspect of the future computing environment, which aims to provide relevant services and information to users, based on their situational conditions. Context is any information that can be used to characterize the situation of an entity. An entity is a person, place or object that is considered relevant to the interaction between a user and an application, including the user and application themselves [5]. Context-aware computing offers many advantages, allowing systems to act more autonomously and take initiatives, being informed by a better model of what their users need and want. Building and deploying context aware systems in open, dynamic environments raises a new set of research challenges on context data acquisition, context interpretation, context data modeling and reasoning etc.

4   Architecture and Sub Systems

Figure 1 illustrates the overview of architecture and sub systems of context Aware Health Monitoring System. In our experiment, we use three wireless medical devices: - Alive Pulse Oximeter, Alive Heart Monitor, Corscience Blood Pressure Monitor 705IT. All three devices are Bluetooth enabled. Patients visit the PHCs and ANMs take their readings using these devices. Health parameters are collected by Mobile phone through Bluetooth and transmitted to the Context Aware System through General packet radio service (GPRS). Inference engine of Context Aware System compares the health parameters against the rules available in the Rule-base to check for any abnormalities. Meanwhile patient's health parameters are stored in the health database. If any abnormalities are detected, SMS (Short Message Service) alert about the health status of the patient will be sent to the medical practitioner who resides in well equipped hospitals in the urban areas. On receiving the alert message, he / she can login to the UbiHealth Web interface from his / her hospital to study the patients' health report to take necessary action.

There are three subsystems in the Context Aware Health Monitoring System: - Mobile Application, Context Aware System and UbiHealth Web interface. Our research focus is towards building the Context Aware System while Mobile Application and UbiHealth Web interfaces are built for showing the capability of Context Aware System.

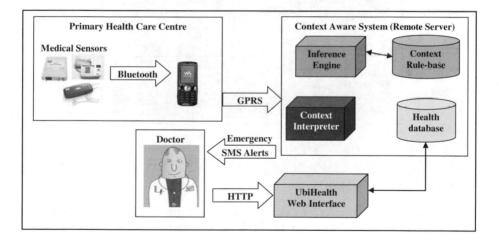

Fig. 1. Architecture and Sub systems of Context Aware Health Monitoring System

4.1   Mobile Applications

It is developed using J2ME which can run on any java enable mobile phones. We have tested on Sony Ericsson and Nokia models. When a patient visits the PHC for the first time, ANM carries out registration for the patient using UbiHealth Web interface and assigns a patient ID. ANM uses the mobile application to record the health parameters of the patients. Mobile Application starts with the authentication screen for ANMs to login. Once the ANM is authenticated, the screen displays the list of wireless medical devices. ANM chooses the device, enters the patient ID and starts taking readings for the patient. These readings are captured by the mobile. After ANM is satisfied with the readings taken, he/she can save the readings. The readings are then transmitted using GPRS to the Context Aware System via a Web server. Mobile applications are developed for wireless devices used in our experiment. If any new wireless device needs to be introduced in the PHC, a J2ME application needs to be developed for data acquisition. Figure-2 shows the mobile application for recording Pulse Oximeter readings.

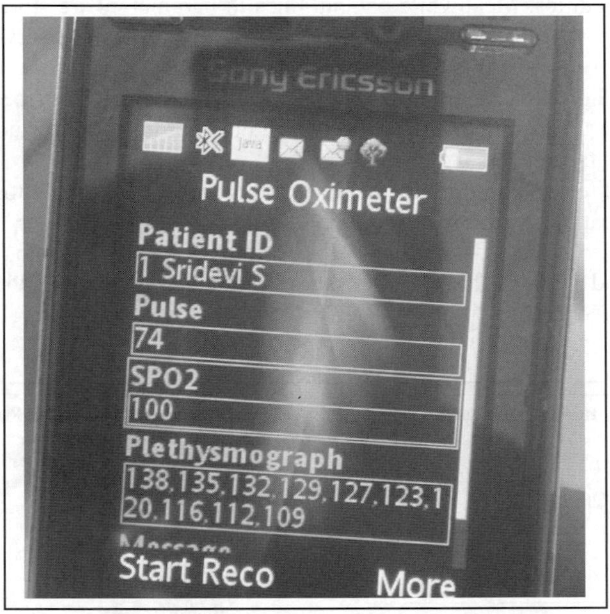

Fig. 2. Mobile Application for Pulse Oximete

4.2   Context Aware System

It is an ontology based middleware developed by our team for Ubiquitous Computing applications, which is being used for health applications, aware home and smart meeting rooms. It acts as a generic middleware between the sensor layer and the application layer. It uses publish-subscribe paradigm and provides event notification service. It manages the various context parameters of various entities such as person, place, smart space artifacts etc. Sensors keep publishing the data in the form of XML using Publish API and Applications subscribe to interested context using Subscribe API. The System maintains the list of subscribers for each context using the Java Hash map. Any change in the context will be notified to the application layer along with the action to be taken. Application will decide on executing the actions. Any new sensor has to be registered in to the Context Aware System using a GUI (Graphical User Interface) to provide the metadata information of the data sent by sensor. Once the sensor is registered, a XML format is provided for the sensor layer to send data. Sensor Layer developers does the data extraction from the sensor based on the datasheet provided along with device and sends the data in the XML format provided by Context Aware system. So any new device can be added without any change in the Context Aware System.

Main sub systems of Context Aware System are the Knowledge base, Context Interpreter, Inference engine and Rule-base. Context Aware System maintains its own knowledge base to store the data for later retrieval in the form of OWL (Web Ontology Language). Knowledge base information is used to compare the historical data with the current context data to make some conclusion. SPARQL, an RDF query language is used to query the knowledge base. Jena, a Java framework for building Semantic Web applications is used for managing the information stored in the Knowledge base. Context Interpreter interprets the XML data received and passed on to the Inference engine. Inference engine compares the data against the rules stored in the Rule-base. Context Rules are added / updated to the Rule-base by the respective domain experts using the GUI provided by the system. Rules can be updated at any time, without disturbing the system. These rules act as the trigger points to send alerts when needed. SMS / Email service is built along with the system. SMS module uses GSM / GPRS modem and Java SMSLib package for sending SMS alerts. Some of the challenges of Context Aware System like Context Interpretation, Context Data Modeling and Reasoning are addressed in our system.

Table 1. Context Rules for Blood Pressure

Systolic	Diastolic	State
210	120	Stage 4 High Blood Pressure
180	110	Stage 3 High Blood Pressure
160	100	Stage 2 High Blood Pressure
140	90	Stage 1 High Blood Pressure
140	90	BORDERLINE HIGH
130	85	High Normal
120	80	NORMAL Blood Pressure
110	75	Low Normal
90	60	BORDELINE LOW
60	40	TOO LOW Blood Pressure
50	33	DANGER Blood Pressure

In the view of Context Aware Health Monitoring System, medical data received using GPRS by the Web server is published in Context Aware System. Health Service Provider is the service running in the application layer which subscribes to the Context Aware System for any change in the health parameters like Blood Pressure, SPO2 (amount of oxygen being carried by the red blood cell in the blood), Pulse rate etc. The Context Aware System receives health parameters in the XML format. These health parameters are parsed by the Context Interpreter and the Inference engine compares the parsed health data against the Rule-base to find if any abnormities are present for the patient records. Knowledge base information will be used at this point to know about past health history. If any abnormality is detected, an event will be triggered and notified to the application layer requesting to send SMS / Email to the domain experts depending on the state of the abnormities. Applications can make use of SMS and Email services that are built along with the Context Aware System. A sample set of rules added in the Context Rule base for Blood Pressure is listed in the Table 1. Trigger points can be set to send alerts.

4.3 UbiHealth Web Interface

It is a web interface developed using JSP / Servlets to provide a web access to the health data for Medical practitioners, ANMs and Patients. Administrator login manages the UbiHealth Website. Patient can login and see their own health records, and past historical health data. The patients need not carry health records along with them. Medical practitioners can login and view the patients assigned for them, their medical records, the medicines given during earlier visits, the state of their health etc. ANMs login to do patient registration, linking doctors with the PHCs, taking statistical report of how many patients visited the centre in a given period, etc.

Currently we have developed interfaces for viewing the health parameters recorded from the Pulse Oximeter, Heart Monitor and Blood Pressure Monitor. Pulse Oximeter

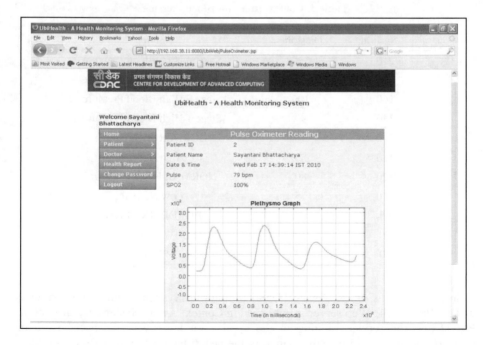

Fig. 3. Screenshot of UbiHealth Web Interface

provides Pulse rate, SPO2 and Plethysmograph. Heart monitor provides heart rate and ECG. Blood Pressure Monitor provides Systolic pressure rate, Diastolic pressure rate and Pulse rate. ECG and Plethysmograph are plotted as graphs in the web interface using Java Applets. If a patient moves to another village, he doesn't need to register again in the PHC of that village. Patient health records are available across all PHC since records are maintained at a centralized location. Figure 3 shows the screenshot of UbiHealth web interface showing the most recently taken Pulse oximeter readings for patient login.

5 Wireless Medical Sensors

In our experiment, we use three wireless medical devices: - Alive Pulse Oximeter, Alive Heart Monitor, Corscience Blood Pressure Monitor 705IT. All three devices are Bluetooth enabled. Figure 4 shows the patient using the Pulse Oximeter to measure his / her pulse and the data being transmitted to J2ME application installed in Sony Ericsson Mobile.

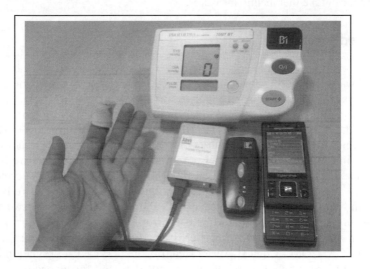

Fig. 4. Alive Pulse Oximeter, Alive Heart Monitor and Corscience Blood Pressure monitor used in our experiment

5.1 Alive Pulse Oximeter

The Alive Pulse Oximeter has a built-in Bluetooth interface, which provides a wireless connection between the patient and host receiving system. SPO2, Pulse, Signal Quality, and Plethysmograph data are transmitted in real time over a Bluetooth connection. When a Bluetooth connection is made to the pulse Oximeter, it starts transmitting data packets automatically.

5.2 Alive Heart Monitor

The Alive Heart Monitor is a data acquisition, wireless transmitter and recording device of patient ECG and acceleration signals. The signals can be stored to an SD memory card on the device and/or transmitted via a Bluetooth connection to a mobile phone, PDA, or PC for real-time display, recording and analysis.

5.3 Corscience 705IT Bluetooth Blood Pressure Monitor

The 705IT BT from Omron is a high-quality meter, which measures blood pressure on the upper arm. The measurement is carried out as in conventional devices with simple and intuitive handling. The big difference is that Corscience has equipped the 705IT BT with a wireless interface, which automatically transfers the measured value to a central archive via a base station. The 705IT BT wirelessly sends the measurement with a time stamp to the base station. The base station can be a Bluetooth-capable cell phone or an approved Bluetooth modem (e.g. BL analog). By means of this automatic progress check, deterioration of the patient's condition can be detected at an early stage and the responsible tele-monitoring center, the physician or the hospital can take therapeutic measures.

6 Related Work

There are many research works related to Context Aware Healthcare. Implicit, Context-Aware Computing for Health Care [6] is a research work which made an ontology-based Context Management System (CMS) that allows the user to define contexts using terms from the medical field. The two most important domain ontologies they have considered are the HL7 reference information model (RIM) and the HL7 clinical document architecture (CDA). A Context Aware Framework for u-Healthcare in a Wearable System [7] is a research work on context aware framework as the middleware between wearable sensor systems and u-Healthcare service entities. HCRBAC – An Access Control System for Collaborative Context-Aware HealthCare Services in Mauritius [8] allow the Mauritian Public Healthcare Service not only to have a secure computerized system for its current services but will also allow it to provide an augmented service that crosses the boundaries of the healthcare institutions. CASCOM - Context-Aware Health-Care Service Co-ordination in Mobile Computing Environments [9] architecture is not designed for a specific application, its primary field of validation is the telemedicine domain and on-the-fly coordination of pervasive healthcare services.

Our Context Aware Health Monitoring System acts as communication bridge between Primary Healthcare Centers in the rural areas and the Medical Practitioners working in the well-equipped hospitals in urban areas of India. It provides alerts at abnormal / emergency conditions to Medical Practitioners to take necessary action to avoid loss of life. Our system is different from other existing context aware systems by utilizing the mobile communication and IT infrastructures which already exist in the PHCs and Hospitals. Our system does not require any additional infrastructure other than the wireless medical devices to be provided at the Primary Healthcare Centres. Currently the system is tested in our laboratory and in next few months we would be deploying in a primary health care centre for testing.

7 Future Plan

As future work, we plan to improve upon our Context Aware Health Monitoring System by alerting the patients and doctors about epidemic diseases spreading in the different areas. Context Rule-base will be updated with rules related to epidemics. When the patient is having symptoms similar to the epidemic disease, the system shall compare with rules and alert the ANMs and Medical Practitioners. Since lot of healthcare standards are being evolved, we are planning to adopt HL7 India standards for managing the health records in our system. The objective of HL7 India is to support the development, promotion and implementation of HL7 standards and specifications in a way which addresses the concerns of healthcare organizations, health professionals and healthcare software suppliers in India [10].

Acknowledgement

This project is funded under National Ubiquitous Computing Research Initiative by Department of Information Technology, Ministry of Communications and Information Technology, Government of India.

References

1. Bricon-Souf, N., Newman, C.: Context awareness in health care: A review. International Journal of Medical Informatics 76(1), 2–12 (2007)
2. Otto, C., Milenkovic, A., Sanders, C., Jovanov, E.: System Architecture of a Wireless Body Area Sensor Network for Ubiquitous Health Monitoring. Journal of Mobile Multimedia 1(4), 307–326 (2006)
3. News Article Healthcare may log 13-14% growth by Ecnomic Times dated (February 16, 2010)
4. http://india.gov.in/citizen/health/primary_health.php
5. Dey, A.K.: Understanding and Using Context. Personal and Ubiquitous Computing Journal (2001)
6. Jahnke, J.H., Bychkov, Y., Dahlem, D., Kawasme, L.: Implicit, Context-Aware Computing for Health Care. In: Proceedings of OOPSLA 2004 Workshop on Building Software for Pervasive Computing (2004)
7. Kang, D.-O., Kang, K., Lee, H.-J., Ko, E.-J., Lee, J.: A Context Aware Framework for u-Healthcare in a Wearable System. In: Post-PC Research Group, Electronics and Telecommunications Research Institute, Daejeon, Korea. IFMBE Proceedings, vol. 14(6)
8. Moonian, O., Cheerkoot-Jalim, S., Nagowah, S.D., Khedo, K.K., Doomun, R., Cadersaib, Z.: HCRBAC – An Access Control System for Collaborative Context-Aware HealthCare Services in Mauritius. Journal of Health Informatics in Developing Countries 2(2), 10–21 (2008)
9. ERCIM News num. 60, CASCOM - Context-Aware Health-Care Service Co-ordination in Mobile Computing Environments, César Cáceres, Alberto Fernández, Sascha Ossowski (URJC)
10. HL7 India Standards, http://www.hl7india.org/

Novel Two-Stage Analytic Approach in Extraction of Strong Herb-Herb Interactions in TCM Clinical Treatment of Insomnia

Xuezhong Zhou[1], Josiah Poon[2], Paul Kwan[3], Runshun Zhang[4], Yinhui Wang[4],
Simon Poon[2], Baoyan Liu[5], and Daniel Sze[6]

[1] School of Computer and Information Technology, Beijing Jiaotong University, Beijing, China
[2] School of Information Technologies, University of Sydney, Sydney, Australia
{josiah.poon,simon.poon}@sydney.edu.au
[3] School of Science and Technology, University of New England, Armidale, Australia
[4] Guanganmen Hospital, China Academy of Chinese Medical Sciences, Beijing, China
[5] China Academy of Chinese Medical Sciences, Beijing, China
[6] Department of Health Technology and Informatics,
Hong Kong Polytechnic University, Hong Kong

Abstract. In this paper, we aim to investigate strong herb-herb interactions in TCM for effective treatment of insomnia. Given that extraction of herb interactions is quite similar to gene epistasis study due to non-linear interactions among their study factors, we propose to apply Multifactor Dimensionality Reduction (MDR) that has shown useful in discovering hidden interaction patterns in biomedical domains. However, MDR suffers from high computational overhead incurred in its exhaustive enumeration of factors combinations in its processing. To address this drawback, we introduce a two-stage analytical approach which first uses hierarchical core sub-network analysis to pre-select the subset of herbs that have high probability in participating in herb-herb interactions, which is followed by applying MDR to detect strong attribute interactions in the pre-selected subset. Experimental evaluation confirms that this approach is able to detect effective high order herb-herb interaction models in high dimensional TCM insomnia dataset that also has high predictive accuracies.

Keywords: Multifactor Dimensionality Reduction, Features Selection, Traditional Chinese Medicine, Insomnia, Network Analysis, Herb Interactions.

1   Introduction

Insomnia is a "nightmare" to many people. It refers to the poor quality of sleep or insufficient sleep duration; it can lead to adverse daytime consequences. Studies have shown that rates of insomnia with associated daytime dysfunction in attendees of general medical practices are high in both the western countries [1] and Asia region [2-4]. It is also common in many different morbid conditions [5] and it could be a risk factor to other diseases, such as depression and anxiety and further studies also find that there are genetic link between insomnia and depression/anxiety [6]. Therefore, insomnia is a

D. Zhang and M. Sonka (Eds.): ICMB 2010, LNCS 6165, pp. 258–267, 2010.

popular, important and vital disorder that needs to be addressed. However, the treatment of insomnia still remains particularly challenging for clinicians because of the lack of guidelines and the small number of studies conducted in patient populations with behavioural and pharmacologic therapies [7]. In the world especially in Asian region, a significant proportion of patients with insomnia consume herbal hypnotics regularly [8, 9]. Also alternative therapies, herbal products and other agents with sedative hypnotic effects are being increasingly sought after by the general population in the world [10-12].

Traditional Chinese medicine (TCM) has a long history and has been accepted as one of the main medical approaches in China [13]. It has also been successfully applied to the treatment of insomnia. Some of these well-known Chinese formulae include Baixia Tang [14], Suan Zaoren Tang [15] and Wendan Tang [16]. There are clinical studies have shown that the TCM herbal treatment for insomnia is effective [17, 18]. Since a classical TCM formula may be modified according to individual patient, it is necessary to find the common and true effective herb patterns (e.g. herb combinations) that actually facilitate the insomnia treatment. Herb prescription (also named formula in TCM field) is one of the main TCM therapies for various disease treatments. A specific formula constitutes of multiple herbs with appropriate dosages. Therefore, when huge number of clinical encounters occurred, they could generate large-scale formulae in the clinical practice. The herb combination patterns used in the herb prescriptions are both theoretically focused and important for effective TCM treatment. It would be significant to find the effective herb combination patterns; we called them *strong herb-herb interactions*, from the large-scale clinical data of herb prescriptions.

We propose in this paper a two-stage analytical procedure by which (1) the subset of herbs that have a high probability in participating in herb-herb interactions is preselected using hierarchical core network analysis (HCNA) method [19], and (2) the set of pre-selected herbs will constitute the input into multifactor dimensionality reduction (MDR) [20] to detect attribute interactions of order even up to 9, 10. The benefits of our approach are two-fold. First, it overcomes the computational bottleneck in applying MDR to detect attribute interactions of a higher order such as 9, 10, against a high dimensional dataset such as those appearing in TCM. Second, by combining the analysis of the hierarchical core sub-network and the result of MDR, "core" attribute interactions that would otherwise be unable to reveal due to the flat nature of the output of MDR can be made known.

2   Methods

2.1   Two-Step Approach

To study interaction, MDR has been demonstrated to be an effective tool to identify/understand the gene-gene and gene-environment epistasis [21]. It is based on the general idea of attribute induction in data mining research to discover possible nonlinear interactions among genes. However, using MDR, it is still time-consuming in analyzing high-dimensionality dataset to discover higher order interactions using exhaustive search method. One way to address this drawback would be to reduce the dimensionality of the dataset prior to the analysis using MDR.

In TCM clinical datasets such as the insomnia that we are studying in this paper, the dimensionality is often high, above 100 dimensions (in our case, we have 261 dimensions). Also, the TCM physicians often prescribe formulae with 10-20 herbs. This means that the higher order interaction may exist in the prescriptions. We need to run over 4-5 months to perform such analysis (9 variable model with 261 dimensions) to get the best models on HP Proliant ML350 server with Intel Xeon(R) CPU (2.33GHz) and 8GB Memory.

Although MDR is the method of choice for detecting higher order interactions (i.e., beyond main effects and pairwise interaction) among herb-herb combinations in the analysis of the insomnia dataset (or in general, TCM dataset), the high dimensionality of the dataset makes it prohibitive to analyze beyond 4-5 attributes interaction. On the other hand, our recent result [19, 22] on analyzing common herb pairs frequently appearing in regular TCM herb prescriptions using HCNA confirmed that there often exists a core herb network, which comprises a subset of herbs appearing in prescriptions treating the disease concerned in the dataset. Herbs that appear in this core sub-network are expected to have a high probability of participating in herb-herb interactions.

MDR is a still a time consuming approach to find the combination patterns with outcomes. The high dimensionality of herbs (e.g. several hundreds) and the lengthy resulting pattern (e.g. more than ten herbs) of herb combination are still hurdles to apply this tool. Since HCNA can generate the important herb combinations, which are mostly smaller than the dimensionality of the whole herbs dataset, we propose to use HCNA as a filter before MDR analysis.

In order to minimize the computational overhead to extract the strong herb-herb interactions, the high dimensionality of herbs in the original prescriptions is first reduced by a hierarchical core sub-network. This analysis can generate a subset of important herbs (20-30 herbs), which is substantially smaller than the original dimensionality. The result from this extraction step will then serve as input to the MDR software in the next step, which can report a list of strong interacting herbs combination. Hence, in our proposal, it is a two-stage algorithm that uses HCNA to pre-select the important herbs and uses MDR to find the strong herb-herb interactions. Figure 1 is the system overview of this data mining process.

2.2 Hierarchical Core Sub-network Analysis

Social network analysis (or complex network analysis) has penetrated to various scientific fields, especially the biology and medical sciences [23]. Scale-free network is one of the distinguished characteristics of complex networks [24], and of which the degree distribution follows as the power law. This further implies that there are nodes that serve as hubs in a large-scale scale-free network. We have shown that from the related work on TCM herb network, the weight distribution of the herb network follows the power law. This indicates that there are common herb pairs frequently used in the regular TCM herb prescriptions. We have developed a complex network analysis system to model and analyze TCM clinical data [25]. A hierarchical core sub-network analysis method is deployed to retrieve the multi-level core herb combination patterns for insomnia treatment.

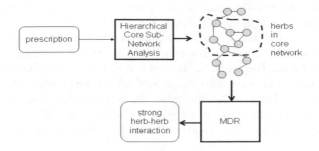

Fig. 1. System overview

The basic formula is usually modified to accommodate individual differences. The main modifications make up of a core herb combination structures with hierarchical levels. In our HCNA, the hierarchical core herb combination structures are discovered based on degree distribution feature of scale-free network. The value of α is assigned near the power law exponent of node weight distribution in herb combination network. The number of hierarchical levels can be manually controlled by researchers. Usually the core sub-networks with 2-4 hierarchical levels will be clinical significant and easily be interpreted. We set the default value of the number of hierarchical levels as 3 to extract the three levels of core herb combination sub-networks.

2.3 Multifactor Dimensionality Reduction (MDR)

MDR [20, 21] has 7 steps and was initially designed to detect and to interpret gene-gene interactions. It utilizes information theory for selecting interesting single nucleotide polymorphisms (SNPs). It is a constructive induction approach to collapse high-dimensional genetic data into a single dimension.

3 Data Set

A clinical data warehouse was developed [22] to integrate and to manage large-scale real-world TCM clinical data. This data warehouse consists of structured electronic medical record from all TCM clinical encounters, including both inpatient and outpatient encounters. There are about 20,000 outpatient encounters of the TCM expert physicians. These encounters included clinical prescriptions for the treatment of various diseases, in which insomnia is a frequently treated disorder.

We have selected and labeled 460 insomnia outpatient encounters. The outcome of each encounter was annotated by TCM clinical experts who went through the changes of the insomnia-related variables over consecutive consultation; these include the sleep time per day, sleep quality and difficulty in falling asleep. The outcomes are then classified into two categories: good and bad. When a treatment was effective, which means that if the patient recovered completely or partly from insomnia in the next encounter, then the prescription of the current encounter would be categorized as 'good'; otherwise, the herb prescription would be categorized as 'bad'.

After labeling these 460 outpatient encounters, there are 68 encounters with bad outcomes in this dataset; in other words, it is an imbalanced dataset to the advantage of the target class. The average good outcome rate (GOR) of the whole data set is 392/460=85.21%. There are 261 distinct herbs in the dataset and there are on average 14 herbs in a formula. The insomnia data set is then transformed into the formats for hierarchical core sub-network analysis as well as for MDR methods. In the transformed dataset for MDR analysis, the value of a herb variable is set to '1' if the current encounter includes the corresponding herb, else it is set as '0'. Also, the value of outcome variable is set to '1' for good outcome and '0' for bad outcome.

4   Results

4.1   Extraction of the Hierarchical Core Herbs

We use the 460 herb prescriptions to construct a herb network with the herbs as nodes and the herb combinations as edges. The result herb network has 8,335 edges. Using the complex network analysis system [25], the HCNA method was applied to extract the top three level core sub-networks. Figure 2 shows the top three levels of core herb combinations. The degree coefficient was chosen as 2.0 to generate these results. A total of 39 core herbs and their corresponding combinations are shown in these figures. It shows that the herbs like *stir-frying spine date seed*, *grassleaf sweetflag rhizome*, *prepared thinleaf milkwort root*, *Chinese angelica*, *oyster shell* and *dragon bone* are used frequently in a combinatorial way for insomnia treatment. The extracted herbs are commonly found in the classical formulae and are frequently used by TCM physicians in insomnia treatment [14, 15, 17, 18]. The result herbs become a filter set with clinical meanings.

4.2   Finding the Strong Herb-Herb Interaction by MDR

We filtered the herb dataset with only 39 herb variables generated by the HCNA step. Then we use the MDR software (MDR 2.0 Beta 6) to find the strong herb-herb interactions for insomnia treatment. Using 10-fold cross evaluation, we select the exhaustive search type and track one top model to get the best model with 9 herb variables. The task took about 83 hours on a PC with Intel Core(TM) 2 Quad 2.66GHz CPU and 4G memory. A tremendous saving in computation cost was obtained with this two-stage approach as it is less than $1/1000^{th}$ of the original effort, i.e. $C(39,9)$ as compared to $C(260,9)$. The training balanced accuracy of the 9 herb variables model is 90.89%. The accuracy with the test instances is 63.13% and the cross-validation (CV) consistency is 3/10. The best model includes the herb variables: VAR33 (*indian bread*), VAR34 (*golden thread*), VAR40 (*grassleaf sweetflag rhizome*), VAR113 (*fresh rehmannia*), VAR117 (*dried tangerine peel*), VAR174 (*Chinese angelica*), VAR196 (*Chinese date*), VAR203 (*white peony root*) and VAR237 (*stir-frying spine date seed*) as the main prediction factors.

Figure 3 depicts the interaction map of the herb variables. The regular boxes are the herbs while the number in each box is its relationship to the outcome; in other words, a positive percentage indicates a positive correlation with the outcome. The edges in this figure represent the interaction, and the percentage on the line is the interaction effect. The thickness of the line is a graphical representation of this effect; the thicker the line, the higher the interaction it is. The figure shows that 5 herbs (enclosed in black circles): VAR235 (*prepared thinleaf milkwort root*), VAR237, VAR196, VAR79 (*cassia bark*) and VAR40 have the independent main effects on the outcome. Furthermore, it shows significant synergic interaction effects between several herb variables which are denoted by the dark thick dotted lines. For example, significant synergic interaction effects between VAR235 and VAR117, VAR76 (*bamboo shavings*) and VAR117, VAR235 and VAR196, and VAR210 (*prepared pinellia tuber*) and VAR117. Also there are strong redundant interactions between several herb pairs like VAR237 and VAR235. The entropy-based interaction map proposes a good approach to show the significant main effect and interaction effect of herbs for outcomes. The discovered strong herb interactions have a significant support for the necessity of multi-herb prescriptions in TCM treatment.

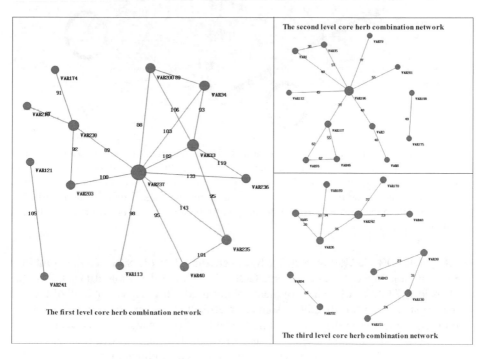

Fig. 2. The top three levels of core herb combination network

To find the final strong herb-herb interaction with good outcome or bad outcome, we could query the whole data set using the significant herb interactions suggested by the result model. The herb pairs with synergic interactions are more useful for finding the final strong effective or *non*-effective herb-herb interactions. Based on the synergic interactions suggested by the interaction map, we have tested the interesting herb combinations with GORs. Table 1 displays the GORs of some of the interesting herb combinations. It clearly shows that there are nonlinear interaction (like 'XOR' function) between VAR235 and VAR196. The single use of each herb has high GOR

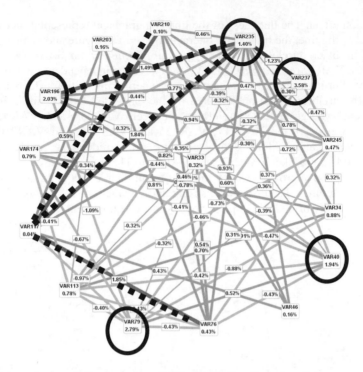

Fig. 3. The interaction graph for the herb variables based on entropy-based measures of information gain. To have a clear view of the graph, we filter out the edges and nodes whose information gain value is smaller than 0.3%. The red lines show positive interaction effects between herb variables. While the blue lines show negative interaction effects between herb variables. The larger of the edge sizes the stronger the interaction between the associated two variables of the edges.

(VAR235:92.80%; VAR196:96.70%), however, the GOR (87.80%) of the combined use is only little higher than the average GOR (85.21%) of the whole data set. Also the GOR is low (74.87%) while both of them are not used. This means that both of the two herbs are significant for effective insomnia treatment, however, there may exist negative interaction between them since while the GOR of combined usage is clearly lower than the GOR of mutually exclusively use of them. It is similar with the interaction between VAR117 and VAR76 except that this pair is a combination to avoid since the GOR of combined usage is largely lower than the average GOR. There are redundant interactions between the herbs, e.g. (VAR235, VAR237) where the second herb covers most of the outcome contributions. This kind of interaction information is both useful for theoretical research and clinical practice of herb combinations for insomnia treatment. From Table 1 (the other records are not listed owing to the page limit), we could find the significant (effective or *non*-effective) herb combinations with strong herb-herb interactions for insomnia treatment, e.g. some herb combinations like (VAR237, VAR203, VAR34) has 97.87% GORs with 46 frequency. However, the herb combination: (VAR46 [*stir-frying immature orange fruit*], VAR76, VAR117, VAR210) has bad outcome, which has a 72.73% GOR with 44 frequency.

Table 1. The example herb combinations with clinical usefulness. The value '01' in the combinations field means the condition that is not use of the former variables in the herb variables field and use of the latter variables.

Herb variables	Combinations	Total frequency	Good outcome frequency	Good outcome rate
VAR235,VAR196	11	41	36	87.80
	01	91	88	**96.70**
	10	125	116	**92.80**
	00	203	152	74.87
VAR76,VAR117	11	62	48	77.42
	01	44	44	**100.00**
	10	25	21	84.00
	00	329	279	84.80
VAR235,VAR237	11	143	132	**92.31**
	01	114	105	**92.11**
	10	23	20	86.96
	00	180	135	75.00

4.3 Comparison of the Different Filters

The HCNA method was included as a filter for herb interaction detection by the MDR. The training balanced accuracy (87.01%) of HCNA filter is slightly better than the other filters (SURF*TuRF, OddsRatioF, ReliefF, x^2F, TuRF, SURF). Although it has lower testing balanced accuracy than several filters, HCNA has the advantages of maintaining the multi-level core herbs in the filtered attribute set. This makes sure that the different levels of core herb variables are retained in the result, which is practically used in TCM herb prescriptions according to the organizational principle of TCM formula theories, such as master-deputy-assistant-envoy (君臣佐使) [26], which indicate that the herbs were used as different roles to form a systematic formula for disease treatment. Thus, HCNA proposes a feasible choice as filter to select the herb variables from the original high dimensional variable set.

5 Discussion and Conclusion

The herb prescriptions in the TCM clinical data consist of multiple herbs and have complicated interactions between different herbs for disease treatment. There are TCM herb combinatorial theories like seven features of herb compatibility (SFHC) [27] to illuminate the combinatorial use of herbs. The SFHC theory says that there are seven kinds of herbs used in prescriptions, namely "going alone", "mutual reinforcement", "assistant", "detoxication", "restraint", "inhibition" and "antagonism". For example, the "inhibition" interaction between two herbs should be avoided in the clinical prescriptions since one herb would inhibit the effect of another herb. Thus, it is significant to find the strong non-linear herb-herb interactions (corresponding to the seven kinds of herb compatibility) in the clinical herb prescriptions, as it would be both useful for drug development and clinical guideline generation.

In this paper, we have proposed a two-stage analytical approach to obtain strong herb-herb interactions of order up to 9 from a TCM clinical dataset of insomnia. Stage 1 pre-selects a subset of the original herbs appeared in the dataset that has a high probability of participating in herb-herb interactions using HCNA method, while stage 2 applies MDR on the pre-selected set of herbs to detect the required attribute interactions. The results show that there exist non-linear herb-herb interactions in the herb prescription for insomnia treatment. Furthermore, the effective or non-effective herb combinations could be found by the proposed two-stage approach and by querying the whole data set. The entropy-based herb interaction map proposes a feasible visualization of the herb compatibility. It is interesting that some types of the herb compatibility, such as the "inhibition", could be recognized as synergic interaction in the model. These kinds of results have very high value to be investigated by further clinical studies.

Experimental results confirmed that our approach is able to detect a 9-order herb interactions with acceptable accuracy, while having the distinct advantage of maintaining the multi-level core herbs in the interaction result when compared to other attribute filtering methods.

Although this paper introduces a computing approach for finding the strong herb-herb interactions and the useful herb combinations in the context of clinical outcomes. Several key issues still exist to be addressed in the future work. Firstly, a straightforward approach to detect the strong herb interactions is needed as the querying of the whole data set could not systematically find the top K strong herb combinations. Secondly, the computing cost of MDR to find the optimal or hyper-optimal herb variable models should be further studied. Thirdly, two important information components, namely herb dosage and clinical manifestation (e.g. symptoms, co-morbid conditions), of the clinical data could be included in the data set. Because it is widely recognized in medical field that the herb dosage has an important effect for treatment and also performs a significant role for herb-herb interactions.

Acknowledgement

This work is partially supported by Program of Beijing Municipal S&T Commission, China (D08050703020803, D08050703020804), China Key Technologies R&D Programme (2007BA110B06-01), China 973 project (2006CB504601), National Science and Technology Major Project of the Ministry of Science and Technology of China (2009ZX10005-019), and the S&T Foundation of Beijing Jiaotong University (2007RC072).

References

1. Sateia, M.J., Nowell, P.D.: Insomnia. Lancet 364, 1959–1973 (2004)
2. Xiang, Y., Ma, X., Cai, Z.: The Prevalence of Insomnia, Its Sociodemographic and Clinical Correlates, and Treatment in Rural and Urban Regions of Beijing, China. A General Population-Based Survey. Sleep 31, 1655–1662 (2008)
3. Cho, Y.W., Shin, W.C., Yun, C.H.: Epidemiology of insomnia in Korean adults: prevalence and associated factors. Journal of Clinical Neurology 5, 20–23 (2009)
4. Doi, Y.: Epidemiologic research on insomnia in the general Japanese populations. Nippon Rinsho. 67, 1463–1467 (2009)
5. Pien, G.W., Schwab, R.J.: Sleep disorders during pregnancy. Sleep 27, 1405–1417 (2004)

6. Gehrman, P.: Heritability of insomnia in adolescents: How much is just depression and anxiety? Sleep 32, A264 (2009)
7. Benca, R.M.: Diagnosis and treatment of chronic insomnia: a review. Psychiatry Service 56, 332–343 (2005)
8. Wing, Y.K.: Herbal treatment of insomnia. Hong Kong Medical Journal 7, 392–402 (2001)
9. Chen, L.C., Chen, I.C., Wang, B.R., Shao, C.H.: Drug-use pattern of Chinese herbal medicines in insomnia: a 4-year survey in Taiwan. Journal of Clinical Pharm Therapy 43, 555–560 (2009)
10. Attele, A.S., Xie, J.-T., Yuan, C.S.: Treatment of insomnia: An alternative approach. Alternative Medicine Review 5, 249–259 (2000)
11. Cuellar, N.G., Roger, A.E., Hisghman, V.: Evidenced based research of complementary and alternative medicine (CAM) for sleep in the community dwelling older adult. Geriatric Nursing 28, 46–52 (2007)
12. Gooneratne, N.S.: Complimentary and alternative medicine for sleep disturbances in older adults. Clinical Geriatric Medicine 24, 121–viii (2008)
13. Tang, J.-L., Liu, B., Ma, K.-W.: Traditional Chinese Medicine. The Lancet 372, 1938–1940 (2008)
14. Sun, H., Yan, J.: Discussion of the 'BuMei' syndrome in Inner Canon of Huangdi. Zhongyi Zazhi, 7 (2004)
15. Sun, H., Yan, J.: Discussion of the 'BuMei' syndrome in Synopsis of Golden Chamber. Lishizhen medicine and material medical research 16, 182–183 (2005) (in Chinese)
16. Fei, Z.D., Wu, X.Q.: Discussion of the effective formula for insomnia: wendang tang. Jiangsu traditional Chinese medicine 26, 39 (2005) (in Chinese)
17. Yu, Y., Ruan, S.: Using cassia bark for insomnia treatment. Zhongguo Zhongyao Zazhi 23, 309–310 (1998) (in Chinese)
18. Cui, Y., Wang, S., Liu, W.: The advancement of Chinese medicine diagnosis and treatment to insomnia. Journal of Henan University of Chinese Medicine 23, 102–104 (2008) (in Chinese)
19. Zhou, X., Chen, S., Liu, B., et al.: Extraction of hierarchical core structures from traditional Chinese medicine herb combination network. In: ICAI 2008, Beijing, China (2008)
20. Ritchie, M.D., Hahn, L.W., Roodi, N., et al.: Multifactor dimensionality reduction reveals high-order interactions among estrogen metabolism genes in sporadic breast cancer. American Journal of Human Genetics 69, 138–147 (2001)
21. Hanh, L.W., Ritchie, M.D., Moore, J.H.: Multifactor dimensionality reduction software for detecting gene-gene and gene-environment interactions. Bioinformatics 19, 376–382 (2003)
22. Zhou, X., Chen, S., Liu, B., et al.: Development of Traditional Chinese Medicine Clinical Data Warehouse for Medical Knowledge Discovery and Decision Support. Artificial Intelligence in Medicine 48(2-3), 139–152 (2010)
23. Wasserman, S., Faust, K.: Social network analysis: Methods and application (1994)
24. Barabasi, A.-L., Reka, A.: Emergence of scaling in random networks. Science 286, 509–512 (1999)
25. Zhou, X., Liu, B.: Network Analysis System for Traditional Chinese Medicine Clinical Data. In: Proceedings of BMEI 2009, Tianjin, China, vol. 3, pp. 1621–1625 (2009)
26. Shi, A., Lin, S.S., Caldwell, L.: Essentials of Chinese Medicine, vol. 3. Springer, Heidelberg (2009)
27. Wang, J., Guo, L., Wang, Y.: Methodology and prospects of study on theory of compatibility of prescriptions in traditional Chinese medicine. World science and technology-modernization of TCM and materia medica. 8(1), 1–4 (2006)

Clinical Usage Considerations in the Development and Evaluation of a Computer Aided Diagnosis System for Acute Intracranial Hemorrhage on Brain CT

Tao Chan

Department of Diagnostic Radiology, The University of Hong Kong
Room 406, Block K, Queen Mary Hospital, Pok Fu Lam, Hong Kong
taochan@hku.hk

Abstract. Acute intracranial hemorrhage (AIH) is a major cause of neurological disturbance or complication of head injury. Its presence dictates different management strategy. In modern medicine, detection of AIH relies on the use of brain computed tomography (CT). But diagnosis of AIH can become difficult when the lesion is inconspicuous or the reader is inexperienced. Therefore it has been envisaged that computer aided diagnosis (CAD) system can help improving diagnosis of AIH, especially those small lesions, and benefit clinicians of variable skill levels.This paper summarizes a previously reported computer aided diagnosis (CAD) system that aimed to help improving diagnosis of AIH on CT by clinicians of variable skill levels. This paper further discusses considerations on clinical usage that affects both the development and evaluation of CAD as exemplified by this application.

Keywords: computer aided diagnosis, computed tomography, brain, acute intracranial hemorrhage.

1 Introduction

Acute intracranial hemorrhage (AIH) literally means recent bleeding inside the confine of the skull. It encompasses intraparenchymal hemorrhage that presents as stroke, or intraparenchymal, subarachnoid, subdural and extradural hemorrhage as complications of head injury. For clinical management of patients presenting with neurological disturbances or head injury, it is of prime importance to identify AIH at the earliest stage as patients with and without bleeding require different management strategies. One major reason that CT has been used as the imaging modality of choice in these clinical scenarios is that it readily depicts AIH as hyperdense regions that show higher signals than normal brain tissues. However, diagnostic errors may still occur when the lesions are inconspicuous, or when the observers are inexperienced. In most parts of the world, brain CT are first read by acute care physicians rather than radiologists. Their skills have been shown to be imperfect in this regard [3]. Even radiology residents may infrequently overlook AIH [4], highlighting the potential benefit of some computerized adjuncts. It was therefore envisaged that CAD can play a useful role for improving early diagnosis of AIH in acute clinical settings, especially for use by clinicians who may not be highly skilled readers.

D. Zhang and M. Sonka (Eds.): ICMB 2010, LNCS 6165, pp. 268–275, 2010.

2   Materials and Methods

2.1   Materials

A total of 186 cases, including all 62 continuous cases that showed AIH not more than 1cm in size obtained during a 6 month period, and 124 randomly selected controls that were obtained during the same period, were retrospectively collected from a 1200 bed acute hospital in Hong Kong. The imaging diagnoses were established by consensus of two experienced radiologists.

2.2   Computer Aided Diagnosis Scheme

A CAD was designed and implemented. The flow chart of the scheme is illustrated in figure 1 and its GUI is shown in figure 2. The system reads and processes standard

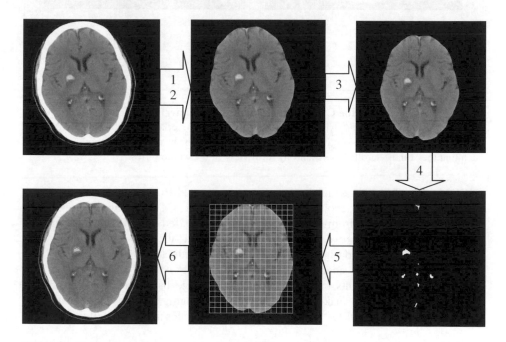

Fig. 1. Individual components and intermediary outputs after successive steps of the algorithm. Basic components of a usual CAD, including image preprocessing, image segmentation, image analysis, and classification are all utilized. 1. intracranial contents segmented using thresholding and morphological operations; 2. preprocessing steps that reduce noise and CT cupping artifacts; 3. intracranial contents aligned by locating mid-sagittal plane and boundaries of the brain; 4. AIH candidates extracted using combined method of tophat transform and left-right comparison; 5. AIH candidates rendered anatomical meaning by registration against a purposely developed coordinate system; 6. genuine AIH distinguished from mimicking variants or artifacts by the rule based classification system, using both image features and anatomical information. The ICH in right basal ganglia is correctly identified as genuine AIH and outlined in red, whilst the mimics are outlined in blue.

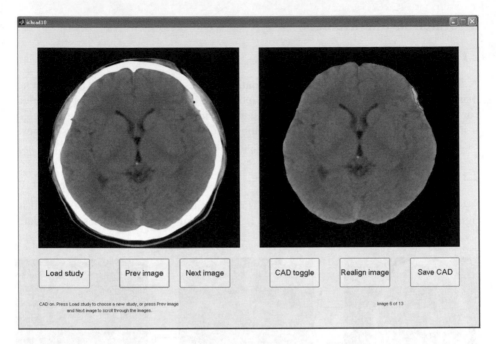

Fig. 2. Screen capture of the CAD system graphical user interface. The original images are displayed on the left window, whilst the output images with overlay of the outlines of AIH are displayed on the right. The original and output images are displayed in stripe mode and linked such that they can be scrolled in synchrony for review of the whole series.

DICOM image files. Intracranial contents are segmented from the CT images, which are then subjected to denoising and adjustment for CT cupping artifacts. AIH candidates are extracted from the intracranial contents based on top-hat transformation and subtraction between two sides of the image about the mid-sagittal plane. AIH candidates are registered against a normalized coordinate system such that the candidates are rendered anatomical information. True AIH is differentiated from mimicking normal variants or artifacts by a knowledge based classification system incorporating rules that make use of quantified imaging features and anatomical information. Some candidates correctly identified as mimics are illustrated in figure 3.

2.3 Observer Performance Test

In the observer performance study, 7 emergency physicians (EP), 7 radiology residents (RR) and 6 radiology specialists (RS) were recruited as readers of 60 sets of brain CT selected from the same 186 case collection, including 30 positive cases and 30 controls. Each reader read the same 60 cases twice, first without, then with the prompts produced by the CAD system. The clinicians rated their confidence in diagnosing each case of showing AIH in both reading modes. The results were analyzed using the multiple-reader, multiple-case receiver operating characteristic (MRMC ROC) paradigm. In addition, the number and correctness in changing diagnosis before and after use of CAD was recorded.

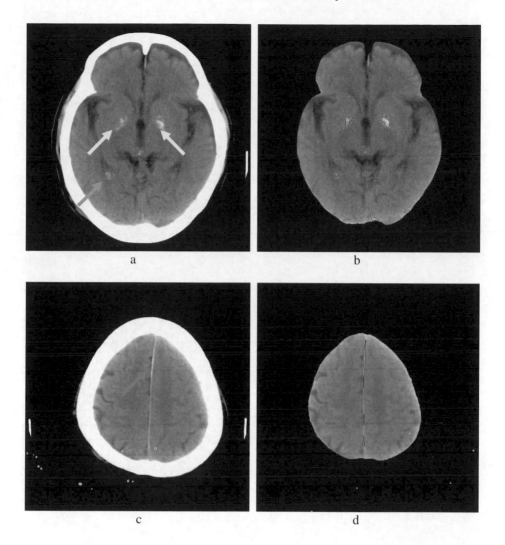

Fig. 3. AIH candidates correctly classified as non lesions by the system. Images *a* and *b* show that the calcifications in basal ganglia, choroids plexus, and pineal gland calcifications, marked by yellow, orange, and green arrows respectively, are correctly classified as mimics of AIH. AIH candidates subsequently discarded as mimics rather than genuine AIH are outlined in blue by the system. They show relatively high attenuations relative to their sizes (image features) and are located at the areas susceptible to calcification deposition (anatomical information). Images *c* and *d* show that the calcifications or ossification of falx cerebri, pink arrow, is also correctly classified, based on their linear configuration, vertical orientation (image features) and central location (anatomical information).

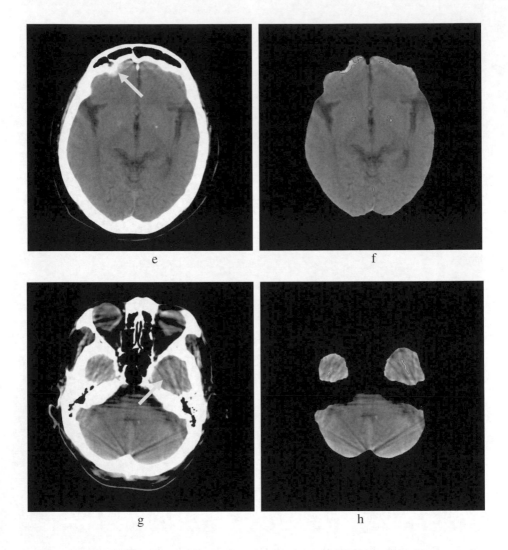

Fig. 3. (*continued*). Images *e* and *f* are the original image and CAD output of an image just superior to the floor of anterior cranial fossa that show partial volume average at the right frontal region. Images *g* and *h* are the original image and CAD output of an image through temporal lobes and posterior fossa that show beam hardening artifact. The artifacts in both cases, yellow arrows, were extracted as AIH candidates, but subsequently discarded by the knowledge base classification, outlined in blue.

3   Results

The CAD algorithm was manually trained using 40 positive and 80 control cases. In the validation test using the remaining 22 positive and 44 control cases, the system achieved sensitivity of 83% for small (< 1cm) AIH lesions on a per lesion basis or 100% on a per case basis, and false positive rate 0.020 per image or 0.29 per case.

The observer performance test showed showed significantly improved performance for EP, average area under the ROC curve (Az) significantly increased from 0.83 to 0.95 without and with the support of CAD. Az for RR and RS also improved, from 0.94 to 0.98 and from 0.97 to 0.98 respectively (figure 4). The EP changed diagnosis correctly 46 times and wrongly 12 times after use of CAD, RR: 29 and 3, RS: 7 and 0.

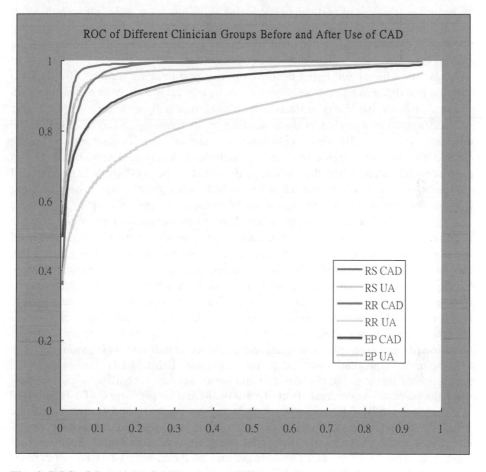

Fig. 4. ROC of detection of AIH amongst different groups of clinicians. EP – emergency physicians. RR – radiology residents, RS – radiology specialists, UA – unaided reading mode, CAD – CAD reading mode.

4 Discussion

It is intuitive that capability to identify however small a lesion is desirable for any CAD. In the initial stage of devising the CAD, the author had conducted a pilot test on the accuracy of two radiology residents (having 1 and 2 years of experience respectively) in detection of AIH using 23 AIH cases of a wide range of sizes, which confirmed that sensitivity reached 100% for AIH above 1cm in size. The finding suggests that finding large AIH can be a technically trivial task for clinicians. This basically defines the goal of the CAD.

It goes without saying that image preprocessing is an important step to produce efficient CAD by facilitating subsequent segmentation and analysis. For each type of images there are specific problems. In the case of brain CT, cupping artefact produced by beam hardening is of particular concern, as this elevates signals subjacent to skull into the range of signals commonly seen in blood clots. Knowledge of the cupping phenomenon hence facilitate implementation of specific strategy to counter act the artefact by suppressing signals close to periphery of the brain.

It was soon found out during the development that imaging features of AIH is so variable that differentiation from other materials, e.g. calcifications, stagnant blood in sinuses, or beam hardening artifacts, which also appear hyperdense on brain CT is very difficult. But the clinical understanding of different appearances between the different types of AIH sheds light into the use of classification system using combination of both imaging features and anatomical location. Such classification is especially important when the system is designed to be used by less experienced clinicians who may not be confident to overrule some false positive outputs from CAD, and in whom use of CAD may actually produce adverse effects [5].

Use of MRMC ROC paradigm rather than simple sensitivity/specificity pairs is popular for evaluation of CAD, because ultimately, benefit of the CAD scheme need to be realized in terms of improved diagnosis made by the reader using the CAD. Indeed, results from MRMC ROC have laid the grounds for FDA approval of some CAD products. It should be recognized that readers may benefit variably from use of CAD according to their own skill level. Therefore it is important to include adequate number of readers including those at the target skill level in the test, such that the results become generalizable to the particular target group. In the current example, inclusion of emergency physicians rather than the usual radiologists is required.

Despite the apparently boosted performance in MRMC ROC, the exaggerated proportion of lesion in the observer test and hence smaller proportion of normal may mask the potential detrimental effect of false positive prompts from CAD. This indeed is the case for AIH. The author had conducted an internal audit from the institute where the brain CT was obtained over a 6 month period, showing that amongst 3341 emergency brain CT performed for initial evaluation of neurological disturbance of head injury, only 279 showed hemorrhage, and of these, only 62 (1.86%, 62/3341) contained blood clot that are defined as small lesion like those used in the development and evaluation of the current CAD. It is readily appreciable that even a small proportion of false positives, if not discarded by the reader as such, can magnify the detrimental effect due to high proportion of normal in clinical setting.

On the other hand, it has been found from during observer performance test that clinicians of higher skill level were more confident in judging if the outputs from CAD were correct or not and change their own judgement accordingly. It is therefore contended that CAD can be tuned to operate at higher sensitivity (hence lower specificity) for such readers.

5   Conclusion

In summary, a CAD system which boasted high sensitivity and low false positive rate has been developed. MRMC ROC study confirmed that it can improve diagnostic performance of clinicians, especially emergency physicians. It is anticipated that such a system can reduce diagnostic errors and improve patient care when it is integrated in the clinical environment for daily operation. Clinical usage considerations are important considerations for the planning, development, and evaluation of CAD, which are crucial for successful of implementation of some useful tools.

References

1. Chan, T.: Computer aided detection of small acute intracranial hemorrhage on computer tomography of brain. Comput. Med. Imaging Graph. 31(4-5), 285–298 (2007)
2. Chan, T., Huang, H.K.: Effect of a Computer-aided Diagnosis System on Clinicians' Performance in Detection of Small Acute Intracranial Hemorrhage on Computed Tomography. Acad. Radiol. 15(3), 290–299 (2008)
3. Schriger, D.L., Kalafut, M., et al.: Cranial computed tomography interpretation in acute stroke: physician accuracy in determining eligibility for thrombolytic therapy. Jama 279(16), 1293–1297 (1998)
4. Wysoki, M.G., Nassar, C.J., et al.: Head trauma: CT scan interpretation by radiology residents versus staff radiologists. Radiology 208(1), 125–128 (1998)
5. Alberdi, E., Povykalo, A., et al.: Effects of incorrect computer-aided detection (CAD) output on human decision-making in mammography. Acad. Radiol. 11(8), 909–918 (2004)

Research on the Meridian Diagnostic System Based on the Measuring Principle of the Electrical Properties of Well, Source and Sea Acupoints

Fengxiang Chang, Wenxue Hong, Jun Jing, Jialin Song, and Chao Ma

Yanshan University, Qinhuangdao, 066004, China
hongwx@ysu.edu.cn

Abstract. Skin impedance at acupuncture points has been used as a diagnostic/therapeutic aid for more than 50 years. Electrical resistance at acupuncture points is normally smaller than that of the surrounding skin. The Well, Source and Sea points are representative in reflection of physiological and pathological states of the human body, according to the theory of Traditional Chinese Medicine. Based on this, we proposed a meridian diagnostic system based on the electrical properties measurement of the Well, Source and Sea points. This paper firstly describes in detail the system's hardware and software designs, and then introduces the calculation methods of the characteristic parameters of the meridian. The visualization analysis results of the meridian data are given at last.

Keywords: electrical property, measurement system, meridian diagnosis.

1 Introduction

The use of electrical devices to detect and monitor acupuncture points has a long and "checkered" history. The first claims for the electrical detection of acupuncture points date to the 1950s, when Reinhardt Voll (Germany) in 1953, Yoshio Nakatani (Japan) in 1956 and J.E.H.Niboyet (France) in 1957, each independently concluded that skin points with unique electrical characteristics were identifiable and resembled traditional acupuncture points. Since then, a number of studies have elaborated the electrical properties attached to these "bioactive" points, which are frequently equated with acupuncture points. These properties include increased conductance, reduced impedance and resistance [1], increased capacitance [2-3], and elevated electrical potential compared to nonacupuncture points [4].

Five Shu points, including Well, Spring, Stream, River and Sea, are five special acupuncture points of the twelve meridians; they locate below elbows, knees of the human body. Qi and blood flowing in the meridians follow certain directions. Meridian Qi starts to flow from the Well points, and end of the Sea points. Though the Source points do not correspond to the five Shu points, the Source points of Yin meridian are the same as the Stream points of the five Shu points, and the Source points locate between the Well points and the Sea points [5].

D. Zhang and M. Sonka (Eds.): ICMB 2010, LNCS 6165, pp. 276–285, 2010.

So we choose the Well points, the Source points and the Sea points as the detective acupuncture points, which could be more comprehensively reflection of the characteristics and changes of the meridians.

The purpose of this study is to develop a meridian diagnostic system based on the electrical properties measurement of the Well, Source and Sea points, which could provide assistance for the diagnosis of disease, especially for early disease detection.

2   Hardware Design of the System

The system can achieve automatic measurement of DC, AC volt-amper characteristics and frequency characteristics, and the frequency range is from 2Hz to 800KHz.

System components and the main work process: the system consists of the signal generator module, the signal overlapped and filtering module, the voltage-controlled current source module, comparison and measurement of the acupuncture points module and the signal gain and phase separation module as shown in Fig. 1. The data acquisition card PCI8335B is employed to control and acquire data, the function generator AD9833 produces a sinusoidal voltage signal, and the signal is overlapped and filtered by the multiplier MC1496 and the filter LTC1560 respectively. Then the processed signal is converted into sinusoidal current signal by the module of voltage-controlled current source (VCCS). The sinusoidal current signal is loaded on the acupuncture points of the human body and the standard resistance R, and then two voltage signals are produced. The difference of the gains and the ratio o0f the phases of the two voltage signals will be outputted by the AD8302 respectively.

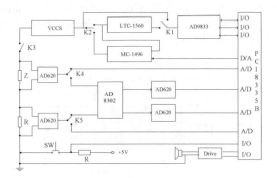

Fig. 1. System hardware schematic chart

2.1   The Signal Generator Circuit

AD9833 is a low power programmable waveform generator capable of producing sine, triangular and square wave outputs. It has three serial interfaces SCLK, FSYNC and SDATA, where SCLK is the serial clock input; FSYNC is the active low control input; SDATA is the serial data input.

The computer sents control word, enable signal and clock signal to three serial interfaces SDATA, FSYNC, SCLK of AD9833 respectively through digital output ports of the data acquisition card PCI-8335B, to control AD9833 producing the desired frequency waveform. As shown in Fig. 2.

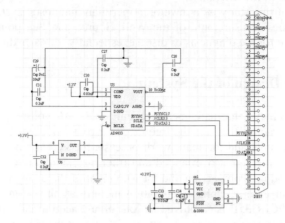

Fig. 2. Signal generator circuit

2.2 The Current Source Circuit

The current source has an important role in the measurement system. It generates a sine stimulating signal according to the required frequency of the measurement, and achieves a constant current output. The impedance values would have great changes when the measuring system has a small error, so the current source must maintain a constant current output. This module can achieve a better constant output by the component LM324. As shown in Fig. 3.

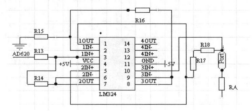

Fig. 3. Current source circuit

2.3 The Gain and Phase Separation Circuit

Fig. 4 shows the gain and phase separation circuit of the system. RF/IF gain and phase detector AD8302 is employed to measure the gain and phase of the system. AD 8302 is a fully integrated system for measuring gain and phase in numerous receive, transmit, and instrumentation applications. It requires few external components and a single supply of 2.7V-5.5V. The ac-coupled input signals can range from -60dBm to 0 dBm in a 50 Ω system, from low frequencies up to 2.7 GHz. The outputs provide an accurate measurement of gain over a ±30 dB range scaled to 30mV/dB, and of phase over a 0°-180°range scaled to 10mV/degree.

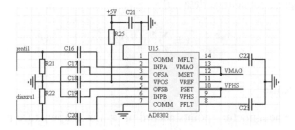

Fig. 4. Gain and phase separation circuit

3 Software Design of the System

The control software was developed using 32-bit windows platform. Data acquisition card PCI-8335B was used as intermedium of communication, and Visual Basic 6.0 was used as software platform. The system acquires the data and then saves and stores the data in a text file with a unique ID code. To begin impedance measurements, the operator selects the sampling frequency, the acupuncture points and the file name in which the data will be saved. Fig. 5 is the flowchart of the data acquisition. Fig. 6 is the system main interface.

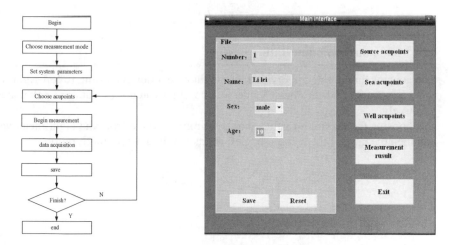

Fig. 5. Flowchart of the data acquisition   **Fig. 6.** System main interface

4 Calculation of the Characteristic Parameters of the Meridian

Taking impedance values of 24 Source acupoints for example, we introduce the calculation method of all characteristic parameters.

4.1 Energy Value

Energy value of the human body is the mean value of all meridians data, which can be gained by averaging impedance data $x(i)$.

$$\bar{x} = \sum_{i=1}^{24} x_i / 24 \tag{1}$$

4.2 Yin-Yang Ratio

Classification the acupuncture points according to the acupuncture points correspond to the meridian is Yin meridian or Yang meridian. In Source points, Taiyuan (lung meridian), Daling (pericardium meridian), Shenmen (heart meridian), Taibai (spleen meridian), Taichong (liver meridian) and Taixi (kidney meridian) belong to "Yin", Yangxi (large intestine meridian), Yangchi (triple energizer meridian), Wangu (small intestine meridian), Yangchong (stomach meridian), qiuxu (gallbladder meridian) and Shugu (bladder meridian) belong to "Yang".

By averaging the data of Yin meridians and Yang meridians respectively, we can achieve the Yin-Yang ratio.

$$\bar{x}_{yin} = \sum x_{yin} / 12 \tag{2}$$

$$\bar{x}_{yang} = \sum x_{yang} / 12 \tag{3}$$

$$r_{yin/yang} = \bar{x}_{yin} / \bar{x}_{yang} \tag{4}$$

4.3 Maximun-Minimum Ratio

The acupuncture points are divided into eight categories according to left-right, hand-foot, Yin-Yang, and the mean values of the eight categories data are calculated respectively.

Choosing the maximum value and the minimum value from the eight mean values, and then the ratio of them could be defined as maximum-minimum ratio.

$$x_1 = \sum x_{left-hand\ yin} / 3 \tag{5}$$

$$x_2 = \sum x_{right-hand\ yin} / 3 \tag{6}$$

$$x_3 = \sum x_{left-hand\ yang} / 3 \tag{7}$$

$$x_4 = \sum x_{right-hand\ yang} / 3 \tag{8}$$

$$x_5 = \sum x_{left-foot\,yin} / 3 \tag{9}$$

$$x_6 = \sum x_{right-foot\,yin} / 3 \tag{10}$$

$$x_7 = \sum x_{left-foot\,yang} / 3 \tag{11}$$

$$x_8 = \sum x_{right-foot\,yang} / 3 \tag{12}$$

$$r_{max/min} = \frac{max[x_1, x_2, \cdots x_8]}{min[x_1, x_2, \cdots x_8]} \tag{13}$$

4.4 Deficiency-Excess Ratio

Subtracting the energy value $\overline{x}$ from the data x_i, the difference is δ_i. If δ_i is positive, then the acupuncture points are classified as "excess", the mean value of these points is $\overline{\delta}_+$, otherwise the acupuncture points are classified as "deficiency", and the mean value of them is $\overline{\delta}_-$, the ratio of $\overline{\delta}_-$ with $\overline{\delta}_+$ is deficiency- excess ratio.

$$\delta_i = x_i - \overline{x} \tag{14}$$

$$\begin{cases} \overline{\delta}_+ = \sum_{i=1}^{m} \delta_i / m, \delta_i > 0 \\ \overline{\delta}_- = \sum_{j=1}^{n} \delta_j / n, \delta_j < 0 \end{cases} \tag{15}$$

$$r = \overline{\delta}_- / \overline{\delta}_+ \tag{16}$$

5 Data Visualization Analysis

Taking DC measurement data of acupoints for example, we analyze the data.

In the static measurement system, the current value is 16uA, and the impedance data can be gained by the formula:

$$R_i = \frac{V_i}{I = 16\mu\,A} \tag{17}$$

The characteristic parameters of the meridian can be achieved by the formulas (1-16).

5.1   The Overall Characteristics Analysis of the Human Body's Meridian

The overall characteristics analysis of human body's meridian is from the overall functions of human body including the energy, the balance of upper and lower, the balance of left and right, the balance of Yin and Yang to interpret the body's health state, it is an expression of the overall characteristics of human body's meridian.

The energy value is 1 to 100, where the value between 1and 20 is deficiency syndrome, between 20 and 40 is relative deficiency, between 40 and 60 is normal, between 60 and 80 is relative excess, and the value above 80 is excess syndrome.

As shown in Fig. 7, the subject's energy value is between 40 and 60, it shows that the body energy is in a normal state, the body's adjustment function is normal, but the body has an unbalance of deficiency and excess. The maximum and minimum ratio is larger, which indicates that the body has higher spirit pressure.

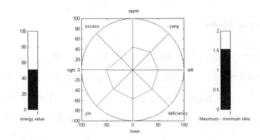

Fig. 7. Overall characteristics analysis of the human body's meridian

5.2   Rader Chart Analysis of the Five Elements Theory

The theory of the five elements explains the physiological characteristics and functions of the five Zang-organs with the corresponding one of the five elements, pairing each of the five Zang-organs with the corresponding one of the five elements. For example, Wood is characterized by free development while the liver prefers freedom to stagnation, so the liver pertains to Wood; Fire is hot and tends to flame up while heart-Yang warms the whole body, so the heart pertains to Fire; Earth receives and generates while the spleen transforms food nutrients and is the source of Qi and blood, so the spleen pertains to Earth; Metal depurates and astringes while lung-Qi maintains inside and normally descends, so the lung pertains to Metal; Water moistens and closes while the kidney stores essence and manages water metabolism, so the kidney pertains to Water.

The human body is an organic whole. The five Zang-organs coordinate with each other physiologically and affect each other pathologically. Under pathological condition, disorder of one organ may be transmitted to another. According to the theory of the five elements, there are two aspects of pathological transmission among the five Zang-organs: one includes over restraint and reverse restraint, the other includes disorder of the mother-organ involving the child-organ and disorder of the child-organ involving the mother-organ.

The Zang-organs and the Fu-organs are connected with each other through the meridians in structure, coordinate with each other and differ from each other in physiology as well as affect each other and transmit to each other in pathology. The relationship between the five Zang-organs and the six Fu-organs, though complicated, can be summarized as "mutual internal and external relationship". The Zang-organs pertain to Yin while the Fu-organs to Yang, Yin controls the internal and Yang manages the external. Such a coordination between Yin and Yang well as the internal and external makes up a special internal and external relationship between the Zang-organs and Fu-organs.

The special mutual internal and external relationship between the Zang-organs and Fu-organs covers five aspects: the heart and the small intestine, the lung and the large intestine, the spleen and the stomach, the liver and the gallbladder as well as the kidney and the bladder. So the gallbladder pertains to Wood, the small intestine pertains to Fire, the stomach pertains to Earth, the large intestine pertains to Metal, the urinary bladder pertains to Water. As shown in Fig. 8, the spleen, the lung and the kidney of the five Zang-organs and the stomach, the large intestine and the bladder of the five Fu-organs are dysfunction.

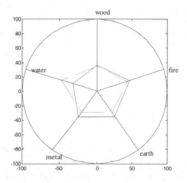

Fig. 8. Rader chart analysis of the five elements (solid line represents Zang-organs, dash line represents Fu-organs)

5.3 The External and Internal Relationships

The twelve meridians can be grouped into six pairs; each pair is internally and externally related to each other. The meridians in external and internal relationship with each other pertain to Yin and Yang respectively. The Yin meridian is the internal and the Yang meridian is the external. The meridians in external and internal relationship with each other are connected with each other at the extremities of the four limbs. The viscera to which the meridians in external and internal relationship with each other pertain are also related to each other externally and internally. In this way the meridians and the viscera in external and internal relationship with each other respectively associate with each other. The Yin meridian pertains to the Zang-organ and connects the Fu-organ with its collaterals while the Yang meridian pertains to the Fu-organ and connects the Zang-organs with its collaterals.

The external-internal relations of the twelve meridians not only strengthen the connection between the external meridians and the internal meridians, coordinate the Zang-organs and Fu-organs that are of mutual external-internal relationship in terms of physiology, but also cause them to have a pathological influence on each other.

Fig. 9 is the radar chart of the external and internal relationships among the twelve meridians. Because the Zang-organs are internal while the Fu-organs external, if the data points fall inside the small round circle, it shows that Zang-organs are relative predominance while Fu-organs relative decline, otherwise, Zang-organs are relative decline while Fu-organs relative predominance. It can be seen from Fig. 9 that the liver and the gallbladder meridians, the kidney and the bladder meridians, the lung and the large intestine meridians are abnormal.

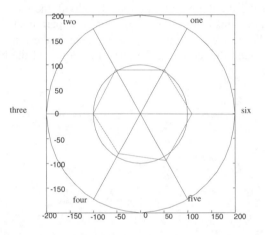

Fig. 9. Rader chart of the external and internal relationships among the twelve meridians (one represents the pericardium and the triple energizer meridians, two represents the heart and the small intestine meridians, three represents the spleen and the stomach meridians, four represents the liver and the gallbladder meridians, five represents the kidney and the bladder meridians, six represents the lung and the large intestine meridians)

6   Conclusions

This paper designed a meridian diagnostic system based on electrical properties measurement of the well, source and sea points. Through rational design and selection of the circuit components and parameters, it performs well and enjoys the advantages of current adjustable, wide output range of voltage. Its function includes the automatic measurement of DC, AC volt-ampere characteristics and frequency characteristics. By analyzing the acupuncture points impedance data and data visualization, we build the system prototype of the analysis and diagnosis of the meridians. The realization of the system could provide assistance for the diagnosis of disease, especially for early disease detection. It also has a positive significance of monitoring the treatment of the disease and evaluating the prognosis of the disease.

Acknowledgments. This paper is sponsored by the National Nature Science Foundation of China (No.60873090).

References

1. Colbert, A.P., Yun, J., Larsen, A.: Skin impedance measurements for acupuncture research: development of a continuous Recording system. eCAM 5, 443–450 (2008)
2. Prokhorov, E.F., Gonzalez-Hernandez, J., Vorobiv, Y.V.: In vivo electrical characteristics of human skin, including at biological active points. Med. Biol. Eng. Comput. 38, 507–511 (2000)
3. Johng, H.M., Cho, J.H., Shin, H.S., et al.: Frequency dependence of impedances at the acupuncture points Quze (PC3). IEEE Eng. Med. Biol. Mag. 21, 33–36 (2002)
4. Ahn, A.C., M.D., M.P.H., et al.: Electrical characterization of acupuncture points: technical issues and challenges. The journal of alternative and complementary medicine 13, 817–824 (2007)
5. Zhongbao, Z., Yuezhong, H., Jinwen, T., et al.: Basic theory of Traditional Chinese Medicion. Publishing house of shanghai university of Traditional Chinese Medicine (2002)

Atrial Fibrillation Analysis Based on Blind Source Separation in 12-Lead ECG Data

Pei-Chann Chang, Jui-Chien Hsieh, Jyun-Jie Lin, and Feng-Ming Yeh

Department of Information Management, Yuan Ze University, Taoyuan 32026, Taiwan
iepchang@saturn.yzu.edu.tw

Abstract. Atrial Fibrillation is the most common sustained arrhythmia encountered by clinicians. Because of the invisible waveform of atrial fibrillation in atrial activation for human, it is necessary to develop an automatic diagnosis system. 12-Lead ECG now is available in hospital and is appropriate for using Independent Component Analysis to estimate the AA period. In this research, we also adopt a second-order blind identification approach to transform the sources extracted by ICA to more precise signal and then we use frequency domain algorithm to do the classification. The strategy used in this research is according to prior knowledge and is different from the traditional classification approach which training samples are necessary for. In experiment, we gather a significant result of clinical data, the accuracy achieves 75.51%.

Keywords: 12-Lead ECG, Atrial Fibrillation, Blind Source Separation, Kurtosis.

1 Introduction

ARIAL fibrillation (AF) is the most common sustained arrhythmia encountered by clinicians [1] and occurs less than 1% of those under 60 years of age and more than 6% in those aged over 60 [2]. The interest in the research and understanding of AF has considerably increased during the last years [3]. The analysis and characterization of AF and other atrial tachyarrhythmias from noninvasive techniques requires the previous estimation of the atrial activity (AA) signal from the standard electrocardiogram (ECG). During AF, the electric waveform arising from the atria appear in a disordered way. In normal situation, steady heart rhythm typically beats 60-80 times a minute, but in cases of atrial fibrillation, the rate of atrial impulses can range from 300 to 600 beats per minute. That is, in ECG, AF is often described by those consistent P-waves that are replaced with fast, irregular fibrillatory waves (F-waves) of variable sizes, shapes and timing [4].

Recently, researches indicate that the dynamics of atrial activity play an important role to characterize and classify atrial tachyarrhythmias. To extract the feature, some approaches base on either QRST cancellation through spatiotemporal method [5], average beat subtraction (ABS) [6] or blind source separation (BSS), e.g. independent component analysis (ICA) [7], Hilbert-Huang transform [8] and time frequency analysis approaches, like wavelet decomposition [9]. Regarding the ECG, these researches observed that the AA signal typically exhibits a narrowband spectrum with a

D. Zhang and M. Sonka (Eds.): ICMB 2010, LNCS 6165, pp. 286–295, 2010.

main frequency of between 3.5-9 Hz [10]. This is an important information we use to classify the AF from AA period in frequency domain.

2 Methodology

2.1 12-Lead ECG Data

In this research, we proposed a blind source separation based approach to classify the atrial fibrillation in 12-lead ECG data. The data in this research is collected from Taoyuan Armed Forces General Hospital in Taiwan. All rhythms were diagnosed and verified by a cardiologist, Doctor Yeh. There are totally 98 12-lead ECG data selected and recorded by Philps TC50. The sampling rate is 500Hz with a 10-second record. According to the result diagnosed by the doctor, there are 25 normal rhythms while the number of AF is 73 and these are the fundamental data we used to compare the accuracy.

Clinical 12-lead ECG data is now available in most hospitals and it includes more detailed information about cardiac disease. The standard 12-lead ECG is composed of six horizontal plane leads called the limb leads, and six horizontal leads, which were also called chest leads [11]. Those leads offer 12 different angles for visualizing the activities of the heart and are named lead I, II, III, aVL, aVF, aVR, V1, V2, V3, V4, V5 and V6, respectively. It is worth noting that ECG complex does not look the same in all the leads of the standard 12-lead system and the shape of the ECG constituent waves may vary depending on the lead.

Atrial tachyarrhythmias are cardiac arrhythmias in normal atrial electrical activation is substituted by continuous activation with multiple wavelets depolarizing the atria simultaneously [4]. AF is one of the most frequent atrial tachyarrhythmias and is characterized by apparently chaotic atrial activation with the irregular and rapid waves, so called fibrillation. Figure 1 shows an example of normal rhythm and AF ECGs.

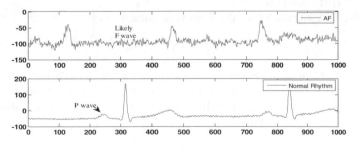

Fig. 1. The examples of Atrial Fibrillation (top) and normal rhythm(bottom) with different atrial activity

In this research, we follow Castells and Rieta's research [7, 12] primarily to process our ECG data but with some different modifications according to our experiment. One of the differences is this study reconstructs the signals before executing Second-Order Blind Identification (SOBI) and another one is we use the Power Spectral Density (PSD) diagram to classify the ECG data belongs to AF or not. The strategy used in this research is according to prior knowledge and is different from the traditional classification approach which training samples are necessary for. The detail flowchart in this research is shown in Figure 2.

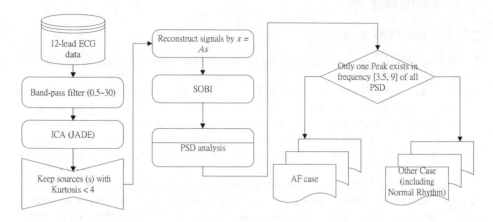

Fig. 2. The overall framework

2.2 Independent Component Analysis (ICA)

Independent component analysis is considered because it has precise and successful work in biomedical engineering. In multichannel signal process such as 12-lead ECG data, ICA especially shows the effective utilization to extract AA and ventricular activity (VA) signal [13].

ICA is a statistical method that seeks nonsingular linear transformations of multivariate data so that the transformed variables have the minimal dependence between each other statistically [14].

The basic ICA approach uses the following linear model:

$$x = As \qquad (1)$$

where s represents d independent sources and A is a nonsingular $d \times d$ linear mixing matrix. The x means the composed result of d observed signals. The goal of ICA is to recover "unknown" original source. In order to get the result, ICA calculates the inversed matrix, $W = A^{-1}$ and the formula can be rewritten as

$$s = A^{-1}x = Wx \qquad (2)$$

General speaking, the key to estimating the ICA model is nongaussianity [15]. A classical measure of nongaussianity is kurtosis. For a signal x, it is classically defined as

$$kurtosis(x) = E\left(x - \mu\right)^4 / \sigma^4 \qquad (3)$$

where μ is the mean of x, σ is the standard deviation of x and E is the expected value. The kurtosis value is three for Gaussian densities and distributions that are more outlier-prone than the normal distribution have kurtosis greater than 3; distributions that are less outlier-prone have kurtosis less than 3. In some literature, definitions of kurtosis subtract 3 from the computed value, so that the normal distribution has kurtosis of 0. This study chooses the formula (3) as our kurtosis calculation.

In ECG data, we can assume that AA and VA are independent and do not present random Gaussian distributions. The assumption makes ICA is possible to reconstruct the AA signals from VA or other, like breath effect and muscular noise in ECG. Before applying ICA on the ECG data, this study uses a band-pass, infinite impulse response (IIR) filter to suppress the noise and then the results of normal rhythm and AF cases are processed by ICA. In general [16] and our experiment, the frequency is range from 0.5 to 30 to keep the appropriate information for further operations. For taking an example, Figure 3 shows a filtered lead I and lead II ECG rhythm in three seconds and compares the result with original one.

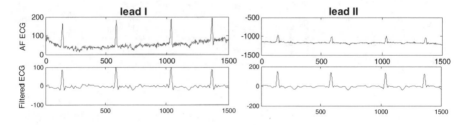

Fig. 3. The result in lead I and lead II after applying an IIR band-pass filter

This study mainly uses the filtered ECG data as input for ICA. In ICA stage, we choose the algorithm named higher order statistics Joint Approximate Diagonalization of Eigen matrices (JADE). ICA can extract all non-Gaussian sources based on the kurtosis measurement. For all the data applied by ICA, the result is resorted by kurtosis value in ascending order. By the experiments, the ventricular sources should show the high kurtosis value while AA is quasi-Gaussian and has smaller kurtosis. This study keeps the source signals with kurtosis value small than four; because of those signals contain more AA information than others. The following Figure 4 shows a result (estimated sources) as an example when JADE is been used and the kurtosis value is taken as a rule for ordering.

As shown in Figure 4, R-wave of ECG data has higher kurtosis value and the rest parts has lower ones, Figure 5 shows the results if we keeps the estimated source with kurtosis value smaller than four and reconstructs the signals and Figure 6 and Figure 7 shows AF and normal rhythm respectively lead II only for detailed comparison. The figures clearly the R-wave and some noise are removed successfully and the AA period are reserved for succeeding processes in next section.

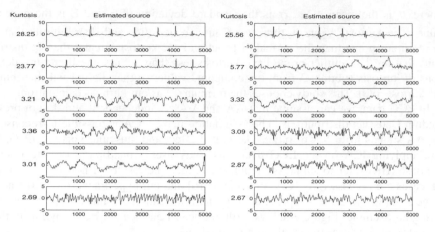

Fig. 4. The estimated source and its kurtosis value

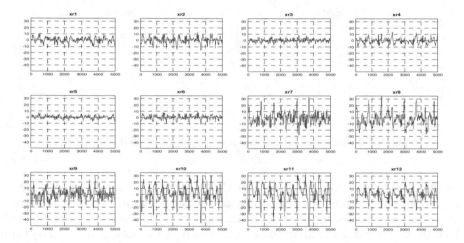

Fig. 5. The reconstructed signals by sources with kurtosis value smaller than 4

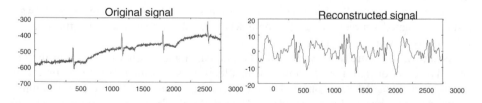

Fig. 6. Original signal (left) and reconstructed signal (right) in lead II (AF)

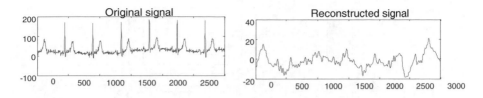

Fig. 7. Original signal (left) and reconstructed signal (right) in lead II (Normal Rhythm)

2.3   Second-Order Blind Identification (SOBI)

The SOBI algorithm is to separating a mixture of uncorrelated sources with different spectral content through a second-order statistical analysis which also takes into consideration the source temporal information [17]. SOBI seeks a transformation that simultaneously diagonalizes several correlation matrices at different lags. SOBI shows a number of attractive features including (1) SOBI only relies on second-order statistics of the received signals. (2) It allows the separation of Gaussian sources that cannot be separated by ICA. (3) The use of several covariance matrices makes the algorithm more robust and unlikely indeterminacies so that this study adopts SOBI to do the second times transformation after applying ICA.

The goal of separating the AA from other sources of interference is equivalent to finding an orthogonal transformation from the observation filtered by ICA. Because AA has a narrowband spectrum, SOBI algorithm is appropriate for estimating the AA.

2.4   Power Spectral Density (PSD) Based Classification

SOBI can separate the mixed uncorrelated sources. An important assumption in this study is that the sources contain vibrated waves in AF cases while normal rhythm should have regular signals. This assumption leads us to use frequency domain technique to classify the cases are AF or not. The procedure is estimated by using Welch's averaged modified periodogram and the window size is 1000. In classification stage, we also only consider the bound of frequency is between 0.5 to 30 Hz because the rest frequencies have a low contribution. The results of AF and normal rhythm after PSD analyses the transformed sources from SOBI. For classification, for all PSD diagrams, if there exists only one peak in one or more PSD diagrams, then the input case will be considered as AF; if there is no peak in any PSD diagrams or a PSD diagram has more than one peak in itself, then the case should be classified as non-AF. The Figure 8 to Figure 11 shows the PSD and fast Fourier transform (FFT) results of AF and normal rhythm respectively.

In Figure 8, because of there exists a diagram with only one peak in frequency 3.5 to 9, hence the case will be regarded as AF, in contrast, the normal rhythm should not have such phenomenon.

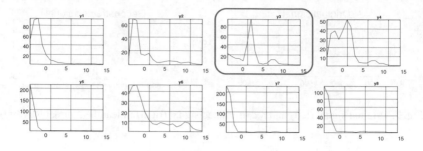
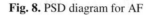

Fig. 8. PSD diagram for AF

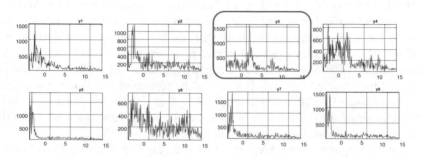

Fig. 9. FFT diagram for AF

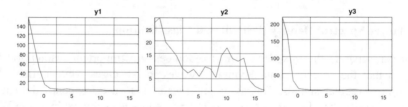

Fig. 10. PSD diagram for Normal Rhythm

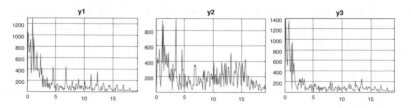

Fig. 11. FFT diagram for Normal Rhythm

3   Experimental Result

The data in this research is collected from Taoyuan Armed Forces General Hospital in Taiwan. All rhythms were diagnosed and verified by a cardiologist, Doctor Yeh.

There are totally 98 12-lead ECG data selected and recorded by Philps TC50. The sampling rate is 500Hz with a 10-second record. According to the result diagnosed by the doctor, there are 25 normal rhythms while the number of AF is 73. For emphasizing the accuracy in our research, we compare our result with the diagnosis from Philps TC50.

The performance for this experiment is measured by three criterions: accuracy, sensibility (SE), giving the ability of true event detection and specificity (SP), and reflecting the probability of a negative test among patients without disease. These performance measures are defined as following [18-19].

$$accuracy = \frac{TP+TN}{TP+FN+FP+TN} \tag{4}$$

$$SE = \frac{TP}{TP+FN} \tag{5}$$

$$SP = \frac{TN}{TN+FP} \tag{6}$$

where TP (True Positive) is the number of matched events and FN (False Negative) is the number of events that are not detected by this approach. FP (False Positive) is the number of events detected by this approach and non-matched to the detector annotations. TN (True Negative) presents as the percentage of events truly identified as not defectives, or normal. In this study, the positive means AF. Table 1 lists the final result.

Table 1. The performance of this study

	Sensibility	Specificity	Accuracy
our approach	68.49%	96.00%	75.51%
Philps TC50	91.78%	20.00%	73.47%

4   Conclusion

According to the literature, it is believed that ICA and SOBI can separate the existed AF signals from AA period and if the ECG is normal rhythm, the PSD morphology should not have any peak information. This study adopts several approaches for different situations. For cancelling the noise simply, the band-pass filter is adopted, for removing VA, a threshold value of four (because of the different definition about kurtosis, the value in Rieta et.al 's research is 1.5, equals to 4.5 in this study) is chosen, as shown in Figure 4, the waveform with the larger value of kurtosis seems contains more VA information. Because of the limitation about Gaussian sources separation in higher order statistics based independent source analysis, the second order statistics based approach can be applied, hence, SOBI is used to separate the Gaussian-like signals from AA. According to the literature and our experiment, the PSD diagram can be used to classify the input case is AF or not because of the peak shall exist in frequency 3.5 to 9 when the AA period contains the vibratory waves.

As shown in Table 1, this study's sensibility is lower than Philps TC's diagnosis result while specificity is much higher than Philps TC50. One main reason in our research is the rules for AF case are very strict; for example, Figure 12 shows an example of the false classification in our rules for AF. For this case, we will decide its result as non-AF because of the approaches in this study cannot successfully separate the AF signals from AA period and original ECG data. These cases occur in our data frequently and lower our sensibility value.

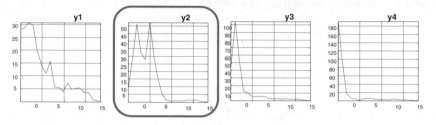

Fig. 12. An example for fault classification

In the future, we will focus on two directions, one is the problem mentioned above, the possible solution including adopting high-pass filter to reduce the lower frequency's contribution, or applying the third signal separation approaches, like wavelet decomposition or Hilbert-Huang algorithm. Another one is how to classify AF or not by some two-class classifier, for instance, maybe Hidden Markov Models used in our previous research [20] can be used to extract the features in the reconstructed ECG data. These two directions should be increase the accuracy in the future research.

References

1. Lin, Y.J., Liu, Y.B., Chu, C.C.: Incremental changes in QRS duration predict mortality in patients with atrial fibrillation. PACE - Pacing and Clinical Electrophysiology 32, 1388–1394 (2009)
2. Hunt, S.A., Baker, D.W., Chin, M.H., Cinquegrani, M.P., Feldman, A.M., Francis, G.S., Ganiats, T.G., Goldstein, S., Gregoratos, G., Jessup, M.L., Noble, R.J., Packer, M., Silver, M.A., Stevenson, L.W., Gibbons, R.J., Antman, E.M., Alpert, J.S., Faxon, D.P., Fuster, V., Jacobs, A.K., Hiratzka, L.F., Russell, R.O., Smith Jr., S.C.: ACC/AHA guidelines for the evaluation and management of chronic heart failure in the adult: Executive summary. A report of the American College of Cardiology/American Heart Association Task Force on Practice Guidelines (Committee to Revise the 1995 Guidelines for the Evaluation and Management of Heart Failure). Journal of the American College of Cardiology 38, 2101–2113 (2001)
3. Alcaraz, R., Rieta, J.J.: A review on sample entropy applications for the non-invasive analysis of atrial fibrillation electrocardiograms. Biomedical Signal Processing and Control (2009)

4. Fuster, V., Rydén, L.E., Cannom, D.S., Crijns, H.J., Curtis, A.B., Ellenbogen, K.A., Halperin, J.L., Le Heuzey, J.Y., Kay, G.N., Lowe, J.E., Olsson, S.B., Prystowsky, E.N., Tamargo, J.L., Wann, S., Smith Jr., S.C., Jacobs, A.K., Adams, C.D., Anderson, J.L., Antman, E.M., Hunt, S.A., Nishimura, R., Ornato, J.P., Page, R.L., Riegel, B., Priori, S.G., Blanc, J.J., Budaj, A., Camm, A.J., Dean, V., Deckers, J.W., Despres, C., Dickstein, K., Lekakis, J., McGregor, K., Metra, M., Morais, J., Osterspey, A., Zamorano, J.L.: ACC/AHA/ESC 2006 guidelines for the management of patients with atrial fibrillation: A report of the American College of Cardiology/American Heart Association Task Force on practice guidelines and the European Society of Cardiology Committee for practice guidelines (Writing committee to revise the 2001 guidelines for the management of patients with atrial fibrillation). Circulation 114, e257–e354 (2006)
5. Stridh, M., Sörnmo, L.: Spatiotemporal QRST cancellation techniques for analysis of atrial fibrillation. IEEE Transactions on Biomedical Engineering 48, 105–111 (2001)
6. Hamilton, P.S., Tompkins, W.J.: Compression of the ambulatory ECG by average beat subtraction and residual differencing. IEEE Transactions on Biomedical Engineering 38, 253–259 (1991)
7. Castells, F., Rieta, J.J., Millet, J., Zarzoso, V.: Spatiotemporal blind source separation approach to atrial activity estimation in atrial tachyarrhythmias. IEEE Transactions on Biomedical Engineering 52, 258–267 (2005)
8. Huang, Z., Chen, Y., Pan, M.: Time-frequency characterization of atrial fibrillation from surface ECG based on Hilbert-Huang transform. Journal of Medical Engineering and Technology 31, 381–389 (2007)
9. Addison, P.S., Watson, J.N., Clegg, G.R., Steen, P.A., Robertson, C.E.: Finding coordinated atrial activity during ventricular fibrillation using wavelet decomposition. IEEE Engineering in Medicine and Biology Magazine 21, 58–61+65 (2002)
10. Langley, P., Bourke, J.P., Murray, A.: Frequency analysis of atrial fibrillation. In: Computers in Cardiology, pp. 65–68 (2000)
11. Garcia, T.B., Holtz, N.E.: Introduction to 12-Lead ECG: The Art of Interpretation. Jones and Bartlett Publishers (2001)
12. Rieta, J.J., Castells, F., Sánchez, C., Zarzoso, V., Millet, J.: Atrial activity extraction for atrial fibrillation analysis using blind source separation. IEEE Transactions on Biomedical Engineering 51, 1176–1186 (2004)
13. Rieta, J.J., Zarzoso, V., Millet-Roig, J., García-Civera, R., Ruiz-Granell, R.: Atrial activity extraction based on blind source separation as an alternative to QRST cancellation for atrial fibrillation analysis. In: Computers in Cardiology, pp. 69–72 (2000)
14. Comon, P.: Independent component analysis, A new concept? Signal Processing 36, 287–314 (1994)
15. Hyvärinen, A., Oja, E.: Independent component analysis: Algorithms and applications. Neural Networks 13, 411–430 (2000)
16. Thakor, N.V., Zhu, Y.S.: Applications of adaptive filtering to ECG analysis: Noise cancellation and arrhythmia detection. IEEE Transactions on Biomedical Engineering 38, 785–794 (1991)
17. Belouchrani, A., Abed-Meraim, K., Cardoso, J.F., Moulines, E.: A blind source separation technique using second-order statistics. IEEE Transactions on Signal Processing 45, 434–444 (1997)
18. Jager, F., Moody, G.B., Taddei, A., Mark, R.G.: Performance measures for algorithms to detect transient ischemic ST segment changes. In: Computers in Cardiology, pp. 369–372 (1992)
19. Kara, S., Okandan, M.: Atrial fibrillation classification with artificial neural networks. Pattern Recognition 40, 2967–2973 (2007)
20. Chang, P.C., Hsieh, J.C., Lin, J.J., Chou, Y.H., Liu, C.H.: Hybrid System with Hidden Markov Models and Gaussian Mixture Models for Myocardial Infarction Classification with 12-Lead ECGs. In: 11th IEEE International Conference on High Performance Computing and Communications, HPCC 2009, Seoul, pp. 110–116 (2009)

State-of-the-Art of Computer-Aided Detection/Diagnosis (CAD)

Hiroshi Fujita[1], Jane You[2], Qin Li[2], Hidetaka Arimura[3],
Rie Tanaka[4], Shigeru Sanada[4], Noboru Niki[5], Gobert Lee[6], Takeshi Hara[1],
Daisuke Fukuoka[1], Chisako Muramatsu[1], Tetsuro Katafuchi[7], Gen Iinuma[8],
Mototaka Miyake[9], Yasuaki Arai[8], and Noriyuki Moriyama[8]

[1] Gifu University, Japan
[2] The Hong Kong Polytechnic University, Hong Kong
[3] Kyusyu University, Japan
[4] Kanazawa University, Japan
[5] Tokushima University, Japan
[6] Flinders University, Australia
[7] Gifu University of Medical Sciences, Japan
[8] National Cancer Centre Hospital, Japan
[9] Kobe University, Japan

Abstract. This paper summarizes the presentations given in the special ICMB2010 session on state-of-the-art of computer-aided detection/diagnosis (CAD). The topics are concerned with the latest development of technologies and applications in CAD, which include brain MR images, fundus photographs, dynamic chest radiography, chest CT images, whole breast ultrasonography, CT colonography and torso FDG-PET scans.

Keywords: Computer-aided detection/diagnosis (CAD), medical imaging, image processing and analysis.

1 Introduction

With the rapid advances in computing and electronic imaging technology, there has been increasing interest in developing computer aided detection/diagnosis (CAD) systems to improve the medical service. CAD is emerging as an advanced interdisciplinary technology which combines fundamental elements of different areas such as digital image processing, image analysis, pattern recognition, medical information processing and management. Although the current CAD systems cannot fully replace human doctors for medical detection/diagnosis in clinical practice, the analytical results will assist doctors to speed up screening large populations for abnormal cases, generate medical records for interdisciplinary interaction on the relevant aspects for proper treatment and facilitate evaluation of treatment for clinical study.

In general, a CAD system consists of four major components: 1) image acquisition and data preprocessing for noise reduction and removal of artifacts; 2) image feature

D. Zhang and M. Sonka (Eds.): ICMB 2010, LNCS 6165, pp. 296–305, 2010.

extraction and representation; 3) detection of region of interest (ROI) by image analysis based on segmentation and matching; 4) evaluation and classification by an appropriate decision-making scheme. The conventional CAD systems are focused on the applications to diagnosis of breast cancer, lung cancer, colon cancer, prostate cancer, bone metastases and coronary artery disease based on the analysis of X-ray or CT images. The extension of the existing medical imaging technologies and their applications were presented in the special session of CAD at ICMB2010 as reported in this short paper. Section 2 highlights these topics which include brain MR images, fundus photographs, dynamic chest radiography, chest CT images, whole breast ultrasonography, CT colonography and torso FDG-PET scans.

2 State-of-the-Art CAD Topics

This section briefly describes the recent research projects and their results on advanced CAD by 7 teams from Japan, Hong Kong and Australia respectively.

2.1 Brain MR Images

The brain check-up which is also referred to as "Brain Dock" is getting popular in Japan because of increasing average longevity. The aim of the Brain Dock is to detect or classify asymptomatic brain diseases in their early stages, e.g., asymptomatic lacunar infarction, unruptured intracranial aneurysms, early dementia, and to prevent such brain disorders. In particular, dementing disorders such as Alzheimer's disease (AD) and vascular dementia are major public health problems in countries with greater longevity, such as in Japan. The process applies the advanced imaging techniques like magnetic resonance imaging (MRI) and positron emission computed tomography (PET) for preventing various neurological conditions including stroke, dementia, etc.

The computer-aided diagnosis (CAD) systems for brain MR images play important roles in the Brain Dock, because it would be laborious for neuroradiologists to detect the lesions among a great number of images or a small number of patients out of many healthy people [1]. Moreover, it could be possible to miss the lesions of patients, because of their heavy workloads. In addition, the number of images, which neuroradiologists have to diagnose, has increased rapidly, because MRI has shifted from two-dimensional (2D) imaging to 3D imaging, and the resolution as well as signal-to-noise ratio has become higher. In neuroradiological field, the CAD systems are needed for not only the Brain Dock, but also the other brain diseases such as multiple sclerosis (MS). Therefore, in recent years, various types of CAD methods have been developed by many researchers including the author's group in the neuroradiology using brain MR images.

Radiologists expect that CAD systems can improve their diagnostic abilities based on synergistic effects between the computer's and radiologist's high abilities by using the information analysis including the medical images. In this presentation, the author described examples of CAD approaches in brain MR images, including detection of abnormalities, e.g., aneurysms in MRA images [2], lacunar infarction [3], Alzheimer's disease [4], white matter hyperintensities [5], MS lesions [6], and concluded with possibilities in the future of the CAD systems for brain diseases in MR images.

2.2 Fundus Photographs

Images of the ocular fundus, also known as images of the retina, can tell us about retinal, ophthalmic, and even systemic diseases such as diabetes, hypertension, and arteriosclerosis. As a non-intrusive method to screen retinopathy, the colour retinal images captured by digital fundus cameras have been widely used in clinical practice. A fully automated segmentation of colour retinal images can greatly help the management of certain diseases, especially the disease which requires the screen of large populations such as diabetic retinopathy.

This presentation consists of two fundamental parts of our work. The first part is on the segmentation of blood vessels. The appearance of blood vessels is a critical feature for diagnosis. Automated segmentation of blood vessels in retinal images can help eye-care specialists screen larger populations for vessel abnormalities which caused by retinal diseases or systemic diseases. The segmentation of retinal vessels is attracting lots of researchers but the effects are still not very satisfied. In this thesis, the major difficulties in vessel segmentation are analysed and two novel vessel segmentation methods are proposed. The second part of this presentation describes the proposed system to segment main regions and lesions of colour retinal images obtained from patients with diabetic retinopathy (DR). There are thousands of retinopathies affecting human being's vision. In this presentation, we focus our research area on DR which affects very large populations. It is one of the most common causes of blindness. Diagnosing DR as early as possible is critical to protecting patients' vision. Our segmentation system is designed to segment the main regions of retina and the major lesions caused by DR. This system is helpful to screen diabetic retinopathy in large populations so that DR can be found earlier.

Retinal vessel segmentation: Many works have been done on vessel segmentation these years [7]. However, automated retinal segmentation is difficult due to the fact that the width of retinal vessels can vary from very large to very small, and that the local intensity contrast of vessels can be weak and unstable. It is still very hard to detect the vessels of variant widths simultaneously and the weak, small vessels effectively. In this presentation, we will present a simple but efficient multiscale scheme to overcome this difficulty by using Multiscale Production of the Matched Filter (MPMF) responses as the multiscale data fusion strategy [8]. Considering that the vessel structures will have relatively strong responses to the matched filters at different scales but the background noises will not, multiplying the responses of matched filters at several scales will further enhance the vessels while suppressing the noise. The vessels of variant widths can be detected concurrently because the MPMF can incorporate the multiscale information. And the MPMF can detect the small and weak vessels which can not be detected by other methods because the weak vessels could be better enhanced (while the noise being significantly suppressed) in the scale production domain. Another difficulty of vessel segmentation is from the affection of lesions. For example, if we need to find the dark lines in an image, the edges of bright blobs will be the major source of false line detection. Consequently, some blobs (bright lesions and the optic disk) in the retinal image may cause false detection of vessels. In this thesis, we propose a modified matched filter to suppress the false detection caused by bright blobs. Instead of subtracting the local mean from the response for removing background and then thresholding to detect vessels as in the traditional matched filter, we first enhance the retinal image by using Gaussian filters and then analyze the local structures of filtering outputs by a double-side thresholding operation. The proposed modified matched filter could avoid responding to non-line edges so that false detection of vessels will be reduced significantly.

DR image segmentation: The objects useful for DR diagnosis include retinal lesions such as red lesions (intraretinal haemorrhages, microaneurysms), bright lesions (hard exudates and cottonwool spots) and retinal main regions such as vessels, optic disk, and fovea. Colour retinal image segmentation to assist DR diagnosis has attracted many researchers these years. But few works have been designed on extracting all above objects in one efficient scheme. The major disadvantages of current colour retinal image segmentation works are (1) the affections among different objects are insufficiently considered so that the false positives of segmentation are relative high; (2) the algorithms are too time-consuming so that the online application is impossible. In this presentation, we propose one efficient scheme to segment all useful objects for DR diagnosis. With sufficiently considering the affections among different objects, our segmentation scheme organized one efficient working flow to segment all objects. This scheme can suppress false positives effectively and improve segmentation speed. And the segmentation speed is further improved by algorithm optimizing and keeping the algorithm as simple as possible.

2.3 Dynamic Chest Radiography

The Background: X-ray translucency of the lungs changes depending on contented air and blood volume. The changes can be observed as changes in pixel value on dynamic chest radiographs obtained with a flat-panel detector (FPD). Thus, the decreased ventilation is observed on dynamic chest radiographs as small changes in pixel values. However, these changes are very slight making their detection by the naked eye difficult, and their correct interpretation requires specialized knowledge of respiratory physiology. Therefore, we have developed a computer-aided diagnosis (CAD) system for dynamic chest radiographs, which quantifies pulmonary ventilation and circulation based on slight changes in pixel value, resulting from respiration and heartbeat. In this presentation, we demonstrate imaging procedures (Exposure conditions, patient positioning, breathing method), quantifiable features from dynamic chest radiographs, image analysis procedures, relationship between the quantified parameters and pulmonary physiology, display methods, and the results in the primary clinical study.

Methods: Dynamic chest radiographs of patients were obtained during respiration using an FPD system. Sixty images were obtained in 8 seconds on exposure conditions as follows: 110 kV, 80 mA, 6.3 ms, SID 1.5 m, 7.5 fps, 2 mm Al filter. Total exposure dose was approximately 1.2 mGy, which was less than those in lateral chest radiography determined as guidance level of the international atomic energy agency (IAEA) (1.5 mGy). The lung area was determined by edge detection using the first derivative technique and iterative contour-smoothing algorithm. For use as an index of respiratory phase, the distance from the lung apex to the diaphragm was measured throughout all frames and then calculated excursion of diaphragmatic movement. Respiratory changes in pixel value in each local area were measured tracking the same area. To visualize changes in pixel values, inter-frame differences were obtained and then mapped on an original image as "Ventilation mapping images" and "Perfusion mapping images". Abnormalities were detected as a deviation from the right–left symmetry of the resulting images. The results were compared with findings in CT, PFT, and pulmonary scintigraphic test.

Results: In normal controls, ventilation and perfusion mapping images showed a normal pattern determined by the pulmonary physiology, with a symmetric distribution and no perfusion defects throughout the entire lung region. Left-right correlation was observed (r=0.66±0.05). In patients, abnormalities were indicated as a reduction of changes in pixel values on the mapping image. In addition, in many abnormal cases, the mapping images lost its symmetry and abnormalities could be detected as a deviation from the right–left symmetry of respiratory changes in pixel value. There was good correlation between the distribution of changes in pixel value and those of radioactivity counts in scintigraphic test (r=0.6±0.2).

Conclusions: Dynamic chest radiography using a FPD combined with a real-time computer analysis is expected to be a new examination method for evaluating pulmonary function as an additional examination in daily chest radiography. Our system can thus aid radiologist to interpret kinetic information on dynamic chest radiographs. In addition, the present method has potential for application to cone beam CT or 4DCT, as an optional function.

2.4 Chest CT Images

CAD has become one of the mainstream techniques in the routine clinical work for detection and different diagnosis of abnormalities in various examinations by using multimodality images. Prototype CAD systems demonstrated by academic institutions and commercial systems are becoming available. However, there are technical problems that prevent CAD from becoming a truly useful tool to assist physicians to make final decisions in the routine clinical work. We will accelerate the innovations in practical CAD systems by blending the outcomes of diverse, rapidly developing basic research in lung microstructural analysis and respiratory lung dynamics, and computational anatomy, and by prompting a smooth bridge for their application on the frontline of medical care. The main target diseases are lung cancers, chronic obstructive pulmonary disease (COPD), and osteoporosis.

The following main technical challenges to be overcome were discussed throughout the presentation:

(a) Technology development for integrating digital diagnostic environment with a large amount of database combination of multimodal radiological images with diagnostic results,

(b) Establishment of computational anatomy based on quantitative analysis of the database,
(c) Development of basic technology to practical CAD system for lung cancers, COPD and osteoporosis,
(d) Development of technology to integrate CAD into the clinical work flow and large-scale, multicenter, prospective clinical trials to prove that CAD system can improve the diagnostic performance substantially.

2.5 Whole Breast Ultrasonography

Breast cancer is one of the leading causes of cancer death in women. The lifetime breast cancer risk for women is about 1 in 9 in Western countries such as the United States, the United Kingdom and Australia, whereas that for Japan and some other Asian countries is significantly lower, but on the rise. Early detection is the key to better prognosis with simpler treatment and lower mortality rate. As we know, mammography is the routine technique used in breast cancer screening. The technique is very effective in screening population of age 50 years or above. However, it is less sensitivity in detecting breast cancer in younger women or women with dense breasts. With the increase in breast cancer incident rate in younger women and in countries where women generally have dense breast, a more suitable modality is needed to provide breast health care for these groups of women.

Breast ultrasonography has a long history in detecting and diagnosing breast diseases. It is widely considered as a valuable adjunct to mammography. In the past, breast ultrasound was limited by poor image quality, low image resolution and limited detail recognition. Recent years have seen many advances in ultrasound technology. Image quality has improved remarkably and the modality has been greatly enhanced. On the issue of detecting breast cancer in dense breasts, breast ultrasound does not have the same limitation as mammography, hence, more suitable to be employed in breast cancer screening in Asian countries, like Japan, where a significant proportion of the screening population has dense breast and in clinics that provide breast health care for younger women.

Currently, breast examination is routinely performed by medical staff who has special training in ultrasonography. During an examination, a small hand-held probe about 4 cm in size is used and the ultraonographer/ ultrasonologist runs the probe over the entire breast or pre-identified regions. The technique can provide very valuable information in the hands of experienced examiners but is generally time-consuming. Results are operator independent and reproducibility is poor. Furthermore, the acquired 2-dimensional scan images contain limited views of sections of the breast. Location and orientation of the scan images are only loosely recorded. Due to the lack of precise location and orientation information, image registration is difficult, if not impossible, hindering longitudinal studies and multi-modalities diagnosis.

In view of the above, an automated breast scanning system that can acquire data of the whole breast systematically (with exact information of location and orientation of the image) and present it in a panoramic view will be of great advantage.

In this presentation, a novel whole breast ultrasound system for auto-acquisition of volumetric breast ultrasound data will be introduced[9] . The main features of the system includes a prototype scanner for automated acquisition of ultrasound images of the whole breast and a image visualization and computer-aided-detection and diagnosis (CAD) platform developed in the Department of Intelligent Image Information, Graduate School of Medicine, Gifu University, Japan [9-14]. The volumetric ultrasound data consists of a stack of two-dimensional images, each depicting a panoramic view of a cross-section of the whole breast. Using the purpose-built visualisation platform, the whole breast data can be displayed in a number of ways such as visualization in axial, saggital and coronal views and synchronized bilateral display to facilitate radiologists' interpretation. The specialized CAD algorithm was purposely developed to handle and analyse ultrasound images acquired with the prototype scanner. The system was able to detect and diagnose suspicious lesions in the whole breast ultrasound data.

2.6 CT Colonography

Computer-aided detection (CAD) in CT colonography (CTC) provides the automatic detection of lesions protruding into the lumen by digitally perceiving the lumen surface in the colorectal tracts[15]. The potential of CAD has been actively investigated in the West countries since the era of single-slice CT, and the utility of CAD systems for detecting colorectal polyps has achieved a clinically applicable level. However, the problem is that the target of CAD for CTC in the West is colorectal polyps because of the respect for the adenoma-carcinoma sequence[16].

Recent recognition of the importance of flat lesions based on development of colonoscopic examination will demand CAD in CTC that can reliably detect the lesions in Japan[17]. Our Center is presently engaged in formal joint research with London University through a British corporation, Medicsight PLC, to develop CAD for CTC. This research focuses specifically on clarifying the characteristics of flat lesions on CTC and developing detection algorithms. We evaluated the current CAD performance regarding 92 early-stage cancers with submucosal (SM) invasion. Colorectal lesions destroying muscularis mucosae and invading the SM layer are more dangerous than intramucosal adenoma or cancer, because these invasive lesions are considered likely to develop into advanced carcinomas. Therefore, we believe that SM cancers are clearly more important than colonic adenomas as a target for early diagnosis.

It is essential to clarify the characteristics of the CTC images and reliably detect them in CTC examinations. Eighty of 92 lesions (86.9%) were detected as well as 100% of protruded lesions and nearly 80% of flat lesions, which was a favorable result beyond our expectation. CAD succeeded in detecting flat lesions that are also challenging for colonoscopy, which suggests the great potential for CTC diagnosis with CAD. To detect a flat lesion, CAD captures some small focal elevations or small nodules in the flat lesion. We are further planning to develop a high-precision CAD algorithm that is applicable to flat lesions to be used clinically for colorectal CTC screening in Japan [18].

2.7 Torso FDG-PET Scans

Objectives: The definition of SUV normality derived from many normal cases will be a good atlas in interpreting FDG-PET scans when suspicious regions with high uptake would appear in high uptake background. The purpose of this study was to develop a new method for determination of SUV of FDG-PET scans to examine the normality by use of a score that is based on the range of SUV on normal cases [19].

Methods: We retrospectively collected normal FDG-PET cases of 143 male and 100 female to determine the normal range of SUV. After all of normal cases in each gender were registered into one model, we configured a normal model which indicates the normal range of SUV voxel by voxel with statistical information of the mean (M) and the standard deviation (SD) that estimate the confidence interval of SUV. The normal models in each gender were stored in iPhone/iPod touch to convert patients' SUV of suspicious regions into abnormal scores defined as deviations from normal SUV where the regions were existed when readers indicate locations on PC and iPhone/iPod touch screen with patients' SUV.

Results: Four hundred thirty two abnormal regions from cancer cases were extracted to measure the SUVs and the abnormal scores. The SUVs in 49 of 432 regions were less than 2.0, but the scores of them were larger than 2.0. Although the SUVmax in 299 of 432 regions were less than 5.0, the scores were larger than 2.0 in 285 of 299 regions.

Conclusions: We have developed a computerized scheme on PC and iPhone/iPod Touch to estimate the abnormalities by the score of SUV using the normal model. Our computerized scheme would be useful for visualization and detection of subtle lesions on FDG-PET scans even when the SUV may not show an abnormality clearly.

3 Conclusions

The presentations given in the special session of CAD/ICMB2010 covered the latest research and technology development in multi-disciplinary areas, which will have long-term significant impact and immediate beneficial to the community on the aspects of medicare, telemedicine and multimedia information processing. These applications will not only increase efficiency and productivity in the business environment, but also enhance the health service for the public.

Acknowledgments

The research team from the Hong Kong Polytechnic University would like to thank the research support from Hong Kong Government under its General Research Fund (GRF) and Innovation Technology Fund schemes for the computer-aided medical diagnosis and monitoring projects. The work presented by the Japanese teams is supported partly by a grant from Japan Government Grant-in-Aid for Scientific Research (C) 19500385.

References

1. Arimura, H., Magome, T., Yamashita, Y., et al.: Computer-aided diagnosis systems for brain diseases in magnetic resonance images. Algorithms 2(3), 925–952 (2009)
2. Arimura, H., Li, Q., Korogi, Y., et al.: Computerized detection of intracranial aneurysms for 3D MR angiography: Feature extraction of small protrusions based on a shape-based difference image technique. Medical Physics 33(2), 394–401 (2006)
3. Uchiyama, Y., Yokoyama, R., Ando, H., et al.: Computer-aided diagnosis scheme for detection of lacunar infarcts on MR images. Academic Radiology 14, 1554–1561 (2007)
4. Arimura, H., Yoshiura, Kumazawa, S., et al.: Automated method for identification of patients with Alzheimer's disease based on three-dimensional MR images. Academic Radiology 15(3), 274–284 (2008)
5. Kawata, Y., Arimura, H., Yamashita, Y., et al.: Computer-aided evaluation method of white matter hyperintensities related to subcortical vascular dementia based on magnetic resonance imaging. Computerized Medical Imaging and Graphics (in press, 2010)
6. Yamamoto, D., Arimura, H., Kakeda, S., et al.: Computer-aided detection of multiple sclerosis lesions in brain magnetic resonance images: False positive reduction scheme consisted of rule-based, level set method, and support vector machine. Computerized Medical Imaging and Graphics (in press, 2010)
7. Hoover, A., Kouznetsova, V., Goldbaum, M.: Locating blood vessels in retinal images by piecewise threshold probing of a matched filter response. IEEE Trans. Med. Imag. 19(3), 203–210 (2000)
8. Zhang, L., Li, Q., You, J., Zhang, D.: Modified matched filter with double-side thresholding and its application to prolifetive diabetic retinopathy screening. IEEE Trans. on Information Technology in Biomedicine 13(4), 528–534 (2009)
9. Takada, E., Ikedo, Y., Fukuoka, D., Hara, T., Fujita, H., Endo, T., Morita, T.: Semi-Automatic Ultrasonic Full-Breast Scanner and Computer-Assisted Detection System for Breast Cancer Mass Screening. In: Proc. of SPIE Medical Imaging 2007: Ultrasonic Imaging and Signal Processing, vol. 6513, pp. 651310-1–651310-8. SPIE, Bellingham (2007)
10. Lee, G.N., Morita, T., Fukuoka, D., Ikedo, Y., Hara, T., Fujita, H., Takada, E., Endo, T.: Differentiation of Mass Lesions in Whole Breast Ultrasound Images: Volumetric Analysis. In: Radiological Society of North America. In: 94th Scientific Assembly and Annual Meeting Program, p. 456 (2008)
11. Lee, G.N., Fukuoka, D., Ikedo, Y., Hara, T., Fujita, H., Takada, E., Endo, T., Morita, T.: Classification of Benign and Malignant Masses in Ultrasound Breast Image Based on Geometric and Echo Features. In: Krupinski, E.A. (ed.) IWDM 2008. LNCS, vol. 5116, pp. 433–440. Springer, Heidelberg (2008)
12. Fukuoka, D., Hara, T., Fujita, H., Endo, T., Kato, Y.: Automated Detection and Classification of Masses on Breast Ultrsonograms and its 3D Imaging Technique. In: Yaffe, M.J. (ed.) IWDM 2000: 5th International Workshop on Digital Mammography, pp. 182–188. Medical Physics, Madison (2001)
13. Fukuoka, D., Hara, T., Fujita, H.: Detection, Characterization, and Visualization of Breast Cancer Using 3D Ultrasound Images. In: Suri, J.S., Rangayyan, R.M. (eds.) Recent Advances in Breast Imaging, Mammography, and Computer-Aided Diagnosis of Breast Cancer, pp. 557–567. SPIE, Bellingham (2006)
14. Ikedo, Y., Fukuoka, D., Hara, T., Fujita, H., Takada, E., Endo, T., Morita, T.: Development of a Fully Automatic Scheme for Detection of Masses in Whole Breast Ultrasound Images. Med. Phys. 34(11), 4378–4388 (2007)

15. Summers, R.M., et al.: Automated polyp detection at CT colonography: Feasibility assessment in the human population. Radiology 219, 51–59 (2001)
16. Morson, B.C.: The polyp-cancer sequence in the large bowel. Proc. R. Soc. Med. 67, 451–457 (1974)
17. Taylor, S.A., et al.: CT colonography: computer-aided detection of morphologically flat T1 colonic carcinoma. European Radiology 18, 1666–1673 (2008)
18. Iinuma, G., et al.: The Challenge: Detection of Early-Stage Superficial Colorectal lesions. Virtual Colonoscopy - A Practical Guid
19. Hara, T., Kobayashi, T., Kawai, K., Zhou, X., Ito, S., Katafuchi, T., Fujita, H.: Automated scoring system of standard uptake value for torso FDG-PET. In: Proc. of SPIE Medical Imaging 2008: Computer-aided diagnosis, vol. 6915, pp. 691534-1 – 691534-4 (2008)

Advances in Detecting Parkinson's Disease

Pei-Fang Guo[1], Prabir Bhattacharya[2], and Nawwaf Kharma[1]

[1] Electrical & Computer Engineering, Concordia University, 1455 de Maisonneuve Blvd.,
Montreal, QC H3G 1M8, Canada
pf_guo@ece.concordia.ca, kharma@ece.concordia.ca
[2] Fellow, IEEE, Computer Science Department, University of Cincinnati, 814 Rhodes Hall,
Cincinnati, OH 45221-0030, USA
bhattapr@ucmail.uc.edu

Abstract. Diagnosing disordered subjects is of considerable importance in medical biometrics. In this study, aimed to provide medical decision boundaries for detecting Parkinson's disease (PD), we combine genetic programming and the expectation maximization algorithm (GP-EM) to create learning feature functions on the basis of ordinary feature data (features of voice). Via EM, the transformed data are modeled as a Gaussians mixture, so that the learning processes with GP are evolved to fit the data into the modular structure, thus enabling the efficient observation of class boundaries to separate healthy subjects from those with PD. The experimental results show that the proposed biometric detector is comparable to other medical decision algorithms existing in the literature and demonstrates the effectiveness and computational efficiency of the mechanism.

Keywords: Feature creation, medical biometrics, genetic programming, the expectation maximization algorithm, Parkinson's disease, medical decision system.

1 Introduction

Symptoms of Parkinson's disease (PD) include muscle rigidity, tremors, and change in speech and gait. Parkinson's causes are unknown but genetics, aging, and toxins are being researched. There is no cure for PD and the prognosis depends on the patient's age and symptoms [1].

Research has shown that the typical Parkinsonian movement disorders (i.e. tremor at rest, rigidity, akynesia and postural instability) considerably reduce when medication is available offering clinical intervention to alleviate symptoms at the onset of the illness [2]. To this aim, studies in medical biometrics on detecting PD in the early stage are under way and have drawn a lot of attention from the biometrics community in recent years [3], [4].

D. Zhang and M. Sonka (Eds.): ICMB 2010, LNCS 6165, pp. 306–314, 2010.

Feature creation refers to the study of creating new feature functions by projecting raw data from ordinary feature space onto a lower dimensional feature space. The purpose is to employ fewer features to represent data without a decline in the discriminative capability [5] [6]. In this paper, a methodology of combining genetic programming and the expectation maximization algorithm (GP-EM) is proposed to create feature functions, based on ordinary features of voice, to address the problem of diagnosing PD. The objectives are:

- to strengthen the detection accuracy by creating the functions of ordinary feature vectors.
- to increase the reliability of medical decision systems by reducing feature dimensionality.

The rest of the paper is as follows: Section 2 presents the general steps of GP; the descriptions of the GP-EM method for the feature function creation are given in Section 3; the experimental results are reported in Section 4 and conclusion is given in Section 5.

2   The Basic Steps of GP

GP initially creates parameters (controlling a run) and a population of computer programs (candidate solutions to a problem) at random. A precondition for solving a problem with GP is that the function and terminal sets must meet the requirement for expressing solutions to the problem. There are two major steps in a run:

(1) the fitness measure represents the degrees of the fitness associated with each computer program;
(2) the new population of offspring is produced to replace the old one after applying genetic operations, such as reproduction, crossover and mutation. This process is repeated until a termination criterion (which is chosen according to a problem domain) is performed to terminate a run.

See further details of GP in [7].

3   Designs of the Biometric GP-EM Detector

On the basis of input ordinary features of voice, the biometric GP-EM detector, as illustrated in Fig.1, is designed to create feature functions that act as medical decision boundaries to discriminate healthy people from those with PD. With GP, the detector is able to search for the space of all possible computer programs (trees), which are composed of functions and terminal arguments. On the other hand, the EM algorithm is used to model a set of the transformed data under the Gaussian mixture, which simplifies the learning task of creating feature functions, resulting in the optimal hypothesis of the k-means problem. Table 1 summarizes the basic GP-EM procedures, and further details are described in the following subsections.

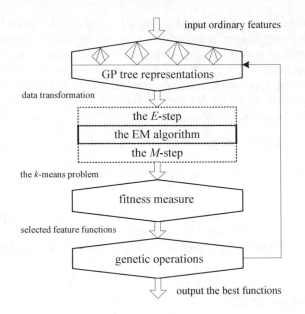

input ordinary features

GP tree representations

data transformation

the *E*-step

the EM algorithm

the *M*-step

the *k*-means problem

fitness measure

selected feature functions

genetic operations

output the best functions

Fig. 1. The process of the biometric GP-EM detector

Table 1. The basic GP-EM procedures

No	Procedures
1	initialize parameters.
2	perform the data transformation from the ordinary feature space to the space of created functions via the GP tree representations, which are composed of the function and terminal sets.
3	complete the EM cycle, the *E*-step and the *M*-step, to search for the optimal hypothesis of the *k*-means problem for the transformed data.
4	evaluate each of the created functions (presented by GP programs) using the fitness measure.
5	vary and create new functions via reproduction, crossover and mutation.

3.1   Created Function Representation via the GP Trees

Computer programs can easily be expressed in the form of parse trees. Each of them can be graphically depicted as a rooted point with internal nodes and external points. In this study, we choose the following mathematical elementary operators to constitute the function set:

$\emptyset$ = [+, -, ×, ÷, *square root, sine, cosine, tan, exponential, absolute, square, negative*];

the terminal set is designed to receive a set of d ordinary features, $\mathbf{F} = \{f\mathrm{d}\}$, from the ordinary feature data environment. An example of the GP parse tree for the rational expression of $\mathbf{Pa} = f3 \times \sin{(f11 + f2)}$ is shown in Fig. 2. The internal nodes of the tree correspond to functions (i.e., mathematical operators '+', '$\times$' and 'sin') and the external points correspond to terminal arguments (i.e., input ordinary features $f2$, $f3$ and $f11$).

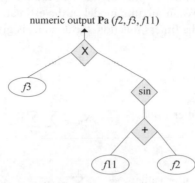

Fig. 2. The parse tree for the rational expression, $\mathbf{Pa} = f3 \times \sin{(f11 + f2)}$

3.2 Searching for Optimal Means of the k Gaussians via EM

In this section, we begin with the input data for the EM algorithm, $\mathbf{C} = \{c_i\}^M{}_{i=1}$ with the k known classes, which are transformed from the ordinary feature space to the space of the created feature functions via GP tree representations. Assuming that the data $\mathbf{C}$ are modeled as a Gaussian mixture in the estimation of the parameters $\{\pi_j, \mu_j, \Sigma_j\}^k{}_{j=1}$. For simplicity, we assume that each of the k Gaussian distributions has the same variance σ^2, and the equal prior probability of $1/k$ [8]. Now, consider fitting this mixture by applying the maximum likelihood approach to the data $\mathbf{C}$ and let us derive the k-means problem via the EM algorithm to estimate the mean values $\{\mu_j\}^k{}_{j=1}$, with the following steps [9]:

$$E - step : E[z_{ij}] = \frac{\exp(-\frac{1}{2\sigma^2}(c_i - \mu_j)^2)}{\sum\limits_{n=1}^{k} \exp(-\frac{1}{2\sigma^2}(c_i - \mu_n)^2)}. \tag{1}$$

$$M - step : \mu_j' \leftarrow \sum\limits_{i=1}^{m} E[z_{ij}]t_i \Big/ \sum\limits_{i=1}^{m} E[z_{ij}]. \tag{2}$$

$E[z_{ij}]$ is the probability that c_i is generated by the jth Gaussian distribution, and z_{ij} is a binary k-dimensional vector in which $z_{ij} = 1$ if and only if c_i arises from the jth Gaussian.

We can see, from (1) and (2), that the EM algorithm breaks down the potentially difficult problem of the feature creation into its two stages, the E-step and the M-step. Each of the steps is simple for the GP-EM implementation. See further details in [9], [10] regarding the derivative of the k-means problem via EM.

3.3 Fitness Measure

The evolutionary GP-EM process is driven by a fitness measure that evaluates how well each individual tree (a computer program in the population) performs in its problem environment. In order to provide the measure for the transformed data $\mathbf{C}$ to fit the corresponding Gaussian mixture model in the estimation of the means of the k Gaussians, the design of the fitness function, the J value, is given by [11]:

$$J = \ln |S_W^{-1} S_B|, \tag{3}$$

with the within – class scatter value:

$$S_W = \sum_{j=1}^{k} \sum_{c \in \mathbf{C}_j} (c - \mu_j)(c - \mu_j)^{T'}, \tag{4}$$

and the between – class scatter value:

$$S_B = \sum_{j=1}^{k} n_j (\mu_j - \mu_T)(\mu_j - \mu_T)^{T'}. \tag{5}$$

$\mathbf{C}_j$: the jth subset of $\mathbf{C}$;
μ_j : the mean of the jth Gaussian;
μ_T: the total mean value;
n_j : the number of samples in the jth class.

4 Experiments

The problem of classification can be formulated in terms of a predetermined measure. For the minimum distance classifier (MDC), the distance is defined as an index of similarity so that the minimum distance is identical to the maximum similarity [11]. Using the classifier MDC, we employ the 10-fold cross-validation approach to assess the capabilities of the GP-EM detector for the implementation of Oxford Parkinson's Disease Database (OPDD) [12]. To avoid unnecessary computation in the experiments, we choose the medium population size of 32 with the maximum tree depth of 5.

4.1 Real Dataset

OPDD contained 195 biomedical voice samples recorded from 31 people, in which 23 were diagnosed with PD. Each datum is 22-dimensional and consisted of certain types of voices ranging from 1 to 36 seconds in length, listed in Table 2. The dataset is divided into two classes according to its "status" column which is set to 0 for healthy subjects and 1 for those with PD. The data set used for this implementation are described in detail in [13], and at the UCI website [12].

Table 2. Descriptions of the features of the dataset OPDD

No	Features	Descriptions
1-3	MDVP (Hz)	ave./max. /min. vocal fundamental frequency
4	MDVP (%)	MDVP jitter as a percentage
5	MDVP (abs)	MDVP absolute jitter in microseconds
6-8	RAP/ PPQ/ DDP	3 measures of variation in frequency
9-10	shimmer / (dB)	2 MDVP local shimmer
11-12	shimmer: APQ	3- /5- point Amplitude Perturbation Quotient
13	MDVP: APQ	11- point Amplitude Perturbation Quotient
14	shimmer: DDA	a measure of variation in amplitude
15-16	NHR / HNR	2 ratios of the noise to tonal components
17-18	RPDE / D2	2 nonlinear dynamical complexity measures
19	DFA	signal fractal scaling exponent
20-21	spread 1 - 2	2 measures of frequency variation
22	PPE	pitch period entropy

4.2 Resulting Created Functions

In the evolutionary explorations, the specification of a termination criterion is usually required to terminate runs. In this study, we ended the evolutionary GP-EM training process when a certain number of iterations are reached, as the result of a run. After running 350 iterations, the resulting created feature function from one of the 10 runs is given by:

$$\mathbf{Pa}_{-\text{detection}} = \cos\{\cos\{-[(f4-f5)/(f4+f16)^2]\times tg[f2+f7-(f5\times f18)^2]\}$$
$$\times[f6+f3-(f18/f21)+tg(f15^2)/(f4-f0)]\},$$

where fd, d = 0,..., 21, is the (d+1)th ordinary feature of OPDD listed in Table 2; normalized in the range of (0, 1), its fitness values is presented in Fig. 3.

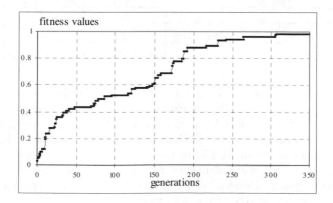

Fig. 3. The convergence results for the created feature function, $\mathbf{Pa}_{-\text{detection}}$

The average time taken for 10 independent runs was about five minutes on a Pentium 4 PC with CPU 1GHz and 1GB of RAM. The Microsoft Visual C++ 6.0 was used for the implementation.

4.3 Detection Performance

In the detection experiments, the resulting feature function, Pa_detection, is employed as the medical decision space to diagnose PD. By using the classifier MDC, the detection performance for 10 runs on the combined training and test sets is illustrated in Table 3. We achieve the average training score of 95.06 % detection accuracy; with only one created function Pa_detection, the average performance over 10 runs shows that 93.12 % of the test data can be classified correctly.

Table 3. Detection accuracy (DA) for 10 runs on OPDD

Performance	The training (DA)	The test (DA)
maximum (%)	97.71	95.00
minimum (%)	91.42	80.00
average (%)	95.06	93.12
Std (%)	1.42	2.86

4.4 Comparison Works

To assess whether a single feature function created by the GP-EM detector would be sufficient for the identification of the Parkinsonian subjects, we compare our algorithm with other methods existing in the literature on OPDD [12]. The comparison works with [13], [14] are given in Table 4.

Table 4. Comparisions of the GP-EM detector and other methods in terms of the detection accuracy (DA) on OPDD; note: the sign '~' means 'not specified'

Methods	DA (%)	No. of features	Time
rough set [14]	100	22	~
feature preselection with an exhaustive search [13]	91.40	10	~
the GP-EM detector	93.12	1	5 minutes

Little *et. al* [13] first applied a feature preselection *filter* that removed redundant measures; using an SVM classifier, an exhaustive search was executed by testing all possible subsets of features in order to discriminate healthy from disordered subjects; consisted of 10 ordinary features, the subset was thus selected that produced the best classification performance of 91.4%. In [14], features sets were analyzed using the rough set approach that mapped feature vectors associated with objects onto the decision support system. Using all of 22 ordinary features, the approach yielded the highest recognition rate of 100% from one example of 10 runs.

Significantly, as shown in Table 4, by using the single created function, Pa_detection, the GP-EM detector achieves the average classification accuracy of 93.12% over 10 runs, indicating the advantages of the reliability and efficiency for the discrimination of healthy subjects from those with PD.

5 Conclusion

In this study, based on ordinary features of voice, an automated GP-EM method of finding feature functions is proposed for the discrimination of healthy people from those with PD. We allow the system to dynamically alter the structures of the functions, which are undergoing adaptation of ordinary feature vectors without any manipulated actions in the evolutionary process. Using the created function, the proposed detector obtains an average detection accuracy of 93.12%, which is a promising improvement.

A major advantage of this biometric GP-EM detector rises from its particular ability to reduce feature dimensionality, making the data representation free from the complication of the multidimensional space. As a result of the improved data representation, the detector reduces computational cost with the improved measures.

The choices of medical biometric methods affect the performance of the design systems. The strategies for integrating GP and EM in this study are:

(1) GP tree structures are conducive to a global search of optimum medical decision boundaries;
(2) via EM, the learning task then changes to output a hypothesis that estimates the means of each of the k Gaussians, thus making it easy to find solutions.

The motivations behind this study are to present the automatic method of finding medical decision boundaries, and to explore the possible use of such a method to support medical decision systems in the biometrics community.

Acknowledgments. Authors thank the source of the database provided by Max Little, University of Oxford, in collaboration with the National Centre for Voice and Speech, Denver, Colorado, who recorded the speech signals.

References

1. Eeden, S.K.V.D., Tanner, C.M., Bernstein, A.L., Fross, R.D., Leimpeter, A., Bloch, D.A., Nelson, L.M.: Incidence of Parkinson's disease: Variation by age, gender, and race/ethnicity. Amer. J. Epidemiol. 157, 1015–1022 (2003)
2. Rahn, D.A., Chou, M., Jiang, J.J., Zhang, Y.: Phonatory impairment in Parkinson's disease: Evidence from nonlinear dynamic analysis and perturbation analysis. J. Voice 21, 64–71 (2007)
3. Cnockaert, L., Schoentgen, J., Auzou, P., Ozsancak, C., Defebvre, L., Grenez, F.: Low-frequency vocal modulations in vowels produced by Parkinsonian subjects. Speech. Commun. 50, 288–300 (2008)

4. Sapir, S., Spielman, J.L., Ramig, L.O., Story, B.H., Fox, C.: Effects of intensive voice treatment (the Lee Silverman voice treatment [LSVT]) on vowel articulation in dysarthric individuals with idiopathic Parkinson disease: Acoustic and perceptual findings. J. Speech. Lang. Hear. Res. 50, 899–912 (2007)
5. Krawiec, K., Bhanu, B.: Visual learning by coevolutionary feature Synthesis. IEEE Trans. Syst., Man, Cybern. Part B 35, 405–425 (2005)
6. Guo, H., Jack, L.B., Nandi, A.K.: Feature generation using genetic programming with application to fault classification. IEEE Trans. Syst., Man, Cybern., Part B 35, 89–99 (2005)
7. Koza, J.R.: Genetic Programming II: Automatic Discovery of Reusable Programs. MIT Press, Cambridge (1994)
8. Raymer, M.L., Doom, T.E., Kuhn, L.A., Punch, W.F.: Knowledge discovery in medical and biological datasets using a hybrid Bayes classifier/evolutionary algorithm. IEEE Trans. Syst., Man, Cybern., Part B 33, 802–813 (2003)
9. Bishop, C.M.: Pattern Recognition and Machine Learning. Springer, New York (2006)
10. Mitchell, T.M.: Machine Learning. McGraw-Hill, New York (1997)
11. Fukunaga, K.: Introduction to Statistical Pattern Recognition, 2nd edn. Academic, Boston (1990)
12. Blake, C., Keogh, E., Merz, C.J.: UCI repository of machine learning databases, Dept. Inform. Comput. Sci. Univ. California, Irvine (1998),
   http://www.ics.uci.edu/~mlearn/MLRepository.html
13. Little, M.A., McSharry, P.E., Hunter, E.J., Spielman, J., Ramig, L.O.: Suitability of dysphonia measurements for telemonitoring of Parkinson's disease. IEEE Trans. Biomed. Eng. 56, 1015–1022 (2009)
14. Revett, K., Gorunescu, F., Salem, A.M.: Feature selection in Parkinson's disease: a rough sets approach. In: Proc. Int. Multiconference on Comput. Science and Information Technology, Mragowo, Poland, pp. 425–428 (2009)

Using Formal Concept Analysis to Visualize Relationships of Syndromes in Traditional Chinese Medicine

Xulong Liu, Wenxue Hong, Jialin Song, and Tao Zhang

[1] Department of Biomedical Engineering, Yanshan University, Qinhuangdao 066004, China
lxl815@88mail.ysu.edu.cn

Abstract. Computer-aided syndrome differentiation is one of most active issues in how to digitize the Traditional Chinese Medicine. The main purpose of this paper is to show the use of Formal Concept Analysis (FCA) to visualize the dependencies of syndromes of Traditional Chinese Medicine (TCM) with Heart-Spleen syndrome serving as an example. First, the syndromes and syndrome factors are defined as the objects and attributes, then their concepts and the hierarchical relationships are described in mathematical language, finally their hierarchy is visualized by concept lattice. This approach may make syndrome's knowledge clearer, and may make data mining of TCM more efficient.

Keywords: Formal Concept Analysis, syndrome, visualization, Traditional Chinese Medicine.

1 Introduction

Traditional Chinese Medicine (TCM) is a range of medical practices used in China for more than four millenniums, a treasure of Chinese people [1]. TCM is featured by such treatment based in pathogenesis obtained through differentiation of symptoms and signs as its essence. It is necessary to differentiate disease and constitutes a special procedure of research and treatment in traditional Chinese diagnosis. Based on such treatment and diagnosis, a traditional Chinese medicine practitioner can prescribe drugs, identify and treat diseases. As a key concept of syndrome differentiation, syndrome of TCM has become the focus of research on TCM itself. It is a kind of pathology of the disease development of a body in a certain stage, including the disease wherefrom, the cause, the feature and the conflicts between healthy energy and evils. It reflects the nature of pathological change at a certain stage, and reveals the intrinsic quality of disease more completely, profoundly, and accurately than symptoms, for example, excessive rising of liver-YANG, damp invasion of lower energizer. Another symptom of TCM should be clarified, such as fever, headache, yellow tongue coat and pulse.

Key issues concerning syndrome of TCM include: 1) multiple-names-one-syndrome, 2) Standardized naming the syndromes, 3) the relationships among symptoms, syndromes and diseases, 4) management and mining mass clinical data.

D. Zhang and M. Sonka (Eds.): ICMB 2010, LNCS 6165, pp. 315–324, 2010.
© Springer-Verlag Berlin Heidelberg 2010

To address these issues, a variety of Computational methods for TCM have been carried out. For example, the standard syndrome-name database aims to deal with the problem that a syndrome was assigned with different names by different TCM practitioners [2]. The Individual Diagnosis and Treatment System (IDTS) aims to mine association rules between Chinese medicine symptoms and the indications of patients [3].Wang et al. [4] combined cluster analysis and Bayesian networks to find causal relationships in TCM in a kidney study. Information Management System of TCM Syndrome was developed based on Prior knowledge Support Vector Machine (P-SVM) [5], which was based on traditional SVM.

Although the above methods bear fruits in mining the relationships between syndromes, the substantial of that has not been illustrated till now. In this paper, we propose the use of Formal Concept Analysis (FCA) [6] as a foundation for visualize relationships between syndromes of TCM which makes the intrinsic relationship vivid and concise. FCA has been successfully developed for more than a decade. Theoretic and practical evidences revealed that its connections to the philosophical logic of human thought became clearer and even consistent connections to Piaget's cognitive structuralism that Seiler elaborated to a comprehensive theory of concepts. Seiler's theory has been proven that there is a closely linking between mind concepts and formal concepts. Therefore, the mathematical of concepts and concept hierarchies used in formal concept analysis opens up the opportunity of supporting mathematically the logical thinking of humans. The support can really take place in the documentary report when apply FCA on it [8-9]. Therefore, FCA will be adopted as the knowledge concept visualization method in this study. In this paper, we view syndromes as objects that possess certain attributes named Symptom Factor [10]. First, the Formal Context of syndromes is established by the theory [6], then using Concept Lattice [6] to express the dependencies of syndromes, finally utilizing a Hasse diagram to visualize Concept Lattice of syndromes and analysis the relationships between syndromes of TCM.

The rest of the paper is organized as follows. Section 2 introduces the above-mentioned concept of Symptom Factors and the theory of FCA. Section 3 presents the system architecture used to visualize and analyze the relationships of Syndromes of Traditional Chinese Medicine. Section 4 introduces a method of visualization for establishing syndrome dependencies using the theory of FCA and demonstrates the use of this method in practice. Finally, Section 5 presents the conclusions and directions for future research.

2 Relation Concepts

2.1 Symptom Factors

As the essence of syndrome of TCM, Symptom Factor [10] is the minimal key elements of syndrome differentiation. It can be divided into two categories. Generally, the one indicates the disease occurrence site (such as lung and heart) and the other indicates disease causes or pathological causes (such as wind, cold, dampness, etc.).A standard syndrome name always includes the above two categories. For example, syndrome of Kidney Yang Deficiency is composed of the Kidney and the Yang Deficiency.

2.2 Formal Concept Analysis

FCA is a mathematization of the philosophical understanding of concept [6]. As a human-centered method to structure and analyze data, FCA is used to visualize data and its inherent structures, implications and dependencies. Following basic definitions have been taken from [6] which has been used throughout the paper.

A formal context K: = (G, M, I) consists of two sets G and M and a relation I between G and M. The elements of G are called the objects and the elements of M are called the attributes of the context.

For $A \subseteq G$ and $B \subseteq M$, define

$A' := \{m \in M \mid g \ I \ m$ for all $g \in A\}$(the set of all attributes common to the objects in A)

$B' := \{g \in G \mid g \ I \ m$ for all $m \in B\}$(the set of objects common to the attributes in B)

The mappings $X \rightarrow X''$ are closure operators.

A formal concept of the context (G, M, I) is a pair (A, B) with $A \subseteq G$, $B \subseteq M$, $A'=B$ and $B'=A$. A is called the extent and B is the intent of the concept (A, B). ς (G,M,I) denotes the set of all concepts of the context (G,M,I).

If (A1, B1) and (A2, B2) are concepts of a context, (A1, B1) is called a subconcept of (A2,B2), provided that $A1 \subseteq A2$ (which is equivalent to $B2 \subseteq B1$). In this case, (A2,B2) is a superconcept of (A1,B1) and we write $(A1,B1) \leq (A2,B2)$. The relation $\leq$ is called the hierarchical order of the concepts. The set of all concepts of (G, M, I) ordered in this way is called the concept lattice of the context (G, M, I) [11].

Formal concepts can be ordered by

$(A1,B1) \leq (A2,B2) : \Longleftrightarrow A1 \subseteq A2$.

The setς(G,M, I) of all formal concepts of (G,M, I), with this order, is a complete lattice, called the concept lattice of (G,M, I).

Every pair of concepts in this partial order has a unique greatest lower bound (meet). The greatest lower bound of (A_i, B_i) and (A_j, B_j) is the concept with objects $A_i \cap A_j$; it has as its attributes the union of B_i, B_j, and any additional attributes held by all objects in $A_i \cap A_j$. Symmetrically, every pair of concepts in this partial order has a unique least upper bound (join). The least upper bound of (A_i, B_i) and (A_j, B_j) is the concept with attributes $B_i \cap B_j$; it has as its objects the union of A_i, A_j, and any additional objects that have all attributes in $B_i \cap B_j$ [6].

A formal context is best understood if it is depicted by a cross table with the formal context about integers form 1 to 10 as an example (see Table 1). A concept lattice is best pictured by a Hasse diagram as the concept lattice of our example context in Fig. 1 As shown in Fig. 1.Consider A = {1,2,3,4,5,6,7,8,9,10}, and B = {composite, even, odd, prime, square}. The smallest concept including the number 3 is the one with objects {3, 5, 7}, and attributes {odd, prime}, for 3 has both of those attributes and {3, 5, 7} is the set of objects having that set of attributes. The largest concept involving the attribute of being square is the one with objects {1,4,9} and attributes {square}, for 1, 4 and 9 are all the square numbers and all three of them have that set of attributes.

Table 1. A formal context about integers form 1 to 10

	composite	even	odd	prime	square
1			√		√
2		√		√	
3			√	√	
4	√	√			√
5			√	√	
6	√	√			
7			√	√	
8	√	√			
9	√		√		√
10	√	√			

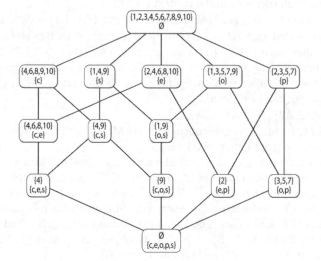

Fig. 1. A concept lattice for objects consisting of the integers from 1 to 10, and attributes composite (c), square (s), even (e), odd (o) and prime (p). The lattice is drawn as a Hasse diagram

3 System Architecture

Our proposal to integrate FCA in an information retrieval and visualization system is based on the architecture presented in Figure 2. It is divided in four main subsystems that solve the data management, human-computer interaction, lattice building and lattice representation tasks. Interactions between the subsystems are represented in the same figure and can be summarized as follows:

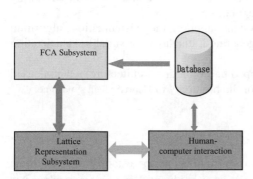

<zhengming syndromes="心脾阳虚[虚寒]证
" therapeutic-methods="温补心脾" prescription="
附子理中汤/桂附理中丸" zhs="4">
<zh>心</zh>
<zh>脾</zh>
<zh>阳虚</zh>
<zh>寒</zh>
</zhengming>
<zhengming syndromes="心脾阳虚，血瘀
水停证" therapeutic-methods="温阳化瘀利水
" prescription="苓理汤" zhs="6">
<zh>心</zh>
<zh>脾</zh>
<zh>阳虚</zh>

Fig. 2. System architecture   **Fig. 3.** Syndrome information compile in XML

1) TCM syndrome information is described by an accredited standard text format, XML (Extensible Markup Language), in database subsystem. XML's design goals emphasize simplicity, generality, and usability over the human-computer interaction. It is a textual data format, with strong support via Unicode for the languages of the world. The database subsystem compiles the domain knowledge rule and relationship by XML syntax as the Fig. 3, which contains the rule of the TCM syndrome, and includes the connecting symptom Factors, treatments and prescription.

2) The subsystem of human-computer interaction plays the role of mixed initiative [11] for visualization and analysis of the relationship between TCM syndromes, involving humans and machines sharing tasks best suited to their individual abilities. The computer performs computationally intensive tasks and prompts human-clients to intervene when either the machine is unsuited to make a decision or resource limitations demand human intervention. For example, we may alter the algorithm parameter to browse results of visualization more efficient using nesting and zooming like Fig. 4.

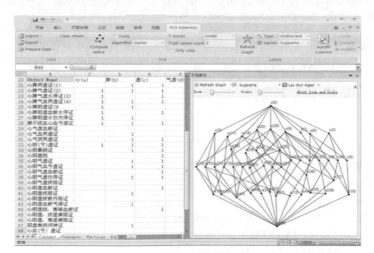

Fig. 4. Nesting and zooming

3) The FCA subsystem builds a formal context using the documents described by XML like Fig. 3, which generates a concept lattice.

4) The Lattice representation subsystem applies the visualization algorithm introduced in next section to the lattice generated, displaying a suitable visualization for user browsing purposes.

Currently, a prototype has been developed based on this architecture, which is used as our tool for visualization and analysis of the hierarchy in Heart-Spleen syndrome.

4 Visualization and Case Study

In this section, we introduce a method of visualization for establishing syndrome dependencies using the theory of FCA. A case study of Heart-Spleen syndrome is presented for demonstrating the effect of visualization in practice.

Heart-Spleen syndrome of TCM includes nine kinds of sub-syndromes, namely Deficiency of Heart and Spleen (DHS), Heart Spleen Qi Deficiency (HSQD), Heart Spleen Qi Deficiency Shuiting (HSQDS), Heart Spleen Qi-Blood Deficiency (HSQBD), Heart Spleen Yang Deficiency (HSYD), Heart Spleen Yang Deficiency Blood Stasis Shuiting (HSYDBSS), Heart Spleen Yang Deficiency Shuiyinneiting (HSYDS) and Failure of Spleen to Control Blood Vessels (FSCBV).

Step 1: Constructing the formal context of Heart-Spleen syndrome as Table 2.

Step 2: Aggregating context and using the extension table (like Fig. 5) to generate all concepts of Heart-Spleen syndrome.

Step 3: After the process by the above steps, a fast algorithm was adopted for building the Hasse diagram of concept lattice. This algorithm was developed by Petko Valtchev [12] and has three steps: the first step generates the intents of all least upper bounds (LUBs), the second finds all maximal sets among the intents and the third step finds all nodes in the partial ordered set represented by the structure after the insertion of the node that correspond to maximal intents and then connects the current node to them. Additional effort is necessary to maintain the list of border elements that is independent from the upper cover computation [13].

Algorithm 1:

```
1: PROCEDURE HASSE
2: INPUT: C = {c₁, c₂... cᵢ}
3: OUTPUT: L = (C, <= l)
4: SORT (C)
5: L ⇐ {c₁}
6: Border ⇐ {c₁}
7: FOR i FROM 2 TO l DO
8: Intents ⇐ ∅
9: FOR ALL c' ∈ Border DO
10: Intents ⇐ Intents ∪ (Int(c') ∩ Int(cᵢ))
11: CoverIntent⇐ MAXIMA(Intents)
12: FOR ALL Y ∈ CoverIntents DO
```

13: c' ⇐ FIND - CONCEPT(Y)
14: MAKE - LINK(c_i, c')
15: Border ⇐ (Border – uppercovers(ci)) ∪ {ci}
16: L ⇐ L ∪ {ci}

The Algorithm 1 first sorts the lattice nodes in C in order to obtain a liner extension of <= L. Then, the nodes are processed in a decreasing order, each time integrating the current node into the partial structure L that finally contains the entire lattice. At each

Table 2. Formal context of Heart-Spleen syndrome

G \ M	Heart	Spleen	Deficiency	Qi Deficiency	Shui Ting	Yang Deficiency	Blood Deficiency	Dampness	Blood Stasis
DHS	X	X	X	X					
HSQD	X	X	X	X				X	
HSQDS	X	X	X	X	X				
HSQDD	X	X	X	X			X		
HYSD	X	X			X				
HSYDBSS	X	X	X		X	X			X
HSYDS	X	X			X	X		X	
FSCBV	X	X	X	X			X		

Step	Extension
1	{DHS,HSQD,HSQDS,HSQDD,HYSD,HSYDBSS,HSYDS,FSCBV}
{Heart, Spleen}	
Deficiency	{ DHS,HSQD,HSQDS,HSQDD, HSYDBSS,FSCBV}
Qi Deficiency	{ DHS,HSQD,HSQDS,HSQDD, FSCBV}
Shui Ting	{ HSQDD ,HSYDBSS,HSYDS,FSCBV},{ HSQDD ,HSYDBSS,FSCBV}
	{ HSQDD ,FSCBV}
Yang Deficiency	{ HYSD,HSYDBSS,HSYDS },{HSYDBSS,HSYDS },{ HSYDBSS }
Blood Deficiency	
Dampness	{ HSYDS }
Blood Stasis	

Fig. 5. Extension table of the formal context in Table 2

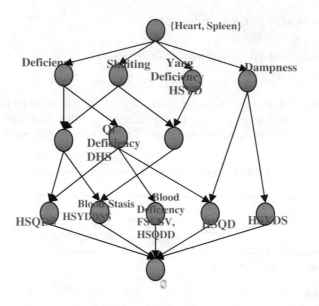

Fig. 6. Hasse diagram of Concept lattice in Table 2

step, the intersections between the current concept intent and the intents of all border concepts are computed and sorted in the Intents set. The maximal elements of Intents are selected and put in Cover-Intents (the MAXIMA primitive). These intents correspond to the most specific concepts that are greater than c_i, the upper covers. Next, for each element of Cover-Intents, the corresponding concept is localized within the partial structure P_i, (FIND-CONCEPT) and the respective link is created. Finally, the upper covers of a concept are excluded from the border of the partial structure and the current element is added [13].

In Fig. 6, each round node represents a concept of Heart-Spleen syndrome, in which blue semicircle represents attributes (symptom factors) and red semicircle represents objects (syndromes). Speaking in human logical terms, by the Basic Theorem of FCA, each concept is represented by a little circle so that its extension (intension) consists of all the objects (attributes) whose names can be reached by a descending (ascending) path from that circle. For instance, the circle with the label Qi Deficiency represents the formal concept with the extent {DHS, HSQDS, FSCBV, HSQDD, HSQD} and the intent { Qi Deficiency, Deficiency, Heart, Spleen}. FSCBV and HSQDD are located in a common node, so they have the same implication in Syndrome Differentiation. The attribute implicate {Qi Deficiency} $\subseteq$ {Deficiency} because there are ascending paths from the circle with the label Qi Deficiency to the circles with the labels Deficiency. There are three kinds of relationships that exist among these concepts, which are independence, intersection, and inheritance. Using the basic principles of FCA to analyze Fig. 6, we can obtain three conclusions:

(1) In the formal context of Heart-Spleen syndrome, the symptom Factors associated with heart and spleen are divided into three hierarchies. The highest level is {Deficiency, Shui-Ting, Yang-Deficiency, Dampness}, the middle level is {Qi-Deficiency}, and the

lowest level is {Blood-Stasis, Blood-Deficiency}. On one hand, the higher level syndrome is located, the better universality can be reflected between the syndrome and Heart-Spleen. On the other hand, the lower level syndrome is located, the better specificity can be reflected between the syndrome and Heart-Spleen.

(2) Deficiency is the most frequent item of the symptom factor nodes in Fig. 5, the most common disease factor in Heart-Spleen syndrome.

(3) DHS and HSYD constitute the foundation of Heart-Spleen syndrome, because the rest of these nodes are derived from two nodes mentioned above.

5 Conclusion and Future Research

In this paper, we have presented a system framework for visualization and analysis TCM syndrome dependencies using formal concept analysis. In this approach, we have described the steps including creation of formal context of TCM syndrome, extraction of concepts, application of the formal concept analysis to establish a set of conceptual relationships and hierarchy of TCM syndrome, and analysis of the established dependencies. Finally, we have demonstrated the usage of our approach by applying it to a case study of the Heart-Spleen syndrome of TCM.

In future research, our work is focused on three main objectives: a) the evaluation of the generated lattices in an interactive setting involving users, b) the research on lattice visualization and c) Improving the TCM concept process model obtained by provide weighted value between concepts. Furthermore, we intend to further investigate the suitability of this approach by applying it to different case studies that relate to different relationships of the syndrome of TCM.

References

1. Suryani, L., Yulan, H., Huic, S.C.: Computational methods for Traditional Chinese Medicine: A survey. Computer Methods and Programs in Biomedicine 88, 283–294 (2007)
2. Wang, X., Qu, H., Liu, P., Cheng, Y.: A self-learning expert system for diagnosis in traditional Chinese medicine. Expert Systems with Applications 26, 557–566 (2004)
3. He, X., Huang, W., Lu, M., Xue, W., Lu, Y.: Research and application of data mining in individual diagnosis and treatment based on Chinese traditional medicine. In: 2006 IEEE Int. Conf. Granular Computing, pp. 417–419 (2006)
4. Wang, M., Geng, Z., Wang, M., Chen, F., Ding, W., Liu, M.: Combination of Network Construction and Cluster Analysis and Its Application to Traditional Chinese Medicine. In: Wang, J., Yi, Z., Żurada, J.M., Lu, B.-L., Yin, H. (eds.) ISNN 2006. LNCS, vol. 3973, pp. 777–785. Springer, Heidelberg (2006)
5. Yang, X., Liang, Z., Zhang, G., Luo, Y., Yin, J.: A classification algorithm for TCM syndromes based on P-SVM. In: Proc. 2005 Int. Conf. Mach. Learn. Cybern., pp. 3692–3697 (2005)
6. Ganter, B., Wille, R.: Formal Concept Analysis: Mathematical Foundations. Springer, Heidelberg (1999)
7. Carpineto, C., Romano, G.: Concept Data Analysis: Theory and Application. John Wiley & Sons Ltd., Chichester (2004)

8. Daey, B.A., Priestley, H.A.: Introduction to Lattices and Order. Cambridge University, Cambridge (2006)
9. Zhu, W.F., Yan, J.F., Huang, B.Q.: Application of Bayesian network in syndrome differentiation system of traditional Chinese medicine. Journal of Chinese Integrative Medicine 4, 567–571 (2006)
10. Anamika, G., Naveen, K., Vasudha, B.: Analysis of Medical Data using Data Mining and Formal Concept Analysis. In: The Fourth World Enformatika Conference, Istanbul, Turkey (2005)
11. Horvitz, E.: Uncertainty, Action and Interaction: In pursuit of Mixed initiative. In: IEEE Computing Intelligent Systems, pp. 17–20 (1999)
12. Valtchev, P., Missaoui, R., Lebrun, P.: A Fast Algorithm for Building the Hasse Diagram of a Galois Lattice. In: Colloquy LACIM 2000, Combinatoire, Informatique et Applications, pp. 293–306 (2000)
13. Pan, T., Tsai, Y.C., Fang, K.: Using formal concept analysis to design and improve multidisciplinary clinical processes. WSEAS Transactions on Information Science and Applications 5, 880–890 (2008)

Wrist Pulse Diagnosis Using LDA

Bo Shen and Guangming Lu

Bio-computing Research Center, Dept. of Computer Science,
Harbin Institute of Technology Shenzhen Graduate School, China
shebo101@gmail.com, luguangm@gmail.com

Abstract. The wrist pulse signal is a kind of important physiology signal which can be used to analyze a person's health status. This paper applies a linear discriminant analysis (LDA) to extract feature and used k-nearest neighbor (KNN) algorithm to distinguish the patients from health. In order to reduce the interference of noise, we first drew a series of pulse data of good quality from the original wrist pulse signal. We then reduced all high dimensional pulse signals to low dimensional feature vectors using LDA. Finally, we used a KNN algorithm to distinguish healthy persons from patients. The classification accuracy is over 83% in distinguishing healthy persons from patients with all kinds of diseases, and over 92% for single specific disease. The experimental results indicate that LDA is an efficient approach in telling healthy subjects from patients of specific diseases.

Keywords: Pulse diagnosis, LDA, KNN.

1 Introduction

Traditional Chinese pulse diagnosis (TCPD) has been successfully used for thousands of years in oriental medicine [1-6]. Traditional pulse diagnosis methods use fingertips to feel the pulse at the wrist, which requires that practitioners have considerable training and experience. Even though the practitioners are very experienced, different practitioners might have different diagnosis result. Fortunately, several kinds of pulse sensors have been invented in recent years, which can digitize and record pulse signals. With the digitized pulse signal, the research of pulse waveform analysis becomes very active.

Pulse waveform analysis is used for variety of pathological, physiological purposes in both TCPD and Western medicine [7], [8]. Some methods have been proposed to analyze the digitized pulse signals [9], [10]. Leonard et al. [9] used wavelet power features and wavelet entropy of the pulse signal to distinguish healthy and unwell children. Chen et al. [10] proposed an auto-regressive (AR) based method to extract the pulse signal features. Although the AR model has achieved encouraging results, the analysis of pulse signal is mainly based on the global domain and usually uses the whole series of pulse signal, which might cause that the local information of

D. Zhang and M. Sonka (Eds.): ICMB 2010, LNCS 6165, pp. 325–333, 2010.

each pulse waveform was ignored. Here, we proposed a local analysis of pulse signal, which just analyze one period of the whole signal.

A pulse signal is a time series, which has many different points, so pulse signals are high dimensional data. In this paper, each of the samples of all pulse signals were cropped to generate single-period data, which contains only one period. It seems that a dimensionality reduction process is still necessary for further reducing the dimensionality of the sample vectors. As an important approach to linear feature extraction, linear discriminant analysis (LDA) can be used to reduce the dimension of pulse signal. The research on LDA feature extraction is very active [11], [12]. There are two very effective LDA methods: Foley-Sammon linear discriminant analysis (FSLDA) [13] and uncorrelated linear discriminant analysis (ULDA) [14]. We note that the Fisher criterion ratio of each FSLDA discriminant vector is not less than that of the corresponding ULDA discriminant vector, so each FSLDA discriminant vector has more powerful discriminant ability than the corresponding ULDA discriminant vector [11]. On the other hand, the fact that ULDA discriminant vectors can produce uncorrelated feature components is a big advantage of ULDA. It has been pointed out that ULDA will be indeed identical to the classical LDA if the classical LDA has no equivalent eigenvalues [18]. Since the classical LDA is computationally more efficient than ULDA, in this paper we adopt the classical LDA shown in Section 3.

2 The Proposed Method

2.1 Wrist Pulse Signal Preprocessing

In this study, the wrist pulse database is from Chen [10]. Figure 1 shows a typical pulse signal from this database. Chen has used auto-regressive (AR) model to extract features from the pulse signal and got very good result [1]. However, Chen's method is based on the whole series of the pulse signal, which contains many periods of pulse waveform. We can see from Fig. 1(a) that the quality of each period may not the same and some periods are seriously interfered by noise. In order to get the pulse waveform of good quality, we manually cropped a piece of pulse signal with only one period of good quality pulse signal, which is shown in Fig. 1(b). However, the length of period is different for different person, so we normalize all the single-period pulse signals to 100-dimensional vectors. For each single-period pulse signal s_i, the normalization process is described as follows:

1. up sample s_i to $100 \times n_i$-dimension vector s_i';
2. down sample s_i' to 100-dimension vector x_i.

Finally, we got a 100-dimensional vector for each sample, which is denoted by x_i $(i = 1,2 \ldots N)$, where N is the sample number.

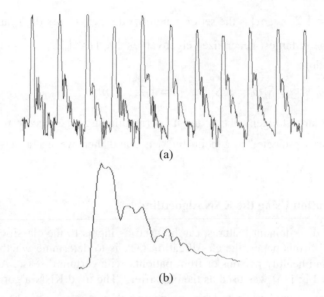

(a)

(b)

Fig. 1. Wrist pulse signal from Chen's database: (a) is original pulse signal; (b) is a single-period pulse signal cut from (a)

2.2 Feature Extraction Using LDA

Suppose $\omega_1, \omega_2, ..., \omega_L$ are L known pattern classes, and let $\{x_1, x_2, ..., x_N\}$ be N d-dimensional single-period pulse signals. Let the between-class scatter matrix be defined as

$$S_b = \sum_{i=1}^{L} N_i (\mu_i - \mu)(\mu_i - \mu)^T \tag{1}$$

and the within-class scatter matrix be defined as

$$S_w = \sum_{i=1}^{L} \sum_{x_k \in X_i} (x_k - \mu_i)(x_k - \mu_i)^T \tag{2}$$

where μ_i is the mean signal of class ω_i, and N_i is the number of samples in class ω_i. If S_w is nonsingular, the optimal projection $W_{opt} = \varphi_1, \varphi_2, ..., \varphi_m$ is chosen as the matrix with orthonormal columns which maximizes which maximize Fisher criterion

$$J(\varphi) = \frac{\varphi^T S_b \varphi}{\varphi^T S_w \varphi} \tag{3}$$

where $\{\varphi_i \mid i = 1,2,\ldots,m\}$ is the set of generalized eigenvectors of S_b and S_w corresponding to the m largest generalized eigenvalues $\{\lambda_i \mid i = 1,2,\ldots,m\}$ of the following eigenequation:

$$S_b \varphi_i = \lambda_i S_w \varphi_i, \quad i = 1,2,\ldots,m \tag{4}$$

We get the largest eigenvalue λ and its corresponding eigenvector w. Then each pulse signal was projected to a point by $x_i^T w$, and then we get a new sample set $Y = \{y_1, y_2 \ldots, y_n\}$.

2.3 Classification Using the KNN Algorithm

The extracted discriminant features, can be used as inputs to the classifier for further classification. In this paper, the classification task is to determine whether the pulse signals are from healthy persons or from patients with certain disease. A KNN algorithm presented in [15] was used as the classifier. The used KNN algorithm worked as follows: for a sample x_q in test set, supposing that $x_1, x_2,\ldots, x_k$, k samples from the set of training samples, are the nearest neighbors of x_q, we assign the most common class label among the k nearest training examples to the test sample x_q.

3   Experiments

We did the wrist pulse diagnosis using Chen's database [10], which is collected by using a Doppler ultrasonic device. In this database, there are totally 320 samples, including 100 healthy persons, 35 patients with appendicitis, 54 patients with acute appendicitis, 54 patients with pancreatitis and 77 patients with Duodenal Bulb Ulcer (DBU). That is to say, there are 5 pattern classes in the database. We first used all the pulse signals to compute the eigenvectors of LDA. We then used the eigenvectors to extract features from pulse signals. Finally, we randomly divided samples of each class into two halves, one half for training and the other for testing and used the KNN algorithm to classify the testing samples.

In this experiment, all the pulse signals were used to compute the eigenvectors of LDA, and then the test samples were reduced to low dimension feature vectors. The distribution of 2-dimensional features that are extracted from single-period pulse signals is shown in Figure 2, and the distribution of 2-dimensional features that are extracted from multi-period pulse signals is shown in Figure 3. Five groups of comparison were done in the experiment: Healthy vs. Appendicitis, Healthy Acute appendicitis, Healthy vs. Pancreatitis, Healthy vs. DBU and Healthy vs. All kinds of diseases. From the distribution, we can see that single-period pulse signals are well divided in both 1-D and 2-D features, but the multi-period pulse signals are not divided so well. That is to say, single-period pulse signals are more dividable than multi-period pulse signals by using LDA.

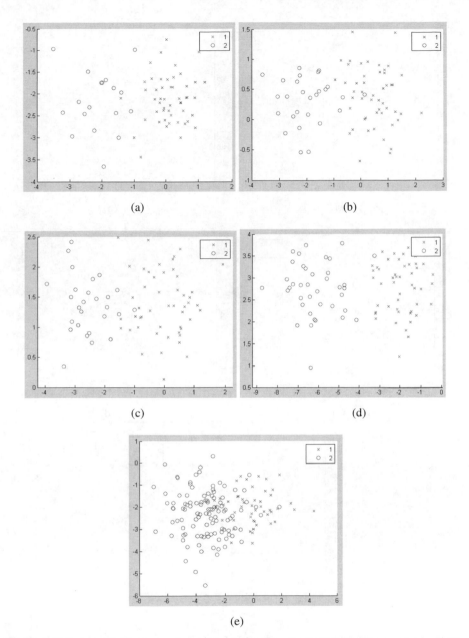

(a)

(b)

(c)

(d)

(e)

Fig. 2. 2-D feature distributions of single-period pulse signals: (a) Healthy vs. Appendicitis, (b) Healthy Acute appendicitis, (c) Healthy vs. Pancreatitis, (d) Healthy vs. DBU (e) Healthy vs. All kinds of diseases

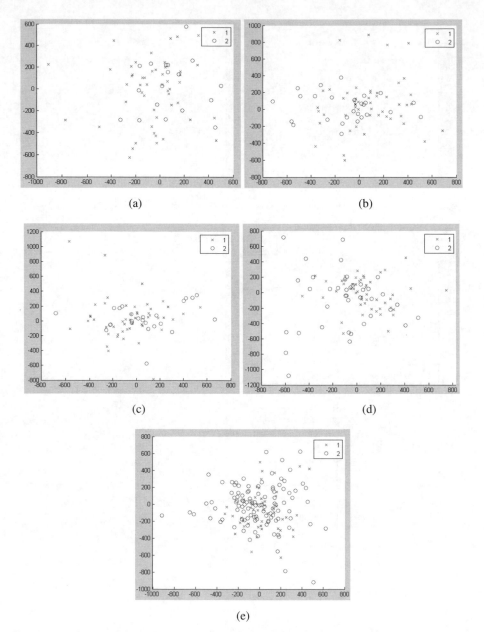

Fig. 3. 2-D feature distributions of multi-period pulse signals: (a) Healthy vs. Appendicitis, (b) Healthy Acute appendicitis, (c) Healthy vs. Pancreatitis, (d) Healthy vs. DBU (e) Healthy vs. All kinds of diseases

It is believed that the accuracy of statistical pattern classifiers increases as the number of features increases, and decreases as the number becomes too large [16]. Fukunaga proposed the optimal dimensionality of feature space is $(L-1)$ for L-class problems [17]. Since the experiment is done to classify 2-class pulse signals, the optimal dimensionality of feature space is 1. As a result, a single-period pulse signal x are projected to a point y using $y = x\varphi$, where φ is the eigenvector.

First, one feature vector is used to extract discriminant features, and the classification results are shown as Table 1. In order to show the efficiency of pulse signal preprocessing, single-period pulse signals are processed using LDA and classified using k-means classifier, and then the results are compared to the corresponding result of multi-period pulse signals. The result of comparison shows that the accuracy of single-period pulse signals is much better than the accuracy of multi-period pulse signals.

Table 1. Experimental results of LDA using 1 eigenvector

Sample class	Sample number		Accuracy (%) multi-period		Accuracy (%) single-period	
Healthy	50	68	78.0	64.7	98.0	95.6
Appendicitis	18		27.8		88.9	
Healthy	50	77	78.0	62.3	98.0	94.8
Acute appendicitis	27		33.3		85.2	
Healthy	50	77	70.0	61.0	90	97.8
Pancreatitis	27		44.4		100	
Healthy	50	89	68.0	66.3	100	97.8
DBU	39		64.1		94.7	
Healthy	50	160	36.0	58.8	72.0	83.7
All kinds of diseases	110		69.1		89.1	

Table 2 shows the result when two feature vectors are used to extract discriminant features. In statistic theory, LDA has only one non-zero eigenvalue in a 2-class problem, and the eigenvector of the zero-eigenvalue is meanness, but the experiment shows that some times additional eigenvector can promote the experimental result, some times bring down the experimental result.

Table 2. Experimental results of LDA using 2 eigenvector

Sample class	Sample number		Accuracy (%) multi-period		Accuracy (%) single-period	
Healthy	50	68	78.0	64.7	98.0	95.6
Appendicitis	18		27.8		88. 9	
Healthy	50	77	78.0	62.3	98.0	94.8
Acute appendicitis	27		33.3		85.2	
Healthy	50	77	70.0	61.0	90.0	93.5
Pancreatitis	27		44.4		96.3	
Healthy	50	89	68.0	66.3	100	97.8
DBU	39		64.1		94,7	
Healthy	50	160	36.0	58.8	70.0	80.0
All kinds of diseases	110		69.1		86.4	

Table 3 gives experimental results that did not use LDA to reduce dimension of pulse signals. When we directly use KNN to classify the pulse signals, the result is much worse compared to the result using LDA.

Table 3. Experimental results without using LDA

Sample class	Sample number		Accuracy (%) multi-period		Accuracy (%) single-period	
Healthy	50	68	44.0	50.0	88.0	72.1
Appendicitis	18		66.7		27.8	
Healthy	50	77	74.0	61.0	84.0	63.6
Acute appendicitis	27		37.0		25.9	
Healthy	50	77	32.0	48.1	86.0	75.3
Pancreatitis	27		77.8		55.6	
Healthy	50	89	56.0	50.6	90.0	78.7
DBU	39		43.6		76.92	
Healthy	50	160	18.0	59.4	48.0	72.5
All kinds of diseases	110		78.2		81.8	

According to the experiment results, when LDA is applied on single-period pulse signals, the results are much better for each group of classification. In addition, increase of feature dimension can not always improve the classification results.

4   Conclusion and Discussion

In this research, we proposed to use a single period of pulse signals to do pulse diagnosis. Since the pulse signal was adopted in a noisy environment where the signal itself is sensitive to the noise, it seems that the single period of pulse signal might be more representative than the whole series of the signal. When we combined LDA and the KNN algorithm to recognize healthy people and patients, the experimental results also showed that single-periodic signal can produce a much higher accuracy than the multi-periodic pulse signal. This result suggests that single-periodic waveform of wrist pulse signal contains a plenty of information that can distinguish healthy people from patients. Moreover, this partially shows that noisy multi-periodic pulse signal might weaken the ability of LDA. Indeed, the noise might cause that the multi-periodic pulse signal has a complex distribution, for which LDA usually cannot perform well [19]. In the future work, we will attempt to draw characteristic data such as the frequency and amplitude of vibration from the pulse signal and to use them as features for pulse diagnosis.

Acknowledgement

This work is supported by the NSFC/SZHK-innovation funds under Contract No.60803090 and SG2008101000 03A in China, and the Natural Scientific Research Innovation Foundation in Harbin Institute of Technology, Key Laboratory of Network Oriented Intelligent Computation, Shenzhen.

References

1. Zhu, L., Yan, J., Tang, Q., Li, Q.: Recent progress in computerization of TCM. Journal of Communication and Computer 3(7) (2006)
2. Lukman, S., He, Y., Hui, S.: Computational methods for traditional Chinese medicine: a survey. Computer Methods and Programs in Biomedicine 88, 283–294 (2007)
3. Wang, H., Cheng, Y.: A quantitative system for pulse diagnosis in Traditional Chinese Medicine. In: Proceedings of the 27th IEEE EMB conference (2005)
4. Lau, E., Chwang, A.: Relationship between wrist-pulse characteristics and body conditions. In: Proceedings of the EM 2000 conference (2000)
5. Hammer, L.: Chinese pulse diagnosis—contemporary approach. Eastland Press (2001)
6. Zhu, L., Yan, J., Tang, Q., Li, Q.: Recent progress in computerization of TCM. Journal of Communication and Computer 3(7) (2006)
7. Li, S.Z.: Pulse Diagnosis. Paradigm, Kent (1985)
8. Amber, R.B.: Pulse Diagnosis: Detailed Interpretations for Eastern & Western Holistic Treatments. Aurora, London (1993)
9. Leonard, P., Beattie, T., Addison, P., Watson, J.: Wavelet analysis of pulse oximeter waveform permits identification of unwell children. Emerg. Med. J. 21, 59–60 (2004)
10. Chen, Y., Zhang, L., Zhang, D., Zhang, D.: Computerized Wrist Pulse Signal Diagnosis Using Modified Auto-Regressive Models. Journal of Medical Systems (2009) (available online)
11. Xu, Y., Yang, J.-y., Jin, Z.: Theory analysis on FSLDA and ULDA. Pattern Recognition 36(12), 3031–3033 (2003)
12. Xu, Y., Yang, J.-y., jin, Z.: A novel method for Fisher discriminant Analysis. Pattern Recognition 37(2), 381–384 (2004)
13. Foley, D.H., Sammon Jr., J.W.: An optimal set of discriminant vectors. IEEE Trans. Comput. 24(3), 281–289 (1975)
14. Jin, Z., Yang, J.Y., Hu, Z.S., Lou, Z.: Face recognition based on the uncorrelated discriminant transformation. Pattern Recognition 34(7), 1405–1416 (2001)
15. Mitchell, T.: Machine Learning, pp. 231–233. McGraw-Hill, New York (1997)
16. Hughes, G.F.: On the mean accuracy of statistical pattern recognizers. IEEE Trans. Inform. Theory 14(1), 55–63 (1968)
17. Fukunaga, K.: Introduction to Statistical Pattern Recognition. Academic Press, New York (1990)
18. Jin, Z., Yang, J.-Y., Tang, Z.-M., Hu, Z.-S.: A theorem on the uncorrelated optimal discriminant vectors. Pattern Recognition 34(10), 2041–2047
19. Xu, Y., Song, F.: Feature extraction based on a linear separability criterion. International Journal of Innovative Computing. Information and Control 4(4), 857–865 (2008)

Computerized Wrist Pulse Signal Diagnosis Using KPCA

Yunlian Sun[1], Bo Shen[1], Yinghui Chen[2], and Yong Xu[1]

[1] Key Laboratory of Network Oriented Intelligent Computation,
Shenzhen graduate school, Harbin Institute of Technology,
518055 Shenzhen, China
[2] Biometrics Research Centre, Department of Computing,
Hong Kong Polytechnic University, Kowloon, Hong Kong
jiam469@163.com, shenbo101@gmail.com,
csychen@comp.polyu.edu.hk, laterfall2@yahoo.com.cn

Abstract. Wrist pulse signals can reflect the pathological changes of a person's body condition due to the richness and importance of the contained information. In recent years, the computerized pulse signal analysis has shown a great potential to the modernization of traditional pulse diagnosis. In this paper, we attempted to use the wrist pulse signals collected by a Doppler ultrasonic blood analyzer to perform wrist pulse signal diagnosis. We first cropped the wrist pulse signal to obtain the single-period waveform, and then employed KPCA to extract features from the waveform. Finally, we used a nearest neighborhood classifier to classify the extracted features. We adopted a wrist pulse signal dataset, which includes pulse signals from both healthy persons and patients. Several experiments on the dataset were carried out and the results show that our developed approach is feasible for computerized wrist pulse diagnosis.

Keywords: Wrist pulse diagnosis, kernel principal component analysis, feature extraction.

1 Introduction

As one of the most important diagnostic methods in traditional Chinese medicine, pulse diagnosis has been practiced for thousands of years in oriental countries [1-6]. In traditional pulse diagnosis, the wrist pulse signals, which are engendered by the fluctuation of blood flow in the radial artery, contain rich and vital information. This information can be used in aid of diagnosing some diseases, such as hypertension, anemia and diabetes [7], because the pathologic changes in human body can be reflected from the variations of the shape, amplitude and rhythm of wrist pulses [7]. As a result, the experienced practitioners can identify the health status of a person by feeling the pulse beating at the measuring position of the radial artery.

With the aid of pulse diagnosis, it is convenient and invasive to deduce the location and extent of pathological changes of a person's body condition. However, the wrist pulse diagnosis is heavily dependent on the practitioner's subjective experience [2]. Different practitioners may give inconsistent diagnosis results for the same patient [2,3]. Therefore, to make pulse diagnosis more accurate, developing computerized pulse signal analysis techniques deserves more in-depth research.

D. Zhang and M. Sonka (Eds.): ICMB 2010, LNCS 6165, pp. 334–343, 2010.

Generally speaking, the computerized pulse signal diagnosis includes three major modules: data collection and pre-processing, feature extraction and pattern classification. Recently, a collection of related techniques have been developed to promote pulse diagnosis. In the first module, Doppler ultrasonic devices have been widely used for collecting the wrist pulse signals because of its noninvasive and painless merit [8]. In the feature extraction module, quite a few methods have been put forward to analyze the digitized pulse signals [9-15]. Leonard et al. [9] found that it was possible to tell healthy and unwell children apart by applying wavelet analysis to pulse signals. In [10], a modified Gaussian model was adopted to fit the pulse signal and the modeling parameters were then taken as features. Useful features were extracted by the use of wavelet transform from carotid blood flow signals in [11]. Chen et al. [12] developed an auto-regressive based method to extract the pulse signal features. Moreover, through the instrumentality of pulse signals, it is capable of determining human sub-health status using linear discriminate classifiers for some researchers [13-14]. Finally, the last module can classify the signals into different groups. A number of statistical methods can be used for the classification task, such as support vector machine (SVM) classifier [15-16] and Bayes classifier [5]. As artificial neural network methods [17] can also be used for the purpose of classification, Chiu et al. used the self-organizing neural network to classify the pulse signals [18].

In this paper, we establish a complete framework for computerized pulse signal diagnosis using KPCA. Firstly, a Doppler ultrasonic blood analyzer (Edan Instruments, Inc.) is used to collect the wrist pulse signals. Then a series of procedures are adopted to pre-process the collected pulse signals. In the feature extraction stage, we propose to use KPCA for feature extraction because of the nonlinearity in this method. Finally, for pattern classification, a nearest neighborhood classifier [19] is employed to tell patients from healthy persons. In our experiments, we use a dataset, which includes 100 healthy persons and 148 patients with different diseases, to measure the performance of the developed approach. Four kinds of diseases are involved in the dataset: 46 patients with pancreatitis (P), 42 with Duodenal Bulb Ulcer (DBU), 22 with appendicitis (A) and 38 with acute appendicitis (AA). The encouraging results show that KPCA is an effective tool to the computerized pulse signal diagnosis.

The remainder of this paper is organized as follows. Section 2 describes the pulse signal collection and pre-processing. Feature extraction and classification is presented in detail in Section 3. Section 4 performs experiments to demonstrate the effectiveness of the proposed method. Finally, we conclude the paper in Section 5.

2 Data Collection and Pre-processing

In our scheme, the wrist pulse signals are collected by a USB-based Doppler ultrasonic blood analyzer. The process of signal acquisition can be divided into three stages. First is to locate a rough area where the fluctuation of pulse at this area is bigger than other areas. Then the ultrasound probe is put on and moved around the rough area slowly and carefully until the most significant signal is detected. To get the most significant signal, the angle of the probe needs to be adjusted against the skin constantly during the movement. Finally, the wrist pulse signals can be recorded and saved.

To make the measurement error reduced, it is necessary to repeat the three steps mentioned above several times, i.e., we collect several measurements of a subject. The best one from these measurements of a subject is selected by manual. We represent the acquired signals in the form of Doppler spectrograms (see Figure 1 (a) [12]). The raw pulse signals are often distorted by noise and baseline drift. Thus it is necessary to pre-process the collected data before extracting features. In the pre-processing, the maximum velocity envelope of each spectrogram is first extracted and normalized for the purpose of reducing the dimension of the signal. Then we remove the low-frequency drift and high-frequency noise from the normalized signals without the phase shift distortion. In our work, a 7-level 'db6' wavelet transform [20] is adopted for de-noising and drift removal. Figure 1 (b) illustrates the maximum velocity envelope of a pulse spectrogram. The result of noise and drift removal is shown in Figure 1 (c).

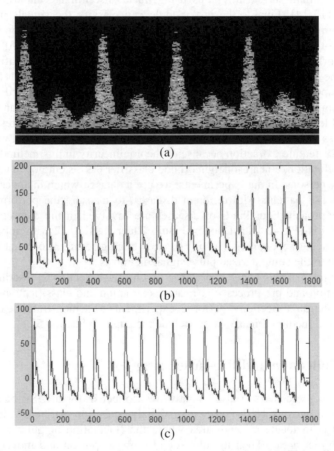

Fig. 1. The Doppler spectrogram before and after pre-processing. (a): The collected Doppler spectrogram. (b): Maximum velocity envelope of the Doppler spectrogram. (c): The signal after de-noising and drift removal.

The wrist pulse signal is not a random process but a semi-periodic signal with regularly occurring systolic and diastolic waves [21], as shown in Figure 1. For the convenience of the following feature extraction, the pulse signal is first partitioned into several single-period waveforms. To handle the partition problem, a segmentation scheme described in [22] is adopted. After the segmentation, each single-period waveform is resized with a fixed length. Figure 2 shows a one-period pulse signal.

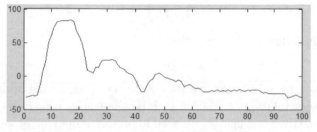

Fig. 2. A typical single-period pulse signal

3   Feature Extraction and Classification

3.1   Feature Extraction Using KPCA

The principal component analysis (PCA) methodology plays an important role in the field of pattern recognition and has been used as the feature extraction procedure of a number of classification problems such as face recognition, handwritten digital recognition and palmprint recognition [19, 23, 24, 25]. Kernel principal component analysis (KPCA) [26, 27] is a nonlinear extension of linear principal component analysis. And the implementation of KPCA is equivalent to the implementation of PCA in the feature space induced by a kernel function [27]. It seems that two procedures are implicitly contained in the implementation of KPCA. The first procedure maps the original sample space into the feature space and the second procedure carries out PCA in the feature space [27].

It is clear that the classification of pulse signals into different groups is not a completely linearly separable problem. In our research, for each single-period pulse signal, we use KPCA to perform feature extraction. In other words, KPCA is expert in dealing with non-linear problems. Now we consider a training dataset $\{x_1, x_2, ..., x_l\}$. Each sample x_i ($i = 1, 2, ..., l$), which is a d-dimensional column vector, represents a one-period pulse signal. Suppose these samples are from two different classes: ϖ_1 and ϖ_2, where ϖ_1 represents the class of signals from health persons, while ϖ_2 is used to indicate the class of signals from patients. l_1 and l_2 denote numbers of samples in the two classes, respectively. Here we give a detailed description of the KPCA feature extraction scheme.

1) Kernel matrix building. Given a suitable kernel function, the kernel matrix can be computed as

$$K_{i,j} = k(x_i, x_j) \quad i, j = 1, \dots, l \ .$$

(1)

where K is a $l \times l$ matrix and $k(x_i, x_j)$ is the kernel function.

2) Kernel matrix centralization. With the kernel matrix computed above, a centering process need to be carried out as follows:

$$\hat{K} = K - 1_l K - K 1_l + 1_l K 1_l, \quad (1_l)_{ij} = 1/l, \quad i, j = 1, 2, \dots, l \ .$$

(2)

3) Solving the following eigen-equation:

$$\hat{K}\alpha = \lambda\alpha \ .$$

(3)

We denote by $\alpha^{(1)}, \alpha^{(2)}, \dots, \alpha^{(m)}$ the m eigen-vectors associated with the first m largest eigen-values $\lambda_1^\alpha, \lambda_2^\alpha, \dots, \lambda_m^\alpha$ of $\hat{K}$. It is clear that m is no larger than l.

4) Projection. Given a sample x, we need to project x onto the m eigen-vectors in order to get the feature vector of x. The feature vector y is computed as

$$y = \left(\frac{\sum_{j=1}^{l} \alpha_j^{(1)} k(x_j, x)}{\sqrt{\lambda_1^\alpha}} \quad \frac{\sum_{j=1}^{l} \alpha_j^{(2)} k(x_j, x)}{\sqrt{\lambda_2^\alpha}} \quad \dots \quad \frac{\sum_{j=1}^{l} \alpha_j^{(m)} k(x_j, x)}{\sqrt{\lambda_m^\alpha}} \right)^T \ .$$

(4)

where $\alpha_j^{(i)}$ denotes the jth component of the vector $\alpha^{(i)}$. It is clear that y is a m-dimensional column vector. After feature extraction, we can obtain l corresponding feature vectors $y_1, y_2, \dots, y_l$.

Given a kernel function, there is at least one kernel parameter. KPCA can be implemented only if a value is assigned to the parameter. It is important to point out that different parameter values can produce different feature extraction results. Therefore, when performing classification task using the extracted features, different classification accuracies will be induced. Except the kernel parameters, m, which is the number of selected eigen-vectors as well as the dimension of feature vectors, is also a parameter. It is clear that different values of m can induce different feature extraction results and then lead to different classification results. In our experiments described in the next section, we will observe the effects caused by different values of the two kinds of parameters. Through the observation, we can determine suitable values for these parameters.

3.2 Classification

After feature extraction, the selected features are then adopted as inputs to the classifier for further pattern classification. The classification aims to determine from these features that whether the pulse signals are from healthy persons or from patients with a certain disease. In this paper, a nearest neighborhood classifier is adopted.

Given a testing sample x, KPCA feature extraction is first carried out for the computation of the corresponding feature vector y. Afterward, the nearest neighborhood classifier with Euclidean distance is employed for classification. The distance between y and y_i is computed as

$$d(y, y_i) = \|y - y_i\| \, . \tag{5}$$

where y_i is the feature vector of the corresponding training sample x_i. We say x belongs to the same class of x_j if $d(y, y_j) = \arg \min_i d(y, y_i)$.

4   Experiments

In our experiments, we use a pulse dataset, which was established by collaborating with the Harbin 211 hospital (Harbin, Heilongjiang Province, China), to test the performance of the proposed method. There are a total of 248 wrist pulse signals included in this dataset. And all these signals were collected by a Doppler ultrasonic blood analyzer from both healthy persons and patients with different diseases, i.e., 100 healthy persons (H), 46 patients with pancreatitis (P), 42 with Duodenal Bulb Ulcer (DBU), 22 with appendicitis (A) and 38 with acute appendicitis (AA). In the following experiments, we split the dataset into a training dataset and a testing dataset. For each of the two groups (healthy persons and patients), half of the samples, which are randomly selected, are used for the training purpose, and the remaining are for the testing use.

In the pre-processing, the collected signals are first cropped to obtain single-period waveforms with a fixed length. Here we set the length to 100. Figure 3 illustrate five single-period pulse signals, which are from healthy persons and patients with four kinds of diseases, respectively. Afterward, the method described in Section 3 is applied to the single-period signals to get feature vectors. The kernel function employed in our experiments is the Gaussian kernel function $k(x, y) = \exp\left(-\|x - y\|^2 / 2\eta\right)$. In KPCA feature extraction, we should determine the values of the two parameters, η and m. To address this problem, we design the following experiment. As was described previously, the training and testing dataset are randomly selected. In consideration of the randomicity, we perform the corresponding experiment more than once for each parameter value. Here the times of performing the experiment is fixed to 20. As a result, there are a total of 20 classification accuracies for each parameter value. Then the maximum, minimum and mean of these accuracies are calculated. Table 1 and Table 2 show the influence of the two parameters on the classification accuracy for healthy people versus patients with pancreatitis. In the two tables, "min", "max" and "mean" denote the maximum, minimum and mean of the observed 20 classification accuracies, respectively. According to the result shown in the two tables, we choose $\eta = 1.5 \times 10^5$ and $m = 20$ for the following experiments.

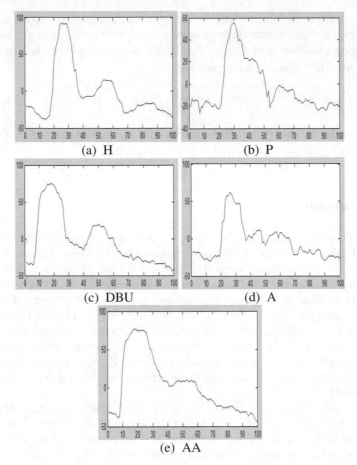

Fig. 3. The single-period signals from healthy persons and patients with four kinds of diseases. (a): Healthy people. (b): Patients with pancreatitis. (c): Patients with Duodenal Bulb Ulcer. (d): Patients with appendicitis. (e): Patients with acute appendicitis.

Table 1. The influence of m on the classification accuracy (%) for healthy people versus patients with pancreatitis (η is fixed to 1.5×10^5)

m	min	max	mean
2	61.64	75.34	69.24
10	79.45	93.15	87.74
20	79.45	94.52	89.38
35	82.19	95.89	89.79
50	83.56	94.52	90.20
70	84.93	94.52	89.59

Table 2. The influence of η on the classification accuracy (%) for healthy people versus patients with pancreatitis (m is fixed to 20)

η	min	max	mean
5.0×10^3	49.32	82.19	67.53
1.0×10^4	72.60	94.52	79.93
5.0×10^4	79.45	95.89	88.63
1.0×10^5	83.56	93.15	88.90
1.5×10^5	83.56	97.26	89.93
2.0×10^5	78.08	95.89	88.50
3.0×10^5	78.08	95.89	88.42

The classification accuracies for healthy people versus patients with four kinds of diseases on the testing dataset are listed in Table 3. The last line of the table illustrates the accuracy for healthy people versus patients with mixed kinds of diseases. In our experiments, we use the wavelet transform method [15], the auto-regressive method [28] and the modified auto-regressive method [10] for comparison. It can be seen from this table that the accuracies for healthy people versus patients with appendicitis and patients with mixed kinds of diseases are no larger than the corresponding accuracies obtained by the modified auto-regressive method. However, it is possible for the accuracy obtained by KPCA to exceed that obtained by the modified auto-regressive method if we adjust the value of η or m to a suitable value. For example, when m is increased to 50, the accuracy of KPCA can be increased to 91.8% which is a higher accuracy than that of modified auto-regressive method. These experimental results show that features extracted by KPCA work well for wrist pulse signal classification.

Table 3. The classification accuracy (%) for healthy people versus patients

Class	WPT[15]	AR[25]	AR&SW[10]	KPCA
H-P	84.8	86.3	90.9	95.9
H-DBU	85.4	82.3	88.0	93.0
H-A	76.1	88.2	91.2	90.2
H-AA	72.4	77.8	80.8	85.5
H-D	80.0	83.7	87.3	82.3

5 Conclusion

The wrist pulse signal carries rich and important information about the health status of a person. Hence, the extraction of this information from wrist pulse signals is significant for the computerized pulse diagnosis. In this paper, we present a framework for computerized pulse diagnosis with the investigation concentrated on the feature extraction and pattern classification. The pulse signals collected by a

Doppler ultrasonic blood analyzer were first pre-processed. Then KPCA was applied to the pre-processed data to extract useful features. Finally, the extracted distinctive features were equipped for a nearest neighborhood classifier for classification. In the experiments, we use a wrist pulse dataset, which includes pulse signals from 100 healthy people and 148 patients, to test the performance of the proposed approach. The experimental results showed that we can achieve an accuracy of over 90% when separating the healthy people from the patients with some particular kind of diseases. And the accuracy of telling healthy persons and patients with mixed kinds of diseases apart can be reached over 80%. These results show that our method is feasible for current computerized pulse diagnosis.

Acknowledgments. This work is partially supported by Program for New Century Excellent Talents in University (No. NCET-08-0156), National Natural Science Foundations of China under Grants Nos. 60803090 and 60902099, 863 Program Project under Grant No. 2007AA01Z195, the CERG fund from the HKSAR Government and the central fund form Hong Kong Polytechnic University for supporting.

References

1. Lukman, S., He, Y.L., Hui, S.C.: Computational Methods for Traditional Chinese Medicine: A Survey. Computer Methods and Programs in Biomedicine 88, 283–294 (2007)
2. Hammer, L.: Chinese Pulse Diagnosis-Contemporary Approach. Eastland Press, Vista (2001)
3. Zhu, L., Yan, J., Tang, Q., Li, Q.: Recent Progress in Computerization of TCM. Journal of Communication and Computer 3 (2006)
4. Wang, K., Xu, L., Zhang, D., Shi, C.: TCPD based Pulse Monitoring and Analyzing. In: Proceedings of the 1st ICMLC Conference (2002)
5. Wang, H., Cheng, Y.: A Quantitative System for Pulse Diagnosis in Traditional Chinese Medicine. In: Proceedings of the 27th IEEE EMB Conference (2005)
6. Lau, E., Chwang, A.: Relationship between Wrist-Pulse Characteristics and Body Conditions. In: Proceedings of the EM 2000 Conference (2000)
7. Shu, J., Sun, Y.: Developing Classification Indices for Chinese Pulse Diagnosis. Complement. Ther. Med. 15, 190–198 (2007)
8. Powis, R., Schwartz, R.: Practical Doppler Ultrasound for the Clinician. Williams and Wilkins, Baltimore (1991)
9. Leonard, P., Beattie, T., Addison, P., Watson, J.: Wavelet Analysis of Pulse Oximeter Waveform Permits Identification of Unwell Children. Emerg. Med. J. 21, 59–60 (2004)
10. Chen, Y., Zhang, L., Zhang, D., Zhang, D.: Computerized Wrist Pulse Signal Diagnosis using Modified Auto-Regressive Models. Journal of Medical Systems (2009), doi:10.1007/s10916-009-9368-4
11. Zhang, Y., Wang, Y., Wang, W., Yu, J.: Wavelet Feature Extraction and Classification of Doppler Ultrasound Blood Flow Signals. J. Biomed. Eng. 19, 244–246 (2002)
12. Chen, Y., Zhang, L., Zhang, D., Zhang, D.: Wrist Pulse Signal Diagnosis using Modified Gaussian Models and Fuzzy C-Means Classification. Medical Engineering & Physics 31, 1283–1289 (2009)
13. Lu, W., Wang, Y., Wang, W.: Pulse Analysis of Patients with Severe Liver Problems. IEEE Eng. Med. Biol. Mag. 18, 73–75 (1999)

14. Zhang, A., Yang, F.: Study on Recognition of Sub-Health from Pulse Signal. In: Proceedings of the ICNNB Conference, vol. 3, pp. 1516–1518 (2005)
15. Zhang, D., Zhang, L., Zhang, D., Zheng, Y.: Wavelet-based Analysis of Doppler Ultrasonic Wrist-Pulse Signals. In: Proceedings of the ICBBE Conference, vol. 2, pp. 589–543 (2008)
16. Burges, C.: A Tutorial on Support Vector Machines for Pattern Recognition. Data Mining and Knowledge Discovery 2, 121–167 (1998)
17. Mitchell, T.: Machine Learning. China Machine Press, Beijing (2008) (in Chinese)
18. Chiu, C., Yeh, S., Yu, Y.: Classification of the Pulse Signals based on Self-Organizing Neural Network for the Analysis of the Autonomic Nervous System. Chinese Journal of Medical and Biological Engineering 16, 461–476 (1996)
19. Bian, Z., Zhang, X.: Pattern Recognition. Tsinghua University Press, Beijing (2000) (in Chinese)
20. Xu, L., Zhang, D., Wang, K.: Wavelet-based Cascaded Adaptive Filter for Removing Baseline Drift in Pulse Waveforms. IEEE Transactions on Biomedical Engineering 52, 1973–1975 (2005)
21. Xia, C., Li, Y., Yan, J., Wang, Y., Yan, H., Guo, R., et al.: A Practical Approach to Wrist Pulse Segmentation and Single-Period Average Waveform Estimation. In: The ICBEI Conference, pp. 334–338 (2008)
22. Yang, W., Zhang, L., Zhang, D., Yang, J.: Computerized Wrist-Pulse Signals Diagnosis using ICPulse. In: The ICBBE Conference, Beijing (2009)
23. Zhang, D., Song, F., Xu, Y., Liang, Z.: Advanced Pattern Recognition Technologies with Applications to Biometrics. IGI Global Press (2008)
24. Xu, Y., Zhang, D., Yang, J.: A Feature Extraction Method for Use with Bimodal Biometrics. Pattern Recognition 43, 1106–1115 (2010)
25. Xu, Y., Zhang, D., Yang, J., Yang, J.: An Approach for Directly Extracting Features from Matrix Data and its Application in Face Recognition. Neurocomputing 71, 1857–1865 (2008)
26. Schölkopf, B., Smola, A., Müller, K.: Kernel Principal Component Analysis. In: Gerstner, W., Hasler, M., Germond, A., Nicoud, J.-D. (eds.) ICANN 1997. LNCS, vol. 1327, pp. 583–588. Springer, Heidelberg (1997)
27. Xu, Y., Zhang, D., Song, F., Yang, J., Jin, Z., Li, M.: A Method for Speeding up Feature Extraction based on KPCA. Neurocomputing 70, 1056–1061 (2007)
28. Chen, Y., Zhang, L., Zhang, D., Zhang, D.: Pattern Classification for Doppler Ultrasonic Wrist Pulse Signals. In: The ICBBE Conference, Beijing (2009)

Framework of Computer-Assisted Instruction and Clinical Decision Support System for Orthodontics with Case-Based Reasoning

Shu-Li Wang[1] and Shih-Yi Yeh[2]

[1] Department of Dental Technology, Central Taiwan University of Science and Technology,
Taichung, Taiwan, 406
slwang591555@gmail.com
[2] Institude of Healthcare Information Management, Chia-Yi, Taiwan, 621
se2412000@hotmail.com

Abstract. Malocclusion is a very common oral disease. People with malocclusion often appeal to orthodontics for treatment. Since the treatment of orthodontics often takes several years, simulation and animation with 3D visual effect provides a good communication vehicle between patients and dentists before the treatment. In this paper, we developed a framework of CAI (computer assisted instruction) system with case-based reasoning for orthodontics. The system collects a lot of past cases of orthodontic and makes a 3D animation for each case, which simulates the moving and morphing from the pre- to the post-treatment. When a dentist encounters a new case, a past case will be selected based on the distance between the new case and the past one. A 3D animation of this most similar case will also be displayed. The system also incorporated pathological rules used for classifying the cases with different symptoms. Based on the rules, the search space for the similar cases can be reduced.

1 Introduction

Recently, information systems have been widely applied to patient education and clinical decision support. The system can help patients understand the medical knowledge and enhance the performance of health care.

Malocclusion is a very common oral disease. Previous studies have pointed out that chewing habit of modern life has caused the ratio of malocclusion to increase, especial the young people in the country whose life style urbanizes more deeply [1]. A variety of empirical studies agreed with this argument that the life style of modernization affects the oral health [2]. For malocclusion, orthodontics is the effective treatment. In the market there are many orthodontic-related information systems, such as dolphin 3D [3], InVivoDental [4], 3DMDvultus [5], which aim at supporting physicians in diagnosis. Those systems are designed in the viewpoint of practitioners: their operations were complex; scope of functions is too gigantic but does not provide educational materials to patients; and interaction between patients and doctors is insufficient. In this paper, we designed a 3D CAI (3-Dimensional Computer-Assisted Instruction) system in the viewpoint of patients, which first explores the rationale of

D. Zhang and M. Sonka (Eds.): ICMB 2010, LNCS 6165, pp. 344–352, 2010.

dentists and displays 3D animations to facility the communication between dentists and patients. Since a variety of malocclusion, like protrusion, overbite, crowding, high canine, spacing, diastema, crossbite, etc, this system adopted the techniques of case-based reasoning (CBR) to find a suitable scenario from the pre-constructed case-based database to demonstrate a 3D animation to patients. Case-based reasoning is an artificial intelligence approach that exploits on past experiences to solve current problems. CBR can be seen as a perspective on human cognition. The current developments in the application of CBR have proven to be efficient, especially for problem solving and decision support in the health sciences [6]. Nevertheless, none of the researchers used CBR in the field of orthodontics. In this paper, we utilized CBR in the field of orthodontics, exploring the degree of malocclusion, which was used as a measure to get the similarity of the new case and existing cases based on a variety of symptoms of the cases. The system could find the most similar cases and display them in 3D animation in accordance with the results of similarity computing.

In addition, this system also provided a clinical knowledge to generate options of treatments automatically while inputting the symptoms of patients. We conducted interviews of the dentists who engaged in the system development. The results show that most of them satisfy the effectiveness in communication, agree with the scenarios generated from the system, and get decision support from the options of treatments provided by the system.

In section 2 we introduce the past researches on orthodontic CAI system. The framework of the system associated with the case-base reasoning methods as well as interview result about the effectiveness of the system is explained in section 3. Finally, a conclusion and the future work are given in section 4.

2   Related Work

Though malocclusion is prevailing recently, there are only a few papers focusing on the orthodontic CAI-related researches. In this paper, we classified these CAI-related researches into three types, namely, dental education system, orthodontic case system, and orthodontic simulation system. The dental education system [7] emphasizes the teaching functions that gather the teaching materials as a library and then organize the library in sections such that the novices and students can easily access the desired knowledge through the classified sections of the library. In addition, the system provides a self-test subsystem, allowing the learners to have an on-line test to verify their learning effectiveness. Such a system is very intuitive for use, similar to the traditional education, so the users can easily adapt themselves to the lessons provided by the system. For professional learners, dental education system is an ideal way to learn the knowledge of dentistry and orthodontics. However, it is not an effective way for patients, novices and students, who cannot learn too abstruse knowledge through the professional system.

The second type of the systems collects a lot of clinical cases to show the treatment of each case associated with the pre- and post-conditions of that treatment [8]. The system is based on the existing orthodontic clinical cases, each of which is annotated with some attributes, such as age, Angel-type, crowding condition, and special features like rotation of tooth, and so on. The system provides educational materials, including diagnoses, malocclusion patterns, photos before and after the treatment, the corresponding treatments, and so on, allowing the students to watch and learn the experiences of the past cases. Such a system is a problem-oriented one that the users can get the related cases after inputting the attributes of desired cases. The system returns the related cases, including photos of a patient's facial appearance, occlusion, tooth model, X-rays of skull and other related photos. This system provided real cases to aid the learning; however, the system must take a lot of efforts to collect many cases for demonstration. In addition, the system needs to get the admission of patients; otherwise, the demonstration will infringe the privacy and confidential of patients.

The above two types of CAI systems are helpful but too instinct that they only provide a static knowledge to the users. It is desired that the patients can have a dynamical explanation to describe the procedure, process and result of the treatment, such as morphing and moving from pre- to post-treatment and possible appearance of options. The third type of the CAI system is a dynamical modeling one. The dentists first scan the tooth model of a patient and then generate a 3D simulation. The dentists can control various variables, or suggested by the patients, to affect the simulation results that the dentists and patients can estimate the outcome of treatment check whether the results can reach their expectation [9], [10]. With the morphing/moving and 3D-display functions, the dentists can build a better communication channel between dentists and patients to increase the satisfaction of patients. Although the simulation can show the processes and outcome of treatments, the system is too tedious to prepare a lot of complicated procedures that the patients may wait for days to get the simulation results.

To reduce the waiting time and tedious procedures of the simulation system, this paper designed an interactive CBIDSS (case-based instruction and decision support system) system that can dynamically and quickly present the process of treatments with 3D animation to facilitate the communication between dentists and patients. The system can also generate the similar scenario automatically after the users enter the physiological attributes of patients. The system can use the rules to weight each physiological attributes and apply case-base reasoning to find a suitable scenario for displaying. With these features, orthodontists can quickly find the most similar case to communicate with patients.

3   Materials and Methods

3.1   Materials and System Architecture

We first collected more than 600 cases of treatments for orthodontics from some hospital in Taiwan from 1995 to 2002. The we filtered the cases and remained 66 cases that belonged to Angle's class II and have a complete treatment history, relevant medical records, and sufficient and clear photos (such as oral photos, cephalometric photos and X-rays). These cases are stored as a knowledge base, and for each case, we draw 3D animation starting from the pre-treatment to the post treatment according to the procedures adapted by dentists.

Based on the knowledge base and the corresponding animations, we construct the system, CBIDSS, which contains knowledge base and diagnostic rules. The diagnostic rules, provided by dentists and extracted from the textbooks, are validated by the cases in the knowledge base. For a new case, CBIDSS will search the rule that matches the symptoms of the case entered by the dentist. The dentists can also enter the new rule of diagnosis into the rule base. The CBR module is responsible for finding the similar cases to users, which can assist patients to listen to the doctors on the content of treatment. The steps as follows (see Figure 1):

Step 1: Initially, the pre-existing diagnosis rules or customized rules are imported to rule base and knowledge base;

Step 2: Doctors diagnose patients and enter clinical records. System accesses clinical records and sends physiological information to knowledge module;

Step 3: This system searches the matching knowledge/rule with the clinical records and physiological information;

Step 4: Knowledge module automatically generates the set of combined symptoms in accordance with patient's condition;

Step 5: The set of combined symptoms are sent into inference engine to find the most similar cases.

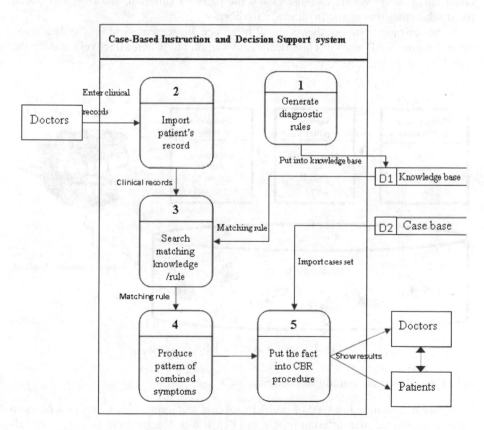

Fig. 1. System Data Flow Diagram

3.2   Rule-Based Reasoning (RBR) to Classify the Symptoms

In our system, rule-based reasoning is used to increase the accuracy and efficiency of inference by pathological parameters. The following describes the concept of reasoning model of pathological logic; the mode expounds how the pathological parameters affect the searching results in CBR procedure, as the following (figure 2):

The patient model in the system consists of skeleton pattern, physiological parameter and symptom type. With the patient model,

First, we have to consider the patient identity, because different physical characteristics produce different results that depict the basic bones. Such as the patient's sex, age, nationality and other information is different, and their bones would have a different profile. This phase derive the appropriate structure of skeleton of patients.

In the basic skeleton has been confirmed, we identified a patient's Angle's class [11] to find the skeletal pattern. However, most of the patients have multiple symptoms at the same time, so the actual situation can not only use Angle's pattern to explain. Therefore, we must combine Angle's pattern and the pattern of symptoms (such as maxillary protrusion, snaggletooth, overbite, etc.) to explain a patient's complete condition. Finally, system can understand the patient's situation, and began to search for similar cases that stored in clinical case library.

In the reasoning process, the results of inference are accuracy even if the data is too large, because RBR used in classification of symptoms which effectively reduce the searching scope and cost.

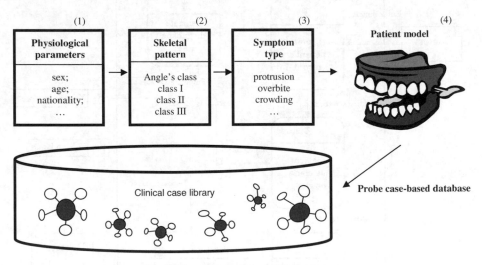

Fig. 2. Flow Chart of the RBR Procedure

3.3   Case-Based Reasoning to Find Similar Case

Case-based reasoning is a type of problem solving technique which uses past cases to solve new, unseen and different problems [12]. It uses the attribute parameter to calculate the value of degree of similarity of cases. This study used the dental parameters and symptom parameters to import the CBR processes, they are described as follows:

1. **Dental Parameters:** Each tooth in accordance with the pathological logic given number, teeth were divided into maxillary area and mandibular area, and then subdivided into the left zone and right zone. From the midline to the distal in sequence number, incisor is numbered as 1; lateral incisor is numbered as 2, and so on.

2. **Symptom parameters:** Because different symptoms have their stress on the specific teeth, each tooth has a different degree of importance. Our system uses symptom parameters to enable the teeth are given different weights.

In the similarity calculation, the weights of attributes are a very important concept, which affect the accuracy of results of inferring; and the next paragraph will explain the relationship between weights and parameters of symptoms.

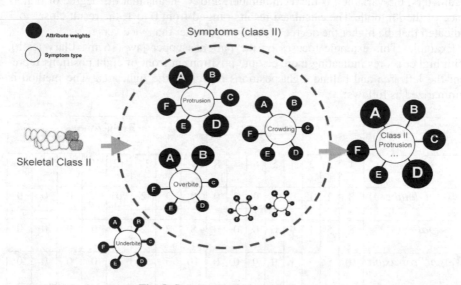

Fig. 3. Symptoms guide attribute weights

Symptoms guide attribute weights: To enable the system to complete the presentation of patient condition, we must base on the type of symptoms to set the weights of all teeth. Figure 3 shows the each feature element that the properties of the different symptoms that have different weight configurations, combination of skeletal pattern and symptom parameters to get a complete model of the symptoms.

Each tooth was seen as independent objects and has different weights; all CBR pre-operation has been completed.

Case-based reasoning model / Similarity model: In order to calculate the similarity values, we sequentially compare the tooth pairs of past and current cases, similarity value is calculated by:

$$SIM(p,c) = \frac{1}{n} \sum_{i=1}^{n} \frac{1}{1 + |p_i - c_i|} \tag{1}$$

The function SIM is used to calculate the degree of similarity between the two cases, n is the number of object attributes that mean the teeth in the cases, p is the case of current patient, c is the old case that exists in the database, p_i is the current patient's tooth numbered as i, and c_i is the tooth numbered as i in the past case. The calculated results ranged from 0 to 1, the result closer to 1 indicated that the higher the degree of similarity. Formula 2 expresses the concept of additional weight of Symptoms.

$$WSIM(p,c) = WSIM([p_1, p_2, ..., p_n], [c_1, c_2, ..., c_n])$$
$$= (\sum_{i=1}^{n} (w_i \times SIM(p_i, c_i))) / \sum_{i=1}^{n} w_i \tag{2}$$

The variable w is the weight configuration that is set in accordance with symptom parameters; the variable i is the teeth number, and w_i means that the degree of importance of the ith tooth. The calculated results ranged from 0 to 1, the result closer to 1 indicated that the higher the degree of similarity, rather than vice versa.

Example: This example focuses on the patient's upper jaw (16 maxillary teeth). With higher scores indicating more ectopic position; in front of eight positions represent the left area and behind eight positions represent the right area. The method is summarized as follows:

	Left zone								Right zone							
i	1	2	3	4	5	6	7	8	9	10	11	12	13	14	15	16
case(in database)	8	2	1	0	0	0	0	0	9	3	2	0	0	0	0	0
patient	8	5	1	0	0	0	0	0	8	2	2	0	0	0	0	0
WeightProtrusion	10	5	2	1	0	0	0	0	10	5	2	1	0	0	0	0

$WSIM(patient, case)$

$$= (\sum_{i=1}^{16} (WeightProtrusion_i \times SIM(patient_i, case_i))) / \sum_{i=1}^{16} WeightProtrusion_i$$

$$= \{10 \times (1+|8-8|)^{-1} + 5 \times (1+|5-2|)^{-1} + 2 \times (1+|1-1|)^{-1} + 1 \times (1+|0-0|)^{-1} + 10 \times (1+|8-9|)^{-1} +$$
$$5 \times (1+|2-3|)^{-1} + 2 \times (1+|2-2|)^{-1} + 1 \times (1+|0-0|)^{-1}\} / (10+5+2+1+10+5+2+1)$$
$$= 0.711538461538462$$

3.4   3D Model

In this study, we use a standard 3D geometric model of the mandible, maxilla, and dental arch of a virtual patient to enable to show vivid treatment. The 3D model of

CBIDSS has the following features: (1) edge-loop based model of a human mouth and dentition; (2) 32 morph targets for any type of mouth expression; (3) polygonal sub-division geometry; (4) anatomically correct full set of 32 teeth; (5) skull with moving jaw, gums, teeth and muscles.

The generated mesh representations are composed of enough number of polygons and vertices to build a model with good degree of accuracy. This model is composed of 9741 vertices (9533 polygons). The state of all geometric meshes generated was kept at an ideal quality, where neither geometric mesh deformations nor mesh nonlin-earities were detected.

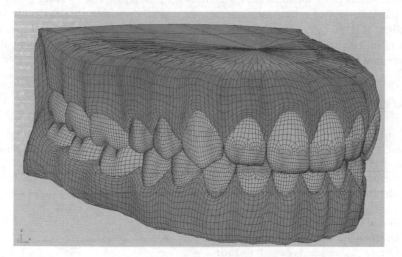

Fig. 4. 3D model: Dental appearance

For a start, system provides an interface to enter clinical records, and generate poss-ible combinations of symptoms of patient. Practitioner can use a combination of au-tomatically generated or manual options to determine the complete status of the pa-tient. The second step, doctors use the interface to set the ectopic values of teeth. Fi-nally, system in accordance with previous operation can find the suitable cases that ex-isted in case database.

Fig. 5. Case-Base Instruction and Decision Support System

4 Conclusions

This paper proposed a new framework of CAI system for orthodontics, which contains CDSS, traditional CAI, knowledge base and clinical data libraries. This paper developed a feasible model for inference of Orthodontic Case with CBR, and adopted RBR to make the results more accurate. The system offers a variety of functional modules so that doctors can quickly find similar cases to demonstrate and help doctors to explain the process and procedure of the treatment to the patient. In addition, this system can effectively improve the quality of communication between doctors and patients, tackling the information asymmetry problem of health care.

In the interview of orthodontic doctors, they give a positive attitude to the system, and admit that the system can effectively improve patient satisfaction and assist practitioners in their work. This system can give more help for new practitioners since of its abundant clinical cases.

References

1. Corruccini, R.S.: Anthropological aspects of orofacial and occlusal variations and anomalies. In: Kelley, M.A., Larsen, C.S. (eds.) Advances in Dental Anthropology, New York, pp. 295–323 (1991)
2. Proffit, W.R., Fields, H.W.: Contemporary Orthodontics, 3rd edn. YearBook Medical Pub., Chicago (2000)
3. Dolphin 3D, http://www.dolphinimaging.com/new_site/3d.html
4. InVivoDental, http://www.anatomage.com/
5. 3DMDvultus,
   http://www.3dmd.com/Products/
   ApplicationSoftware.asp#3DMDvultus
6. Bichindaritz, I., Marling, C.: Case-based reasoning in the health sciences: What's next? Artificial Intelligence in Medicine 36(2), 127–135 (2006)
7. Grigg, P., Stephens, C.D.: Review Computer-assisted learning in dentistry A view from the UK. Journal of Dentistry 26(5-6), 387–395 (1998)
8. Komolpis, R., Johnson, R.A.: Web-Based Orthodontic Instruction and Assessment. Journal of Dental Education 66(5), 650–658 (2002)
9. Makia, K., Ogawaa, N., Nakajimaa, K., Ogura-Abea, Y., Miyazakia, Y., Kubotaa, M., Kosekib, M., Inoub, N.: Biomechanical simulations for orthodontics: 3D FEM based on Cone Beam X-ray International Congress Series. Computer Assisted Radiology and Surgery 1281, 1191–1195 (2005)
10. Rodrigues, M.A.F., Gillies, D.F., Neto, M.E.B., Ribeiro, I.M.M.P., Silva, W.B.: An interactive simulation system for training and treatment planning in orthodontics. Computers & Graphics 31(5), 688–697 (2007)
11. Angle, E.H.: Treatment of malocclusion of the teeth and fractures of the maxillae: Angle's system, 6th edn. SS White dental Mfg Co., Philadephia (1990)
12. Liua, C.H., Chenb, L.S., Hsuc, C.C.: An association-based case reduction technique for case-based reasoning. Information Sciences 178(17-1), 3347–3355 (2008)

Design of an Intelligent Reporting System for Cardiac Catheterization

Fan Wu and Hsiao-Hui Li

Department of Management of Information System, National Chung Cheng University, Taiwan
miswu@ccu.edu.tw, xiasohui@gmail.com

Abstract. Cardiovascular diseases (CVDs) are the no.1 killer in the world. Since the procedure of cardiac catheterization is very complex, the reporting system for cardiac catheterization, as we knew, has not yet well developed in most of the hospitals until now. Our approach is building a system fit into physician workflow and integration the PACS and HIS. The experimental results show that we can reduce operating costs and increase healthcare efficiency by using this system. The implementation of the system paves the way toward a comprehensive EPR of healthcare industry.

1 Introduction

Because fast food culture is expanding worldwide and people do less exercise than before, cardiovascular diseases become the most dangerous killers in the world [1], [2]. According to the statistics [1], the people with cardiovascular diseases are getting younger. In recent years, minimally invasive surgery has been well developed. Cardiovascular patients can receive cardiac catheterization, instead of open heart surgery, which is more risky and painful. Since the procedure of cardiac catheterization is very complex, the reporting system for cardiac catheterization, as we knew, has not yet well developed in most of the hospitals until now. Some hospitals use the plain text editor, like Microsoft Word, to delineate the process and results of cardiac catheterization instead. It is clear that this plain text is difficult to illustrate the positions and degrees of lesions precisely; in addition, the unstructured document cannot be smoothly integrated with the patient records for browsing and interchange.

Electronic patient record (EPR) is a goal pursued by the worldwide healthcare institutions, which provides linkage with information systems among specialists and sites in uniform core data elements [3]. According to the report of the institute of Medicine (IOM), EPR can contain six formats, namely, text, graphics, images, number, sound and video [4], and the contents of EPR should include diagnoses, laboratory results, reports from exam departments, and other information [5]. US Department of Health and Human Services (HHS) hopes to achieve paperless patient record by 2010. Similarly, Taiwan Department of Health (DOH) is pushing to computerize medical records for all hospitals from 2002. Every hospital wants to carry out the policy, but an appropriate

D. Zhang and M. Sonka (Eds.): ICMB 2010, LNCS 6165, pp. 353–362, 2010.

system for cardiac catheterization that can combine the text report and full-motion images is very difficult to develop.

A computer-based system allowing doctors and medical staffs directly enter orders is an efficient way for improving practitioner performance and making patient care more safely and effectively [6], [7], [8]. A machine-readable reporting system can help practitioners to remember the needed actions, recall the available options, and recognize out-of-range values or dangerous trends while making the decisions [6]. In the past, no well developed reporting system associated with the instruments of cardiac catheterization of major manufactures, like PHILIPS, GE, Toshiba, is provided. This paper developed a reporting system, which combines the text and full-motion video for cardiac catheterization. The system efficiently manages all the medication history of patients as well as related medical images and provides a convenient and user-friendly interface to reduce the operation time. In developing the system, we first standardize the format and contents of the reports of cardiac catheterization and Percutaneous Transluminal Coronary Angioplasty (PTCA). Taking advantages of Picture Archiving and Communication System (PACS), we embedded the medical images in the system in order to facility patient, family and physician in communicate activities. In sum, the system offers a tool to facilitate data entry of physicians with minimum efforts, integrate the workflow into the system to reduce the entry error, and provide medical diagnosis accuracy and quickly.

2   Background

Cardiovascular diseases (CVDs) are the no.1 killer in the world [1]. According to the statistics [1], there are about 17.5 million people died from this disease in 2005, accounting for 30% of mortality in global. CVDs include several diseases that damage the structures or the functions of the heart, such as coronary heart disease, cerebrovascular disease, hypertension, peripheral artery disease, rheumatic heart disease, congenital heart defect and heart failure [1], [2]. The cause of CVDs is from lack of exercise, a poor diet, smoking, physical inactivity, and an unhealthy diet [1]. It is estimated that about 20 million people will die from CVDs by 2015 [1] and that over 40% of the deaths caused by CVD were due to coronary artery disease (CAD). CAD is a disease since of hardening of coronary arteries on the surface of the heart.

Early diagnoses and therapies of CADs are very important. The best method in diagnosing CAD is through cardiac catheterization [9]. Cardiac catheterization (also called cardiac cath or coronary angiogram) is an invasive imaging procedure that allows physicians to evaluate the function of the heart. During the test, a thin, flexible tube (called catheter) is inserted and threaded through the artery or vein in the groin (femoral or iliac), neck (carotid) or forearm, elbow or wrist (radial). The catheter reaches the coronary artery in the heart, where it can measure blood pressure within the heart and how much oxygen is in the blood [15]. The results of cardiac catheterization help physicians to determine treatment options such as medications, surgery, PTCA or stent implantation. In the past, the patient needs to receive general anesthesia for open heart surgery, if medication treatment is not good. Now, cardiovascular patients can receive PTCA or stent implantation, instead of open heart surgery, which is more risky and painful [13].

PTCA/stent procedure is similar to cardiac catheterization, utilizing the same local anesthesia. In PTCA, it is used to open up a narrowed coronary artery. The balloon catheter is then removed. However, a sent is usually placed at the narrowed section during a PTCA procedure [14]. A major advancement in the cure of coronary angioplasty is the introduction of coronary stents in the early 1990s [10]. A Stent is a wire mesh tube used to prop the artery open after PTCA. During the procedure of cardiac catheterization, angiograms or images can be recorded.

The major advantages of cardiac catheterization are that the patient does not need general anesthesia to reduce the risk, in addition the less pain, the shorter length of stay and the shorter operating time. However, the most worried shortcoming is that the postoperative patient may face the artery (or vein) to narrow again after 3 to 6 months of PTCA [11]. For following-up the progress, it is needed to record the patients' medical history in a system.

However, there is no integrated reporting system for cardiac catheterization that provides the function or characteristics of uniform terminologies, graph editing, faster data retrieval, and tracing of the progress. The next section describes the development of our system for resolve the deficiency of current reporting system of cardiac catheterization.

3   System Description and Results

We design the system according to the workflow of physicians in order to reduce the time and costs of the physicians and increase the patient safety simultaneously. We first constructed a heart model that contains the arteries and their all possible lesions. The information of lesions contains their possible positions, narrow grade, lesion type, lesion detail, TIMI flow, blush grade and dissection. Then we collect the possible phrases used by practitioners to describing the reports. These phrases are organized into different categories, and can be indexed.

The system integrates with HIS (hospital information system) bi-directionally: one direction is that the HIS can pass the patient demographic data and note of exams from HIS for reducing the clerical work load; the other direction is that the reporting system can upload the report to HIS (or EPR) when the report is done. Note that the video of cardiac catheterization still resides in the reporting system, not propagating to HIS since of its large volume of data. However, the videos stored as independent files can be indexed through the reporting system or HIS. Since the images from medical report were showed the traditional and reasonable content of patient records in general cardiac catheterization procedure. All the operations typically performed before, during, and after the cardiac catheterization are included on the system can help physicians trace of the progress, make clinical decisions, and prevent the errors that may occur. Besides, by using the system, the manager can compile statistics about the result of cardiac catheterization and keep track of how many patients visit the cardiac catheterization room. For the purposes of this article, we have called this medical report the cardiac catheterization Report (CathReport).

Next, we describe the design, implementation, and evaluation of a CathReporitng system.

3.1 Requirement Assessment and Design Process

The design of the CathReport took place in several stages. First, in order to better understand the workflows and clinical processes, we were communicating with physician. The second phase is to collect the paper-based report which physician hand-writing. Finally, we use Spiral Model to implement the system. The prototypes were targeted with user feedback.

The CathReport is a reporting tool which was implemented with Java. Runs on the Microsoft windows XP operating system, a MS SQL 2000 database and is implemented in a networked environment consisting of a server and several clinical workstations. The Server with Intel Core Duron XP 1.60G Hz processor and 1GB main memory, and the workstation with Intel Pentium 4 processor and 512MB main memory. Figure 1 shows the system architecture. CathReport and database are designed in a single server. The workstations work through CathReport Server with HIS Server and PACS Server.

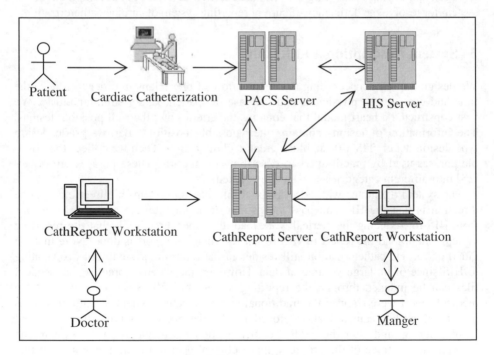

Fig. 1. System architecture

During the analysis phase, we usage unified modeling language (UML) to identify and partition the system functionality. CathReport has been realized as three main components: System Maintain, File Management, and File Analysis. The DBA administrator can setup user security to avoid illegal access. The physicians can add, depict, modify, query, and utilize information about specific patient topics. The manager of hospital can to compile statistics patient report efficiency by use the system. Figure2 shows the use case diagram of the CathReport system.

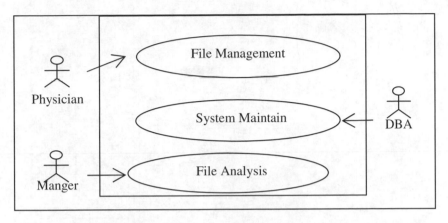

Fig. 2. Use case Diagram

3.2   System Function

As previously mentioned, PACS and HIS are integrated into the CathRepot system, allowing physicians to view the full-motion videos/images and patient history concurrently. The system Framework is divided into three parts: File Management subsystem, System Maintain subsystem, and File Analysis subsystem, it describes as follows:

3.2.1   File Management Subsystem

When a physician is desires to edit a new cardiac catheterization report. From the main menu, click on the "new" icon to create a new report. In the first, physician need to entry a patient's unique identification number(ID) in order to download his/her personal information about this surgical from the HIS-database. Diagnose medical reports is tedious work because most patients symptoms are similar. In order to improve data access, we offer some smart function to use in this system.

Physician can use these functions to editing report: (1) pulling list (2) a pop-up window (3) put a mark on image. In the function, we use Geographic Information System (GIS) conception to design. Computerized information input, storage, display which to cope with geographically referenced spatial data and the corresponding attribute information [12]. We use vector model to help physician mark lesions. In this example when physician is marking up the medical images, the CathReport will positioning automatically and presenting the blood vessel name. By clicking the "click to choose detail" button, various pre-defined items can be choosing. Physician can use the pop-up windows to choose lesion, then the lesion type will gauge by the system automatically. These functions are shown in Fig.3.

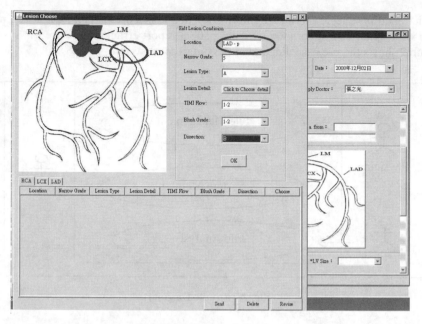

Fig. 3. Lesion choose

The system also may produce some suggestions according to different circumstances (Fig. 4). Then, physician need review each of those items and agree to some resolutions. Use the system, not only support decision making, but also decrease the workload.

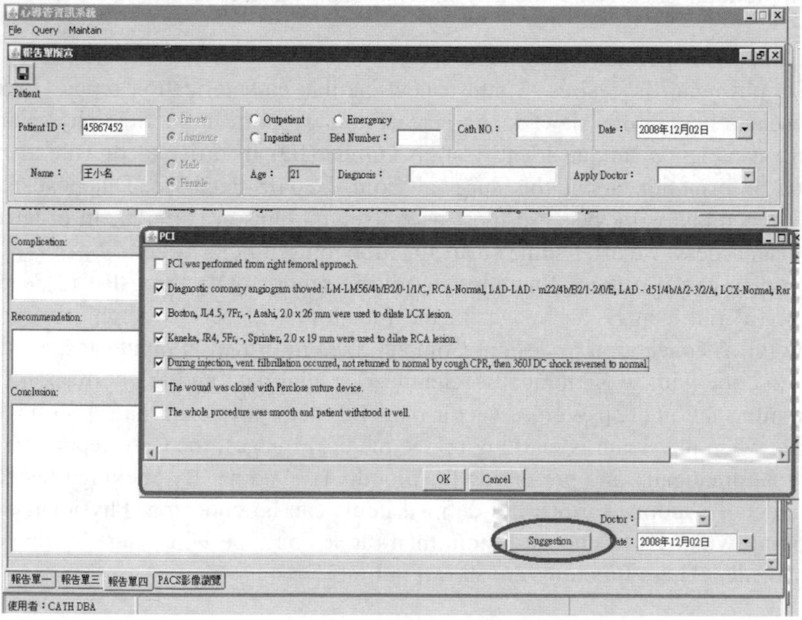

Fig. 4. System Suggestion

3.2.2 File Analysis Subsystem

In order to analyze and inquire the results of cardiac catheterization, the CathReport system offers the analysis functions. Manager can choose the parameters such as lesion type, location, duration and so on. Then the analysis findings will display in two kinds format (bar-chart and pie-char) that allows decision maker to make important decisions effectively and efficiently (Fig.5).

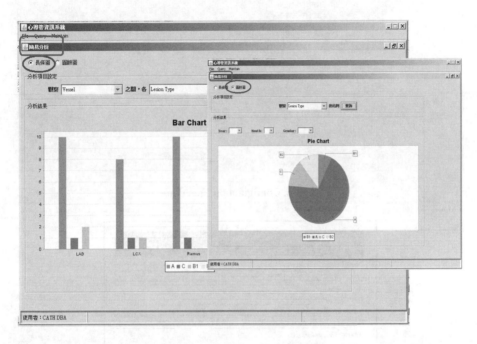

Fig. 5. A bar chart and a pie chart

3.2.3 System Maintain Subsystem

The Maintain subsystem can divide three classifications: user manager, configuration, and report manager. In the "Configuration" Function, system manager can deal with relevant equipment setup without programming, including install path and server setting (Fig.6). The user Manager is an authentication function includes editing user profiles, and adding, disabling or removing accounts. In the "Report Manager" interface, physicians can pre-defined phrases which in common to use, reduce operating time.

4 Evaluation

The CathReport system is designed, developed, and implemented under support of regional hospital in Taiwan. The reporting tool is linkage and integrate with HIS and

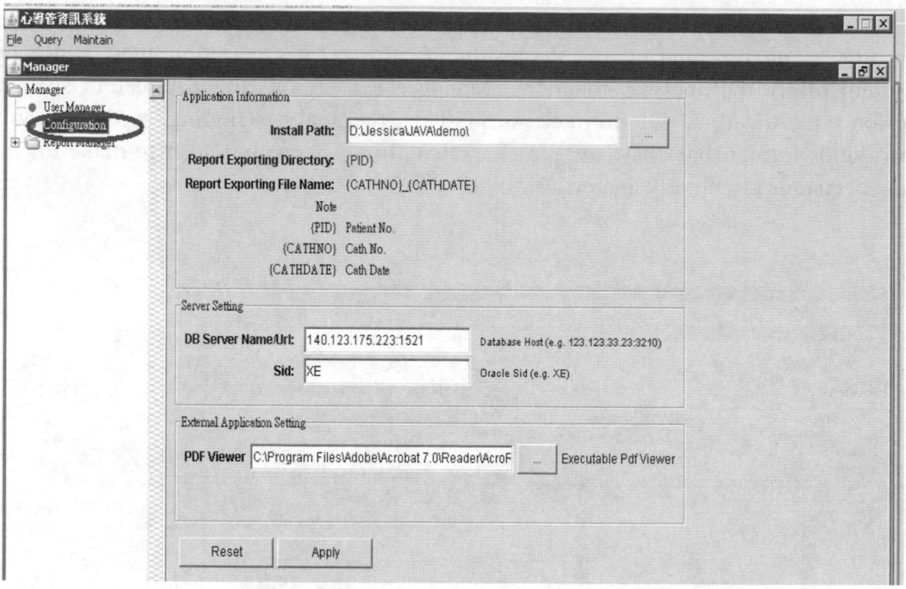

Fig. 6. Configuration Function

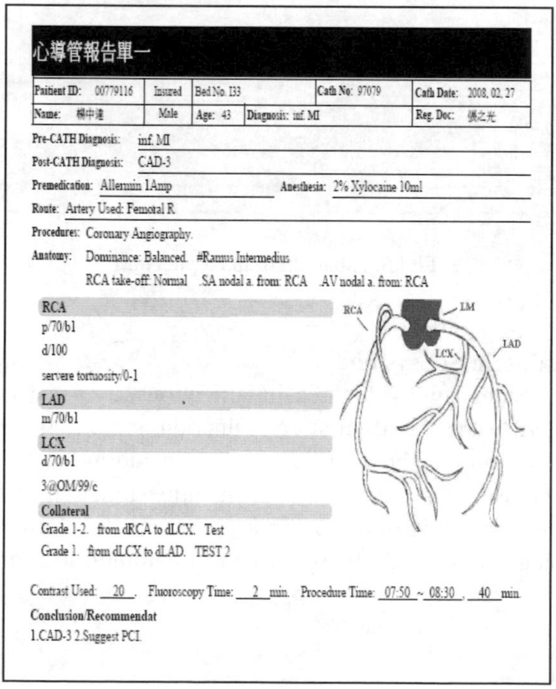

Fig. 7. Cardiac catheterization report

PACS bi-directionally, physician can collection all kinds of medical history from the system. It offers some intelligent functions, medical staffs can save 50% time when use the system. Via the CathReport we can print artistic reports (figure 7). It also can improve patient impression of hospital.

The benefits of hospital by using the CathReport system are below:

- Facilitate patient record edit: The system offers some smart function to help patient record writing; it can reduce time spent on routine tasks and avoid input error.
- Patient care: The system can help physicians trace of the cardiac catheterization progress making patient care more safely.
- Diagnosis suggestion: It provides some suggestion about report of cardiac catheterization that can reduce the operation time and improve staff productivity.
- Management Requirement: In the past, the managers need to compile statistic all kinds of data by hand. When use the system, manager only need using mouse to click. All kinds of data analysis and reporting will produce automatically.
- Promote the development of EPR: Since the procedure of cardiac catheterization is very complex, the reporting system for cardiac catheterization, as we knew, has not yet well developed in most of the hospitals until now. To building the reporting system can promote the development of EPR.

5 Conclusion

This paper demonstrates the procedures and experiences in developing the reporting system for cardiac catheterization in a regional hospital in Taiwan. The results show the system can help physicians to diagnose more precisely and quickly. The system is designed to fit workflow, following the principle of least astonishment to decrease the unintended consequences while using the system. The evaluation results show that the system can reduce a half of operating time and provide a vehicle that the manager and searcher can perform management activities and retrospective researches of the healthcare resources. In addition, the implementation of the system paves the way toward a comprehensive EPR of healthcare industry.

References

1. http://www.who.int/cardiovascular_diseases/en/
2. http://www.heartdiseasesguide.com/search/label/Cardiovascular%20Research
3. Dick, R.S., Steen, B., Detmer, D.E. (eds.): The Computer-Based Patient Record: An Essential Technology for Health Care, Revised edn. National Academy Press, Washington (1997)
4. Ebadollahi, S., Coden, A.R., Tanenblatt, M.A., Chang, S., Syeda-Mahmood, T., Amir, A.: Concept-based electronic health records: opportunities and challenges. In: Proceedings of multimedia 2006, pp. 997–1006. ACM, New York (2006)
5. Computerized patient records goal of new group. Hospitals 20, 48–52 (1991)

6. Joans, K.: The extent and importance of unintended consequences related to computerized provider order entry. Journal of the American Medical Informatics Association, 415–423 (2007)

7. Kaushal, R., Shojania, K.G., Bates, D.W.: Effects of computerized physician order entry and clinical decision support systems on medication safety: a systematic review [see comment]. Arch. Int. Med. 163, 1409–1416 (2003)

8. Garg, A.X., Adhikari, N.K.J., McDnald, H., et al.: Effects of computerzed clinical decision support systems on practitioner performance and patient outcomes. JAMA 293, 1223–1238 (2005)

9. Michaels, A.D., Chatterjee, K.: Angioplasty versus bypass surgery for coronary artery disease. Circulation 106, e187–e190 (2002)

10. Shah, P.B., Lilly, C.M.: Interventional therapy for coronary artery disease. American Journal Respiratory Critical Care Medicine 166, 791–796 (2002)

11. Hwang, J.-J.: Cardiac catheterization inspection and treatment, vol. 5, pp. 221–225 (2005)

12. Aronoff, S.: Geographic Information Systems: A Management Perspective. WDL Publications, Ottawa (1989)

13. http://www.gandau.gov.tw/nurse/inout%20nurse/heart/h010.htm

14. http://www.webmd.com/heart-disease/angioplasty-with-stent-placement-for-heart-attack-and-unstable-angina

15. http://www.americanheart.org/presenter.jhtml?identifier=4491

Presenting a Simplified Assistant Tool for Breast Cancer Diagnosis in Mammography to Radiologists

Ping Zhang[1], Jenny Doust[1], and Kuldeep Kumar[2]

[1] Faculty of Health Sciences and Medicine, Bond University
Gold Coast, Australia
[2] Faculty of Business, Technology and Sustainable Development, Bond University
Gold Coast, Australia
{pzhang,jdoust,kkumar}@bond.edu.au

Abstract. This paper proposes a method to simplify a computational model from logistic regression for clinical use without computer. The model was built using human interpreted featrues including some BI-RADS standardized features for diagnosing the malignant masses. It was compared with the diagnosis using only assessment categorization from BI-RADS. The research aims at assisting radiologists to diagnose the malignancy of breast cancer in a way without out using automated computer aided diagnosis system.

Keywords: breast cancer diagnosis, mammography, logistic regression, simplified model.

1 Introduction

Mammography is considered the most reliable method of early detection. However, in the earliest stage the visual clues are subtle and varied in appearance, making diagnosis difficult, challenging even for specialists. The benefits of early detection generate a powerful motivation for researchers to develop automated detection and diagnosis systems to assist specialists and radiologists.

A historical review of computer-aided diagnosis (CAD) has been given by Doi [1], indicated that CAD has become a part of the routine clinical work for detection of breast cancer on mammograms at many screening sites and hospitals in the United States. The requirement of high sensitivity and high specificity is difficult to meet for researchers who are developing computer algorithms for detection of abnormalities on radiologic images. It would not be possible for most advanced countries to employ the current computer results for automated computer to make the diagnosis decision [1].

A number of investigators have reported positive findings on the usefulness of CAD in detecting various lesions, including clustered microcalcifications [2] and masses [3] in mammograms, based on the evaluation with ROC (receiver operating characteristic [4]).

D. Zhang and M. Sonka (Eds.): ICMB 2010, LNCS 6165, pp. 363–372, 2010.

Recently, the effectiveness of computer assisted detection systems have been assessed for clinical use. In both the studies of Taplin et al [5] and Ko et al [6], CADs were used as the assistants to expert radiologists. Despite the differences between the studies, they both indicated the possible benefit and some of the issues with the clinical use of current CAD technology. Taplin et al used the ImageChecker M2 1000 system (version 2.2, R2 Technology) [7, 8] as the assistant to the radiologists. They compared the reading from an expert radiologist with or without CAD following the BI_RADS[1] criteria. The conclusion drawn from their study was that the CAD did not affect overall sensitivity, but its effect differed for visible masses that were marked by CAD compared with those were not marked by CAD. The study also showed that the CAD had a greater effect on both specificity and sensitivity among radiologists who interpret more than 50 mammograms per week. Ko et al [6] prospectively assessed the clinical usefulness of CAD in the interpretation of screening mammograms. The CAD used in their study was iCAD MammoReader [9]. This study showed that use of CAD can increase cancer detection rate by at least 4.7% and sensitivity by at least 4% with significant increased recall rates and not significant effect on positive predictive value for biopsy.

The effect of CAD on double reading of paired screen-film and full field digital screening mammograms was evaluated by Skaane et al [10]. The CAD (Image Checker, version 8.0 R2 Technology) showed the potential to increase the cancer detection rate for both FFDM (full field digital mammography) and screen-film mammography in breast cancer screen performed with independent double reading. Georgian-Smith et al [11] compared the CAD with a blinded human second reader for detection of additional breast cancer not seen by a primary radiologist. The research aimed to compare the practice of a human second reader with a CAD reader for the reduction of the number of false-negative cases resulting from review by a primary radiologist. A conclusion from their small sample sized investigation was that a human second reader or the use of a CAD system can increase the cancer detection rate, but no statistical significant difference between the two.

Measuring actual improvement of cancer detection is not a simple task and the results of these studies are dependent on the performance of the CAD itself, the population of cases to which it is applied, and the reviewers who use it [6]. CAD has been used as a complementary tool to mammography, prompting the reader to consider lesions on the mammogram that may represent cancer. However emerging evidence and improved CAD technology are likely to help define its role in breast screening [12].

As we know, the commercialized CADs have not been world widely accepted yet, for example, there have not CAD used in Australia. However, radiologists would like the ability of CAD to change the contrast of a mammogram [13]. In this situation, we would like to present a simple model to assist radiologists for their diagnosis without having to use computer.

---

[1] ACR stands for the American College of Radiology. BI-RADS stands for Breast Imaging Reporting and Data System, which contains a guide to standardized mammographic reporting including a breast-imaging lexicon of terminology, a report organization and assessment structure and coding system.

2 Methodology

Various classification techniques have been applied for breast abnormality classification including neural network, KNN (K-neariest neighbour), logistic regression, and support vector machines. In our previous study [14, 15], we also tested various classification techniques including discriminant analysis, logistic regression and neural networks, and compared the different set of features for breast cancer diagnosis in mammography. Our study showed that logistic regression performs very well for breast abnormality classification in mammography, though the theory is not complicated.

2.1 Logistic Regression

Logistic regression is useful for situations in which we want to predict the presence or absence of a characteristic or outcome based on values of a set of predictor variables. It is similar to the linear regression model but is suited to models where the dependent variable is dichotomous [16]. It is widely used in medical research since many studies involve two-category response variables [17].

Logistic regression model for a binary dependent variable can be written as

$$E(y) = \frac{\exp(\beta_0 + \beta_1 x_1 + \beta_2 x_2 + \ldots + \beta_k x_k)}{1 + \exp(\beta_0 + \beta_1 x_1 + \beta_2 x_2 + \ldots + \beta_k x_k)}$$

where in this research, y=1, if the patient has malignant tumor; y=0, if the patient has benign tumor.

$E(y)= P$=probability of the breast area is malignant

$x1, x2\ldots xk$ are quantitative or qualitative independent variables.

2.2 Features for Classification

Researchers have extracted different features from the mammograms for cancer diagnosis, such as region-based features [18], shape-based features [19], image structure [20], texture based features [21] and position related features [22]. In our previous research, we extracted 25 features including computer extracted features and human interpreted features [14, 15]. The human extracted features are the features interpreted from the radiologists including the ones interpreted following the BI-RADS lexicon. The computer extracted features are the commonly used which are statistically calculated based on the grey-level value distribution from the mammograms. These features were used to build different classification models for breast cancer diagnosis. Our research result showed that human extracted features contributed to the diagnosis more significantly than the computer extracted features[23].

The BI-RADS was developed to standardize mammographic reporting, to improve communication, to reduce confusion regarding mammographic findings, to aid research, and to facilitate outcomes monitoring [24, 25]. The assessment categorization from the BI-RADS has been recommended as a predictor of malignancy [26]. Lo et al [27, 28] developed the computer-aided diagnostic techniques that took consideration the description of lesion morphology according to BI-RADS. The models were reported that performed at a level comparable to or better than the overall performance of the expert mammographers who originally interpreted the mammograms.

In this paper, we present a simplified model derived from a logistic regression equation built on human extracted features to assist radiologists for breast cancer diagnosis.

3 Materials

Digital Database for Screening Mammography (DDSM) [29] from University of South Florida was used for this research. In DDSM, the outlines for the suspicious regions are derived from markings made on the film by at least two experienced radiologists. DDSM provides some information related to the marked masses using the ARC BI-RADS definition. Such as assessment, density, mass shape and mass margin are all described using the standard terminology in BI-RADS. Patient age and some others information involved in the digital mammogram, such as subtlety, are also provided in related files. These information are considered that can be important for distinguishing the malignant and benign masses or calcifications.

A total of 200 mass suspicious areas were extracted from DDSM for the experiments. It includes 100 malignant case and 100 benign ones. The mammograms used in this study are all scanned on the HOWTEK scanner at the full 43.5 micron per pixel spatial resolution.

Seven human interpreted features related to mass suspicious areas are patient age, breast density, mass shape, mass margin, assessment, subtlety, and calcification association. Except "calcification association" (Calc-asso) was created by us in our research, others are obtained from DDSM. Calc-asso describes how the mass is related to a calcification, for example, some of the masses are also marked as calcifications. In this research, we simply categorize calc-asso as 'yes' or 'no' for having the association or not having the association. Some of BI-RADS features are described with text in the database. We assign them to the different categories with numbers. For example, "1" represents "round" and "2" is used to represent "oval" for the shape feature. These variables are treated as categorical (nominal) variable for building the models.

4 Implementation

4.1 Cross Validation for Comparison of the Models

The assessment categorization from the BI-RADS as a predictor of malignancy was tested by Orel et al [26]. In their research, 1312 localization (mass areas) were investigated. The PPV (positive predictive value) was 0% for assessment category 2 (refer to BI-RADS), 2% for category 3, 30% for category 4 and 97% for category 5. In our research, we tested logistic regression model built with all 7 human extracted features. This model is compared with the model built with only 6 features (out of 7 features excluding "assessment"), and with using only assessment categorization from the BI-RADS for prediction of malignancy.

In this study, the logistic regression models were constructed based on the set of features with iteratively reweighted least squares (IWLS) method using R (version 10.1.0), a free software environment for statistical computing and graphics [30]. 5-fold cross validation was conducted to compare the performance between the models using 7 features, 6 features or assessment categorization alone. Table 1 shows the validation results from the models using different sets of features. In 5 splits of the datasets, every split used the different 40 cases out of the whole 200 cases as testing and the rest for training.

Table 1. Areas under the ROC from different Models

Dataset (split)	7-features*		6-features**		"assessment" only	
	training	testing	training	testing	training	testing
1	0.989	0.893	0.983	0.886	0.841	0.863
2	0.972	0.998	0.969	0.988	0.835	0.880
3	0.972	0.988	0.971	0.973	0.844	0.850
4	0.977	0.970	0.973	0.968	0.848	0.830
5	0.972	0.991	0.969	0.981	0.845	0.843
Mean	0.977	0.968	0.973	0.959	0.843	0.853
Standard deviation	0.007	0.043	0.006	0.042	0.005	0.019

*7-features: age, density, shape, margin, subtlety, calcification association and assessment;
**6-features: age, density, shape, margin, subtlety and calcification association

4.2 Models to Be Recommended to Radiologists

From table 1, we can see that with 7 features combining the 6 human extracted features and the assessment categorization from the BI-RADS predict better than using only 6 human extracted features or using only the assessment categorization from BI-RADS ($P<0.05$, from t-tests). We used the whole set of 200 cases to build a final logistic regression model with these 7 features. It is shown below:

$log(p/(1-p))= -41.9+C1*age+C2*density+C3*mass\_shape$

$+C4*margin+C5*subtlety +C6*Cal\_asso +C7*assessment$

p is the predicted probability of the mass being malignant. *C1* to *C7* are the coefficient values for the different variables in the model. They take different values when the variables are in different categories. See table 2 for the coefficient values in the final logistic regression model. The coefficients of the independent variables were simplified to an integer weight in order to be easily managed by radiologists for assisting the diagnosis. For example, a 55 years old patient has the mammogram showing a mass in "round" shape, obscured margin, subtlety 1, is seen as also calcifications and is assessed as category 3. The total diagnostic weight will be 2+2+0+1+15+0=19.

Table 2. Logistic Regression Model Built with the Whole Dataset

Variable	Category	Coefficient (C1-C7)	Simplified Weight
Age	Over 50	2.41	2
Density	1	1.99	2
	2	1.97	2
	3	1.54	2
Mass shape	Round	1.68	2
	Lobulated	1.44	1
	Irregular	1.80	2
Mass margin	Circumscribed	16.00	16
	Microlobulated	17.97	18
	Ill_defined	17.23	17
	Spiculated	22.34	22
Subtlety	1	1.25	1
	2	1.36	1
	3	0.98	1
Cal-asso	Yes	15.33	15
Assessment	4	18.20	18
	5	20.19	20

Note: the variable values are interpreted following the BI-RADS standard or recoded in DDSM database.
Assessment categories from BI-RADS: 1- negative, 2- benign, 3- probably benign, 4- suspicious abnormality, 5- highly suggestive malignancy.

5   Results and Discussion

Figure 1 shows the ROC curves produced from the simplified model, and is compared with logistic model with all 7 features and with assessment categorization alone for diagnosis. We can see the area under the simplified model is obviously higher than that from only assessment categorization (0.961 vs 0.843).

Table 3 shows the diagnostic performance from the simplified model built with our 200 cases and the result from using only assessment value for diagnosis. It can be used as the second diagnosis reference on top of the diagnosis with assessment categorization alone. For instance in our above example, though the assessment category is 3 which represent only 13% specificity for malignancy, the model from the logistic regression shows the high probability (specificity 100% with total weight score 19) of it being malignant.

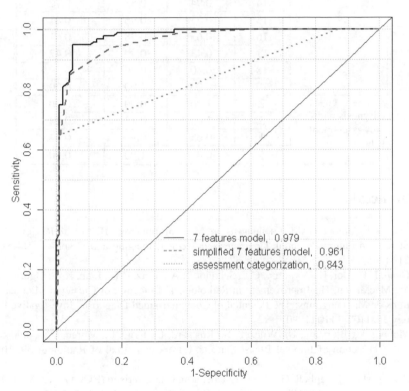

Fig. 1. ROC and the areas under the curves from the different models

From the results we can see, especially when the assessment category falls to 3 (probably benign), the proposed model will give a reference for the diagnosis. Need to mention, the model proposed in the paper is just a preliminary model built based on only 200 masses which do not cover all the possible features of the mammograms. More variety of the dataset will be used in our future research and aim to build a clinical evaluated model for practical use.

Table 3. Performace of Logistic Regression Model Built with the Whole Dataset and using Assessment Categorization Alone

	Total Weight score	Sensitivity	Specificity	Youden index
Final Diagnostic Model Result	9	1.00	0.02	0.02
	10	1.00	0.09	0.09
	11	1.00	0.24	0.24
	12	1.00	0.39	0.39
	13	0.99	0.62	0.61
	14	0.94	0.83	0.77
	15	0.85	0.96	0.81
	16	0.63	0.98	0.61
	17	0.43	0.99	0.42
	18	0.30	0.99	0.29
	19	0.00	1.00	0.00
Diagnosis with only assessment value	2	1.00	0.00	0.00
	3	1.00	0.13	0.13
	4	0.65	0.99	0.64
	5	0.00	1.00	0.00

References

1. Doi, K.: Computer-Aided Diagnosis in Medical Imaging: Historical Review, Current Status and Future Potential. Computerized Medical Imaging and Graphics 31(4-5), 198–211 (2007)
2. Chan, H.P., Doi, K., Vyborny, C.J., Schmidt, R.A., Metz, C.E., Lam, K.L., Ogura, T., Wu, Y., MacMahon, H.: Improvement in Radiologists' Detection of Clustered Microcalcifications on Mammograms: The Potential of Computer-aided Diagnosis. Investigative Radiology 25, 1102–1110 (1990)
3. Moberg, K., Bjurstam, N., Wilczek, B., Rostgard, L., Egge, E., Muren, C.: Computer Assisted Detection of Interval Breast Cancers. European Journal of Radiology 39, 104–110 (2001)
4. Zweig, M.H., Campbell, G.: Receiver-Operating Characteristic (ROC) Plots: A Fundamental Evaluation Tool in Clinical Medicine. Clin. Chem. 39(4), 561–577 (1993)
5. Taplin, S.H., Rutter, C.M., Lehman, C.D.: Testing the Effect of Computer-Assisted Detection on Interpretive Performance in Screening Mammography. Am. J. Roentgenol. 187(6), 1475–1482 (2006)
6. Ko, J.M., Nicholas, M.J., Mendel, J.B., Slanetz, P.J.: Prospective Assessment of Computer-Aided Detection in Interpretation of Screening Mammography. Am. J. Roentgenol. 187(6), 1483–1491 (2006)
7. Wang, S.-p.: Computer-aided Dagnosis System and Method. United States Patent Application No: 6434262, L.A, CA (2002)
8. Wang, S.-p.: Computer-aided Diagnosis Method and System. United States Patent Application No: 6266435, Los Altos, CA, 94022 (2001)

9. Rogers, S.K.B.: Use of Computer-aided Detection System Outputs in Clinical Practice.United States Patent Application No: 6970587, ICAD, Inc, (Nashua, NH, US), OH, US (2005)

10. Skaane, P., Kshirsagar, A., Stapleton, S., Young, K., Castellino, R.A.: Effect of Computer-Aided Detection on Independent Double Reading of Paired Screen-Film and Full-Field Digital Screening Mammograms. Am. J. Roentgenol. 188(2), 377–384 (2007)

11. Georgian-Smith, D., Moore, R.H., Halpern, E., Yeh, E.D., Rafferty, E.A., D'Alessandro, H.A., Staffa, M., Hall, D.A., McCarthy, K.A., Kopans, D.B.: Blinded Comparison of Computer-Aided Detection with Human Second Reading in Screening Mammography. Am. J. Roentgenol. 189(5), 1135–1141 (2007)

12. Houssami, N., Lord, S.J., Ciatto, S.: Breast Cancer Screening: Emerging Role of New Imaging Techniques as Adjuncts to Mammography. eMJA The Medical Journal of Australia 190(9), 493–497 (2009)

13. Rangayyan, R.M., Ayres, F.J., Leo Desautels, J.E.: A review of computer-aided diagnosis of breast cancer: Toward the detection of subtle signs. Journal of the Franklin Institute 344(3-4), 312–348 (2007)

14. Zhang, P., Kumar, K., Verma, B.: A Hybrid Classifier for Mass Classification with Different Kinds of Features in Mammography. In: Wang, L., Jin, Y. (eds.) FSKD 2005. LNCS (LNAI), vol. 3614, pp. 316–319. Springer, Heidelberg (2005)

15. Zhang, P., Verma, B., Kumar, K.: Neural vs. Statistical Classifier in Conjunction with Genetic Algorithm Based Feature Selection. Pattern Recognition Letters 26, 909–919 (2005)

16. Norusis, M.J.: SPSS Advanced Statistics User's Guide. SPSS Inc., Chicago (1990)

17. Everitt, B.S.: Statistical Methods for Medical Investigations, 2nd edn. Edword Arnold, London (1994)

18. Woods, K.S., Doss, C., Bowyer, K., Solka, J., Priebe, C., Kegelmeyer, P.: Comparative Evaluation of Pattern Recognition Techniques for Detection of Microcalcifications in Mammography. International Journal of Pattern Recognition and Artificial Intelligence 7(6), 80–85 (1993)

19. Arbach, L., Bennett, D.L., Reinhardt, J.M., Fallouh, G.: Classificaiton of Mammographic Masses: Comparison between Backpropagation Neural Network (BNN) and Human Readers. In: Sonka, M., Fitzpatrick, M.J. (eds.) Proceedings of SPIE, Medical Imaging: Image Processing, vol. 5032, pp. 810–818. SPIE Press (2003)

20. Chitre, Y., Dhawan, A.P.: ANN Based Classification of Mammographic Microcalcifications Using Image Structure Features. State of the Art in Digital Mammographic Image Analysis 7(6), 167–197 (1994)

21. Varela, C., Karssemeijer, N., Tahoces, P.G.: Classification of Breast Tumors on Digital Mammograms Using Laws' Texture Features. In: Niessen, W.J., Viergever, M.A. (eds.) MICCAI 2001. LNCS, vol. 2208, pp. 1391–1392. Springer, Heidelberg (2001)

22. Antonie, M.L., Zaiane, O.R., Coman, A.: Application of data mining techniques for medical image classification.. In: The Second International Workshop on Multimedia Data Mining (MDM/KDD 2001), in conjunction with ACM SIGKDD Conference, pp. 94–101 (2001)

23. Zhang, P., Kumar, K.: Analyzing Feature Significance from Various Systems for Mass Diagnosis.. In: International Conference on Computational Intelligence for Modeling Control and Automation and International Conference on Intelligent Agents Web Technologies and International Commerce (CIMCA 2006), p. 141 (2006), doi:10.1109/CIMCA.2006.46

24. Amerian College of Radiology: BI-RADS: American College of Radiology Breast Imaging Reporting and Data System (BI-RADS), 3rd edn. American College of Radiology, Reston (1998)

25. Eberl, M.M., Fox, C.H., Edge, S.B., Carter, C.A., Mahoney, M.C.: BI-RADS Classification for Management of Abnormal Mammograms. The Journal of the American Board of Family Medicine (JABFM) 19, 161–164 (2005)
26. Orel, S.G., Kay, N., Reynolds, C., Sullivan, D.C.: BI-RADS Categorization as a Predictor of Malignancy. Radiology 211, 845–850 (1999)
27. Baker, J.A., Kornguth, P.J., Lo, J.Y., Williford, M.E., Floyd, C.E.J.: Breast Cancer: Prediction with Artificial Neural Network Nased on BI-RADS Standardized Lexicon. Radiology 196, 817–822 (1995)
28. Lo, J.Y., Markey, M.K., Baker, J.A., Floyd, C.E.J.: Cross-institution Evaluation of BI-RADS Predictive Model for Mammographic Diagnosis of Breast Cancer. AJR 178, 457–463 (2002)
29. Heath, M., Bowyer, K.W., Kopans, D.: Current Status of the Digital Database for Screening Mammography. In: Digital Mammography, pp. 457–460. Kluwer Academic, Dordrecht (1998)
30. Development Core Team, R: A Language and Environment for Statistical Computing. R Foundation for Statistical Computing, Vienna (2008)

Feature Based Non-rigid Registration Using Quaternion Subdivision

Fahad Hameed Ahmad[1], Sudha Natarajan[1], and Jimmy Liu Jiang[2]

[1] School of Computer Engineering, Nanyang Technological University, Singapore
[2] Institute for Infocomm Research, Agency for Science Technology and Research, Singapore

Abstract. A new weighted quaternion based non-rigid registration is presented in this paper. Strong crest points derived from principal curvatures provide the most robust features for image registration. Crest point strengths are based on their principal curvatures and the number of scales a particular crest point is detected at. Geometric features are extracted which are invariant to rotation, translation and scaling by using neighborhood crest points only as other voxels are susceptible to deformation. The neighborhood size is adjusted according to scale adaptively using a fixed k nearest neighbor to make the extracted feature scale invariant. Statistical properties are used to measure the distribution of these geometric invariant features. The scale and rotation invariant feature points are then used to establish a point to point correspondence between the template crest points and the subject image crest points. A multi-scale feature based subdivision scheme is employed for registration where a weighted quaternion matrix provides a quaternion transformation based on the corresponding points to obtain the best rotation for global as well as local sub-blocks.

1 Introduction

Non-rigid registration is a medical imaging technique used for image analysis and surgical guidance and planning and is well established in neuro, otolaryngology, maxillo-facial and orthopaedic surguries. Problem areas in the patient undergo non-rigid deformations due to operating environment or due to inherent nature of the problem area. A large variation could exist between different images obtained at varying time intervals or between pre-operative and intra-operative scans. The need for registering medical images therefore arises which would improve and aide surgeons and doctors to better diagnose the problem area. A brief overview on soft-tissue modeling and deformable medical image registration and its importance to Image Guided Intervention is provided in [1].

Registration could be classified into land mark based, intensity based and feature based methods. Feature based methods search medical images for distinctive features which could be used as matching points for registration. Feature

D. Zhang and M. Sonka (Eds.): ICMB 2010, LNCS 6165, pp. 373–382, 2010.
© Springer-Verlag Berlin Heidelberg 2010

based methods are comparatively fast as they limit data size to just a few feature points and can handle complex deformations better. The success of the registration is highly dependent on the accuracy of these features to locate and distinguish important anatomical and functional features in an image. Registration is generally achieved by using Iterative Closest Point (ICP) algorithms [2]. Closed form solutions also exist which require prior knowledge of point to point correspondence between template features and subject image features [3].

Many different models exist for non-rigid registration. Some techniques use complex equations describing the underlying physical properties of the tissue [4]. Free-form Deformation techniques also exist for non-rigid registration which can be modeled using thin-plate splines, B-splines or other higher order polynomials as basis functions. Sufficient statistical modeling and training needs to be done before hand which may not be suitable if the technique is being used in a generic sense and its accuracy may vary from case to case. Another class of technique uses subdivision of one or both images into sub-blocks and treats non-rigid registration as a local rigid body registration problem [5]. This technique is inherently simpler to execute and more generic as non-rigid registration is reduced to a rigid registration problem. The HAMMER algorithm [6] uses a multi-grid subdivision approach using Geometric Moment Invariant features. Another subdivision algorithm developed [7] uses MI as a similarity measure for registering individual sub-blocks. The draw back for such schemes is that they are iterative in nature and use fixed sub-blocks which may not be suitable to represent elastic registration. Such techniques could easily lead to local minimum solutions and require successive shifting and calculation of similarity which is computationally intensive.

Our main contribution in this paper is that we extract novel scale, translation and rotation invariant features based on crest points which are capable of distinguishing important voxels and are capable of establishing a point to point correspondence between template and subject image feature points even after a significant deformation has taken place. A new multi-scale feature based subdivision registration using weighted quaternion registration is also proposed which is a better alternative for finding the local minimum in closed form for non-rigid sub-blocks. The subdivision scheme is loosely based on the HAMMER algorithm and uses a multi-scale approach to successively move from global to local registration. Weighted quaternion fitting techniques are then applied to register images once a point to point correspondence is obtained. The number of computations are drastically reduced by using quaternion fitting and the weighted registration scheme provides robustness to any outliers present in the feature set.

The paper is organized as follows. Section 2 describes the two step feature extraction technique. Our proposed weighted quaternion registration and multi-scale sub-division is described in section. 3. Section. 4 explains interpolation and the construction of deformation fields. Experiments and their results are described in section. 5 followed by discussion and conclusion in last section.

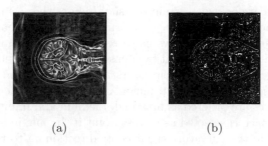

(a)   (b)

Fig. 1. (a) shows principal curvatures extracted from Brain MRI with $\sigma = 2$, (b) shows Crestlines extracted from Brain MRI with $\sigma = 2$

2   Feature Point Extraction

Feature extraction is a very important first step, which helps us to locate relevant features which are treated as landmark or reference points for our image registration step. Different features such as arteries, veins, bony structures or organ surface boundaries are important reference points that are very useful in the registration step. It must be noted that the features used should be invariant when seen from any imaging technology, viewing angle or deformation. Subdivision registration schemes generally use windowing methods to extract invariant features [7]. Different types of windows and different statistical invariants are used. Work has already been done on many geometric and algebraic invariants [8], [9]. The window used is again a limiting factor since it is a rigid window which cannot adaptively change according to the complex deformation of the underlying image. This introduces an error as statistical information would be affected if the sub-block is deformed. Also, if statistical information is not limited to important and more registration relevant features then irrelevant and redundant features might affect the accuracy of the description. If important features move out of a window then accuracy of the invariant feature descriptors for a particular region would change drastically.

Based on the above limitation we deviate from the common window based similarity techniques and propose a two step feature extraction step. In the first step we locate interest points which would be robust to non-rigid deformation and in the second step we extract relevant rotation and scale invariant geometric features from those interest points only. We use differential geometry [10] to extract surface features of anatomical and physiological structures. Similarly crest points could be used to identify ridges and valleys on iso-surfaces in brain MRI scans. Many fast implementations are currently available for extracting principal curvatures which also incorporate multi-scale [11].

A 3D surface is described by two principal curvatures and their corresponding orthogonal principal directions. A 3D surface or boundary region separating different regions in 3D based on their intensity form iso-surfaces and the zero-crossing point of these iso-surfaces are described by crest points. These features are not invariant under affine transformations so to improve robustness we reject crest points which are below a certain threshold under the assumption that high curvature crest points would survive deformations better than low curvature crest points. Crest point also allows us to reduce the huge amount of 3D data and make registration more efficient by focusing on a limited number of interest points. Three eigenvalues and eigenvectors are obtained from the Weingarten matrix [10], corresponding to three principal curvatures and their directions. The principal curvatures are ordered as maximal, medium and minimal curvature. A point in the 3D space is selected as a crest point if the maximum curvature c is locally maximum in the curvature eigenvector direction d. To reduce noise only those crest points are selected that are connected to a certain number of other crest points. Isolated crest points that are not connected to any other crest point or that are connected to only a few crest points are ignored. The crest points join to form crest lines and unwanted noisy features are hence removed. We use multi-scale techniques to separate the features obtained according to their resolution and importance which will then be used for subdivision registration. The multi-scale principal curvature values for each voxel and the crest points obtained after thresholding are shown in Fig. 1.

2.1   Geometric Invariant Feature Extraction

Crest points need to be uniquely identified hence further rotation and scale invariant features should be extracted to differentiate between crest points. Since crest points are the most robustly identifiable feature in a deformed image hence any invariant feature must be extracted using crest points only. We consider the remaining voxels susceptible to deformation variation. Geometric invariant primitives [9] which consist of distance D between two arbitrary points, the area A between three arbitrary points, the dot product d_p between two vectors $\boldsymbol{p_i}$ and $\boldsymbol{p_j}$ are invariant to rotation and translation. The dot product is the only property which is invariant to scale as well. Area A and distance D can be made invariant to scale by normalizing them to 1.

$$D(x_i, x_j) = \sqrt{(x_i^1 - x_j^1)^2 + (x_i^2 - x_j^2)^2 + (x_i^3 - x_j^3)^2} \tag{1}$$

$$A(\boldsymbol{p_i}, \boldsymbol{p_j}) = \frac{1}{2} \parallel \boldsymbol{p_i} \times \boldsymbol{p_j} \parallel \tag{2}$$

$$d_p = \boldsymbol{p_i} \cdot \boldsymbol{p_j} \tag{3}$$

$$\theta_i = \cos^{-1}\left(\frac{\boldsymbol{p_i} \cdot \boldsymbol{p_j}}{|\boldsymbol{p_i}||\boldsymbol{p_j}|}\right) \quad \text{where} \quad j = 1, 2, \cdots, k \tag{4}$$

$$\theta_i^k = [\theta_{i1}, \theta_{i2}, \cdots, \theta_{ik}] \tag{5}$$

$$D_i^k = [D_{i1}, D_{i2}, \cdots, D_{ik}] \tag{6}$$

Since we limit our invariant features to crest points hence a k nearest neighbor search provides us with k nearest crest points for each interest point in the image. Principal curvature eigenvectors $\boldsymbol{p}$ are taken as feature points and dot products are calculated between each k nearest neighbor $\boldsymbol{p_j}$ with the crest point principal curvature eigenvector $\boldsymbol{p_i}$. This gives us a set θ_k consisting of angles which the

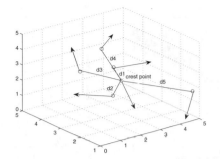

Fig. 2. Crest point surrounded by five nearest neighbor crest points is shown, where $\{d1, d2, d3, d4, d5\}$ are the distances between five nearest neighbor points with the crest point. Arrows represent the direction of principal curvature for the five nearest neighbor crest points

principal curvature eigen-vector makes with the surrounding crest point neighborhood as shown in Fig. 2. Moment invariant features are then extracted from this set θ_k consisting of mean μ, variance σ^2 and kurtosis values μ^4/σ^4 (or fourth standardized moment) which represent low order moment invariants which are invariant to rotation and translation. Similarly moments are calculated for principal curvature eigenvalues e_1, e_2 and e_3 (maximal, medium and minimal direction) for k nearest neighbor crest points, which are also robust to rotation, translation and scaling. A normalized distance distribution D_i^k is also generated between k nearest neighbor crest points and low order moments are extracted from the distribution. These features would help us determine the similarity for point to point correspondence between the subject and template image.

The k nearest neighbors are fixed quantities and they should be sufficiently large to meet statistical requirements. Since multi-scale features would be more spread out at larger scales and more compactly located at finer scales hence a fixed value of k would automatically increase or decrease the size of the region from which neighboring crest points are extracted. The advantage of using such a scheme involving k nearest neighbors is that there is no fixed window being used. Some areas would have more feature compared to other areas where features would be sparse, hence k nearest neighbor would automatically adapt the area being covered. The limitation associated with area based methods especially for regions where features are less is hence removed. A second advantage would be that feature matching is not limited by rigid sub-blocks. If a feature point in one sub-block of the template image matches with a feature from another sub-block of the subject image then these points are paired. Such a situation might occur due to deformation.

3   Registration Using Quaternions

Quaternion fitting techniques based on the Iterative Closest Point (ICP) algorithm provides a closed form solution for a feature based registration problem [3].

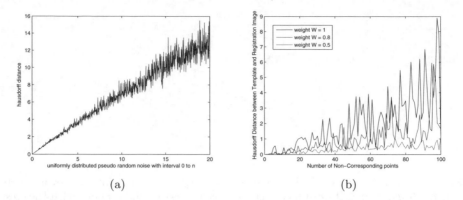

Fig. 3. (a)This graph represents the effect of adding noise to the performance of quaternion registration, (b)This graph represents the effect of increasing the number of non corresponding features on the performance of quaternion registration. Less similar or non-corresponding features are assigned weights $w(x, y, z)$ to increase the accuracy of the registration process.

Quaternion fitting techniques provide the closest match between corresponding registration points in R^3 in a least squares sense. Successive shifting and rotation is eliminated during subdivision registration which saves computation cost. We choose quaternion based ICP algorithm as quaternion multiplication takes sixteen multiplications and twelve additions, compared with orthogonal rotation matrices [12] which require twenty-seven multiplications and eighteen additions.

A coordinate (x_n, y_n, z_n) for a feature point p_n in R^3 is represented as the imaginary part $\mathbf{v_n}$ in quaternion space with real part equal to zero. A quaternion with real part equal to zero is called a pure quaternion. $L_q(v)$ rotates the vector $\mathbf{v_n}$ through an angle θ about $\mathbf{q}$ as the axis of rotation where $\mathbf{q}$ is the imaginary part of q. The imaginary part of pure quaternion $\boldsymbol{v_n}$ is represented by vector $\mathbf{v_n}$. Unit quaternion operator q does not change the length of the vector $\mathbf{v_n}$ and if the direction of q is along $\mathbf{v_n}$ then it is left unchanged. Quaternion rotation implementation has been done using matrices as shown in Eq. (8).

$$L_q(v) = q\boldsymbol{v_n}q^* \tag{7}$$

$$qvq^* = \begin{pmatrix} q_1 & -q_2 & -q_3 & -q_4 \\ q_2 & q_1 & -q_4 & q_3 \\ q_3 & q_4 & q_1 & -q_2 \\ q_4 & -q_3 & q_2 & q_1 \end{pmatrix} \begin{pmatrix} q_1 & -q_2 & -q_3 & -q_4 \\ q_2 & q_1 & q_4 & -q_3 \\ q_3 & -q_4 & q_1 & q_2 \\ q_4 & q_3 & -q_2 & q_1 \end{pmatrix} \begin{pmatrix} 0 \\ x_n \\ y_n \\ z_n \end{pmatrix} \tag{8}$$

Faugeras and Hebert [13] and Horn [3] have proposed a quaternion registration technique to register a set of points $S^1 = \{P_i^2\}$ with a another corresponding set of points $S^2 = \{P_i^1\}$. The condition for registration is that each point must correspond precisely with the same point in the rotated and noisy version. This would present a problem during registration of normal images as feature points extracted are not indexed and a one to one correspondence needs to be extracted. This sort of correspondence could be extracted by finding similarity in a rotation, translation and scale invariant feature space.

The unit quaternion representing the best rotation is obtained by extracting the eigenvector of the 4×4 matrix given in Eq. (9).

$$\begin{pmatrix} S_{xx} + S_{yy} + S_{zz} & S_{yz} - S_{zy} & S_{zx} - S_{xz} & S_{xy} - S_{yx} \\ S_{yz} - S_{zy} & S_{xx} - S_{yy} - S_{zz} & S_{xx} + S_{xx} & S_{xx} + S_{xx} \\ S_{xx} - S_{xx} & S_{xx} + S_{xx} & -S_{xx} + S_{yy} - S_{zz} & S_{xx} + S_{xx} \\ S_{xx} - S_{xx} & S_{xx} + S_{xx} & S_{xx} + S_{xx} & -S_{xx} - S_{yy} + S_{zz} \end{pmatrix} \quad (9)$$

where $S_{aa} = \sum_{i=1}^{n} \acute{a}_i^2 \acute{a}_i^1$ where $\acute{a}_i^j = \acute{x}_i^j, \acute{y}_i^j,$ or $\acute{z}_i^j$ and $\acute{x}_i^j, \acute{y}_i^j, \acute{z}_i^j$ being the centered or mean subtracted coordinates of the points P_i^j. We have simplified translation by simply subtracting the mean from the subject image feature points and centered it on the template image point. A weight $w(x, y, z)$ is introduce for each feature point. This weight corresponds to the degree of mismatch between feature points and the number of scales that a feature point is present at. Experiments were conducted to analyze the performance of a weighted quaternion fitting technique [14]. A rigid registration was carried out in an artificial set of data where the number of non-corresponding points were increased. Since both template image and subject image undergo different types of deformation hence a number of feature points with no correspondence exist. If template feature points are searched in the subject feature space then a number of close matches would be found which may not be a perfect match. A degree of uncertainty would exist between the two matched points based on the euclidean distance similarity. Threshold would then be applied to each point based on the importance of the feature points (detection scale) and its similarity with its corresponding point. Fig. 3b shows a comparison of different weights assigned to different number of non-corresponding points. Hausdorff similarity measure is then used to assess the registration result. It is clear from Fig. 3b that as the weight for non-corresponding points decreases, then registration results improve. We would adapt this technique for our case as well and apply a threshold which will determine the degree of mismatch between corresponding points. The weights used would be adaptable. The fitting quaternion obtained from Eq. (9) would have to obey a global limitation for the angle by which it rotates a particular feature sub-block. This feature sub-block must rotate within a pre-described limits of pi/4 radians. Any rotation beyond that would lead to a deformation which would be unrealistic for a real case. For weighted quaternion fitting S_{aa} could be rewritten as shown in Eq. (10).

$$S_{aa} = \sum_{i=1}^{n} w(i) \acute{a}_i^2 \acute{a}_i^1 \quad \text{where} \quad \acute{a}_i^j = x_i^j, y_i^j \text{ or } z_i^j \quad (10)$$

3.1   Subdivision Registration Using Unit Quaternion

The subdivision registration scheme differs from previous subdivision schemes as it only partitions feature points according to the 3D space but does not create rigid sub-volumes. The shifting and rotation performed would be limited to the feature points. Since feature points are few in number hence rotation and shifting would be much faster. We also use a multi-scale approach which would limit the registration step to important features at higher scales and reduce the number of feature points by removing finer scale descriptors. As subdivision proceeds to smaller sub-blocks, more and more fine scale feature points are introduced to increase the accuracy of registration of smaller blocks. A 3D gaussian kernel is used with standard deviation $\sigma = 1, 2$ and 4.

4 Deformation Field and Interpolation

We obtain a displacement vector after registration which represents a 3-D displacement for each of our feature points. Since these displacements are only for a few selected feature points in the 3-D space hence these values need to be interpolated for the other points. We therefore interpolate these values for each of the 3 dimensions by using tri-cubic interpolation methods [15] to get 3, 3-D matrices X_d, Y_d, Z_d where each point represents a displacement for the point in the original 3-D image. A similar deformation field is constructed in [7]. We use these deformation fields to re-map each pixel in the original 3-D image $L_{deformed}(x, y, z)$ to its new location $L_{registered}(\bar{x}, \bar{y}, \bar{z})$ by adding the displacements in (X_d, Y_d, Z_d) to its current coordinates (x, y, z).

$$L_{registered}(\bar{x}, \bar{y}, \bar{z}) = L_{deformed}(x + X_d(x, y, z), y + Y_d(x, y, z), z + Z_d(x, y, z)) \tag{11}$$

The resulting matrix $L_{registered}(x, y, z)$ could be sparse with many missing data values due to stretching and deformation caused by (11), hence a second interpolation is performed after relocating data points to get these values. Gaussian smoothing windows are not necassary for smoothing out boundary shifts as the first interpolation provides a smooth boundary between sub-blocks. By using just a few feature points in a sub-volume and then creating deformation fields makes sub-volume translations less rigid with lesser discontinuities as could be seen in Fig. 4c.

5 Experiments

We use brain scans from the Internet Brain Segmentation Repositary (IBSR). The data is provided by the Center for Morphometric Analysis at Massachusetts General Hospital and is available at *"http://www.cma.mgh.harvard.edu/ibsr/"*. The experiment is divided into two steps. In the first step we deform MR images by a random deformation field. A deformed image is show in Fig. 4b. In the second step we register and quantitatively measure the registration of this deformed Brain MR to the original MR under varying deformations.

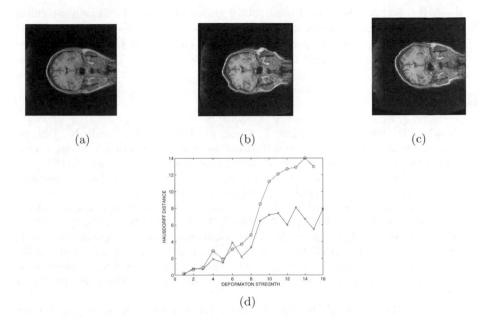

(a)   (b)   (c)

(d)

Fig. 4. (a) Original Image, (b) Image deformed using a uniformly distributed deformation vector, (c) Results obtained after registering deformed image (b) with (a), (d) measures the registration accuracy using hausdorff similarity measure against different deformation strengths where blue line represents the results using our technique and red line represents the results obtained using ELASTIX registration software

A uniform random deformation field is created with random shifts for each feature vector. The deformation is controlled by changing the maximum number of shifts possible for each pixel. By increasing shifts we increase the random deformation which takes place and then try to register image. The random shifts are stored in X_d, Y_d and Z_d. Similarity is measured using hausdorff distance to quantitatively measure the performance of our registration scheme to different strengths of deformation. Fig. 4d shows the performance and robustness of our algorithm under different deformation strengths compared with a popular registration software ELASTIX (*http://elastix.isi.uu.nl/*).

6   Conclusion

We have presented a robust registration method based on geometric invariant features using multi-scale weighted quaternion subdivision. The geometric invariant features are based on selective crest points which are robust to deformation. The k-nearest neighbor approach combined with multi-scale provides a new adaptive method for collecting features. The neighborhood increases size if lesser features are present and decreases size when features are abundant. This is an improvement over previous techniques where lesser features in some parts of the MR image would result in less robust moment invariants. The quaternion fitting approach also provides a much quicker way for registering during subdivision registration, and the associated feature weights help remove any outlier effect on the registration result.

References

1. Hawkes, D., Barratt, D., Blackall, J., Chan, C., Edwards, P., Rhode, K., Penney, G., McClelland, J., Hill, D.: Tissue deformation and shape models in image-guided interventions: a discussion paper. Medical Image Analysis 9, 163–175 (2005)
2. Besl, P., McKay, H.: A method for registration of 3-D shapes. IEEE Transactions on pattern analysis and machine intelligence 14, 239–256 (1992)
3. Horn, B., et al.: Closed-form solution of absolute orientation using unit quaternions. Journal of the Optical Society of America A 4, 629–642 (1987)
4. Pluim, J., Maintz, J., Viergever, M.: Mutual-information-based registration of medical images: a survey. IEEE transactions on medical imaging 22, 986–1004 (2003)
5. Van den Elsen, P., Pol, E., Viergever, M.: Medical image matching-a review with classification. IEEE Engineering in Medicine and Biology Magazine 12, 26–39 (1993)
6. Shen, D., Davatzikos, C.: HAMMER: hierarchical attribute matching mechanism for elastic registration. IEEE Trans. Med. Imaging 21, 1421–1439 (2002)
7. Walimbe, V., Shekhar, R.: Automatic elastic image registration by interpolation of 3D rotations and translations from discrete rigid-body transformations. Medical Image Analysis 10, 899–914 (2006)
8. Taubin, G., Cooper, D.: Object recognition based on moment (or algebraic) invariants. In: Geometric invariance in computer vision, pp. 375–397 (1992)
9. Xu, D., Li, H.: Geometric moment invariants. Pattern recognition 41, 240–249 (2008)
10. Do Carmo, M., Do Carmo, M.: Differential geometry of curves and surfaces. Prentice-Hall, Englewood Cliffs (1976)
11. Zhou, Y., Chellappa, R.: Computation of optical flow using a neural network. In: IEEE International Conference on Neural Networks 1998, pp. 71–78 (1988)
12. Horn, B., Hilden, H., Negahdaripour, S.: Closed-form solution of absolute orientation using orthonormal matrices. Journal of the Optical Society of America A 5, 1127–1135 (1988)
13. Faugeras, O., Hebert, M.: The representation, recognition, and locating of 3-D objects. The international journal of robotics research 5, 27 (1986)
14. Rusinkiewicz, S., Levoy, M.: Efficient variants of the ICP algorithm. In: Proceedings of the Third Intl. Conf. on 3D Digital Imaging and Modeling, Citeseer, pp. 145–152 (2001)
15. Lekien, F., Marsden, J.: Tricubic interpolation in three dimensions. Int. J. Numer. Methods Eng. 63, 455–471 (2005)

A Sparse Decomposition Approach to Compressing Biomedical Signals

Lu Bing and Shiqin Jiang

School of Electronics and Information Engineering, Tongji University, Shanghai, P.R. China
betty20006@163.com

Abstract. An approach of compressing biomedical signals was studied in this paper. First of all, we constructed an over-complete dictionary according to characters of compressing signals. Using the orthogonal matching pursuit (OMP) algorithm, sparse decomposition of biomedical signals was performed based on the dictionary. In this work, we used the optimized results of genetic algorithm (GA) as preliminary particles, and the best atoms were found by local search with particle swarm optimization (PSO). With this genetic hybrid particle swarm (GAPSO) approach, the convergence rate (CR) and the root-mean-square error (RMSE) were improved along with less distortion. For MCG signals in mid-length, simulation results showed that the standard error was 2.78%, when the compression ratio was 15%.

Keywords: Sparse Decomposition, Bio-medicine, Signal Compressing.

1 Introduction

In order to solve the storage and transmission problems of biomedical signals, it is necessary to study an effective signal compression technology. Sparse decomposition is developed recently for signals of compact representing [1]. For example, David Donoho [2] and Emmanuel Candes [3], developed a new signal decomposition method, compressive sensing (CS). G. Monaci, et al, reported learning multi-modal dictionaries [4]. FOCUSS Method was improved by Zhaoshui He, et al. [5]. Chenlu Qiu, et al., reconstructed the real-time dynamic MRI by compressive sensing technology [6]. Chunguang Wang, et al., put sparse decomposition into use of ECG compression [7]. The principle of sparse decomposition can be briefly stated as the following. Once a certain signal only has several non-null entries in time or other transformation domains, it means that the power of the signal is focally concentrated and it can be compactly represented in the corresponding domain with only a few non-null entries [2, 4]. Signal sparsity can be measured with $l_p (p \leq 1)$ norm or the decaying-to null speed of the sorted absolute signal entries [2-4]. With the current algorithms, signals are decomposed sparsely in an over-complete dictionary. Because decomposition in an over-complete dictionary is not unique, it has great freedom for the basis selection.

D. Zhang and M. Sonka (Eds.): ICMB 2010, LNCS 6165, pp. 383–391, 2010.

Moreover, when the over-complete dictionary is large enough, the signal should have sparse representation, if the sparse constraints are emphasized [1].

At present, there are many approaches to sparse decomposition, among which the matching pursuit (MP), developed by Mallat and Zhang, is the popular one [1]. MP decomposes signal into sparse representation iteratively in an over-complete dictionary. And during the decomposition procedure, the optimal basis (atom) for signal representation can be adaptively selected according to the signal characteristic. MP technology has smaller distortion than that of the wavelet translation at the same compression ratio.

In this paper, Orthogonal Matching Pursuit (OMP) developed from MP, which is applied for compressing biomedical signals. First of all, we constructed an over-complete dictionary based on sparse decompression theory. Then, using OMP algorithm, sparse decomposition of biomedical signals was performed based on the dictionary. Because OMP spent most of the optimized computation in finding the best atoms, we proposed a genetic hybrid particle swarm optimization (GAPSO) approach to find the best atoms. Compared with the genetic algorithm [8], the simple particle swarm optimization [9] algorithm has no crossover and mutation operation. However, it is easy to fall into the inferior local optimum, and multi-parameters should be adjusted. The GAPSO algorithm can bring into advantages of two algorithms, i.e. the hybrid heuristic algorithm can give attention to two aspects of efficiency and accuracy. The preliminary optimized result with genetic algorithm was set as particles, and then, a local search by means of PSO algorithm was used to accelerate the convergence speed for finding the best atoms. Our simulation results demonstrated that the presented GAPSO compression algorithm was feasible, and the performance indexes of signals compression were well, which included the algorithm convergence rate, signal compressing ratio and root-mean-square error (RMSE).

The GAPSO approach is introduced in Section 2 and simulation results are shown in Section 3. In the end, conclusions are given.

2   Methods

2.1   Signal Decomposed with OMP

MP decomposes signal into sparse representation iteratively in a dictionary Φ [1]. In iteration, a certain atom with the maximum inner product between itself and the residual signal is selected as the optimal one. After several iterations, when the convergence condition is satisfied, a sparse decomposition of signal in the corresponding dictionary is achieved. The iterative MP algorithm can be briefly stated as follows:

(1) *Initialization*: Set $k = 1$, $S^{(0)} = 0$, $R^{(0)} = S$, where S is the signal to be decomposed, R is the residual signal during iterations; the superscript is the current iteration number; initializing the stopping error ε with a small positive number; initializing the coefficients adders for all the atoms with 0.

(2) *Selection of atom and coefficient*: Calculate the inner product between current residual signal and each atom in the over-complete dictionary, and find the maximum inner product c_k and the corresponding optimal ϕ_{opt} as follows:

$$c_k(\phi_{opt}) = \max_j \langle R^{(k-1)}, \phi_j \rangle$$

Where ϕ_j is the j th atom in the dictionary and $\langle \cdot \rangle$ is the inner product, j varies from 1 to the size of dictionary.

(3) *Update*: Accumulate the current maximum coefficient c_k to the adder of the current optimal atom; add the weight of the selected atom to the decomposed signal $S^{(k)}$ and remove the corresponding contribution of this atom from the residual signal $R^{(k)}$ as

$$S^{(k)} = S^{(k-1)} + c_k \phi_k \tag{1}$$

$$R^{(k)} = S - S^{(k)} \tag{2}$$

(4) *Convergence judgment*: If $\left\| R^{(k)} \right\| \leq \varepsilon$, stop the iterative procedure; else k=k+1, jump to step 2 and go on.

For OMP algorithm, in the kth step of MP, $R^{(k)}$ is projected to the atom ϕ_{jk}, which promise $R^{(k+1)}$ and ϕ_{jk} orthogonal. ϕ_{jk} is not orthogonal with selected $\{\phi_{jp}\}_{0 \leq p \leq k}$ prior. New composition is introduced in the direction of $\{\phi_{jp}\}_{0 \leq p \leq k}$ when projection of $R^{(k)}$ is subtracted. The orthogonality problem is solved by projecting $R^{(k)}$ onto selected orthogonal family $\{v_p\}_{0 \leq p \leq k}$ which calculated from $\{\phi_{jp}\}_{0 \leq p \leq k}$.

2.2 Genetic Hybrid Particle Swarm Optimization

Genetic hybrid particle swarm algorithm is applied to find the optimal atoms during the procedure of OMP iterations. In the first stage of the GAPSO algorithm, we adopt genetic algorithm to make full use of its randomly-search features and global convergence for getting initial search space in multi-objective optimization. In the second stage, particle swarm optimization is adopted, because it has strong local-search ability and fast convergence. The procedures of the GAPSO algorithm are as follows.

(1) Determine the parameters. Set atomic parameters as a parameter group to be optimized, where, $A = \left\| \langle R^{(k-1)}, \phi_j \rangle \right\|$ is a fitness function.

(2) Encoding each candidate solution by a string.

(3) Choice of parent body, according to the fitness function value of all individuals and a certain probability.

(4) Cross-exchange variations according to a certain probability.

(5) Calculation the fitness of offspring, the fitness value is added to pro-group to constitute a new generation, then repeat the procedure until the optimal conditions are met.

(6) Set a number of groups of optimized solution of the genetic algorithm as initialized particles swarm and determine strengths and weaknesses of individual. In PSO, particles cooperate to find the best position (best atom) in the search space (solution space). Each particle moves according to its velocity. The particle movement is computed as follows:

$$v_i^{(t+1)} = wv_i^{(t)} + c_1\gamma_1(p_i^{(t)} - y_i^{(t)}) + c_2\gamma_2(p_g^{(t)} - y_i^{(t)}) \tag{3}$$

$$y_i^{(t+1)} = y_i^{(t)} + v_i^{(t)} \tag{4}$$

In equation (3) and (4), $y_i^{(t)}$ is the position of particle i at time t; $v_i^{(t)}$ is the velocity of particle i at time t; $p_i^{(t)}$ is the best position found by particle i itself so far; $p_g^{(t)}$ is the best position found by the whole swarm so far; w is an inertia weight scaling the previous time step velocity; c_1 and c_2 are two acceleration coefficients that scale the influence of the best personal position of the particle $p_i^{(t)}$ and the best global position $p_g^{(t)}$; γ_1 and γ_2 are random variables between 0 and 1.

(7) Evaluate and calculate the fitness of each particle of each particle. Map the best position $p_i(i = 1,2,\cdots,M)$, M is the number of particles, found by particle i and the best position p_g found by the whole swarm so far.

(8) If satisfying the termination condition $(\phi_{(k-1)} - \phi_{(k)})/\phi_{(k)} \leq \varepsilon$ or $k = N_{max}$, output the optimal solution p_g and the corresponding parameters of the position, otherwise return to (6). Update the signal or change the form of residual signal by the formula:

$$R^{(k)} = S - S^{(k)} \tag{5}$$

Repeat the process many times and achieve sparse signal decomposition.

Using the GAPSO approach to biomedical signal compression, we only needs to storage very sparse solution vectors (storage non-zero value in solution vectors and its position) and the over-complete dictionary, instead of the original signal. The flow chart of this algorithm in biomedical signal compression is shown in Fig. 1.

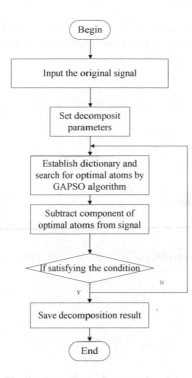

Fig. 1. Flow chart of sparse signal decomposition

3 Simulation Results

3.1 One-Dimensional Signals Compression

Four groups of MCG signals are selected randomly, with the sampling rate of 1k Hz and the length of $N = 1133$. Compression ratio (CR) is defined as $L/N \times 100\%$. Where L is the number of non-zero value in solution vectors. Set the original vector as S, the reconstructed vector as $\hat{S}$. Root-mean-square error (RMSE) of the compression algorithm is expressed as

$$RMSE = \sqrt{\sum_{i=1}^{N}(S(i) - \hat{S}(i))^2 \Big/ \sum_{i=1}^{N}(S(i))^2} \times 100\%$$

In simulation, parameters of the genetic algorithm are chosen as follows: size 60, crossover probability 0.5, mutation probability 0.03, maximum number of iterations 70; and parameters of the particle swarm optimization include the maximum iteration number 25; weight $\varepsilon = 0.8$, $c_1 = 1.89$, $c_2 = 2.50$, $v_1^{max} = 0.77 \times M$, $v_2^{max} = 0.59 \times N$, $v_3^{max} = 0.21 \times M$, $v_4^{max} = 0.13 \times N$. Simulation results are shown in Fig. 2. RMSE of four groups of signals are shown in Table 1. It can be seen that four groups of reconstructed signals with different values of L and CR are well-accorded with the original signals.

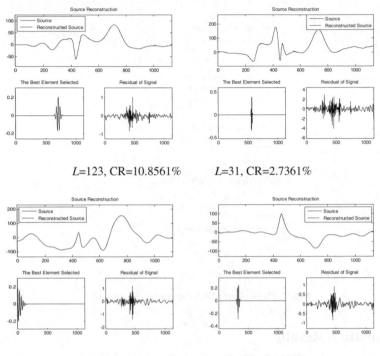

L=123, CR=10.8561% L=31, CR=2.7361%

L=127, CR=11.2092% L=60, CR=5.2957%

Fig. 2. Compression results of MCG signals in mid-length

Table 1. RMSE of four groups of signals (compression ratio 15%)

Group number	1	2	3	4
RMSE (%)	2.50	1.69	4.12	2.88

With the parameters mentioned above in our approach, when the CR is 15%, the root-mean-square errors in Table 1 are all relatively small, along with smaller distortion. We also use the wavelet transform and the OMP approaches for the same MCG signal in mid-length when their average CR achieves 16%, the average RMSE are 7.13% and 11.69%, respectively. When the similar CR is applied, the RMSE of our approach is the smallest, and with the least distortion, followed by original OMP. Wavelet transform is the worst.

Next, we select two groups of MCG signal in long-length, where N=14999. Simulation results are shown in Fig. 3. It can be seen that this algorithm also performs an effective reconstruction for large-scale data. The dictionary constructed can reflect the characteristics of signals, and make vector of signal decomposition sparser. However, at the same time, there is distortion in some peak of mutations.

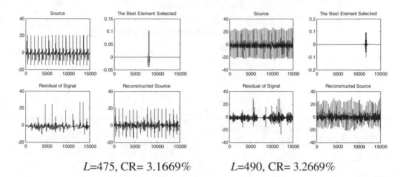

$$L=475, CR= 3.1669\% \qquad L=490, CR= 3.2669\%$$

Fig. 3. Long signal compression of MCG

3.2 Two-Dimensional Signals Compression

Two-dimensional signals are obtained from biomedical image library: Images.MD [10], in which an oblique spin-echo MRI in a transverse four-chamber plane is taken. The image size is 120×120 pixels. Fig. 4 shows the signal compression results of one line which is taken from original image randomly. Fig. 5 shows the original and reconstructed image by means of our algorithm. The RMSE of whole image is shown in Fig. 6. Where X-axis represents 120 lines of image, and Y- axis represents reconstructed RMSE of every line. Relationship between the number of iterations and the RMSE of the image reconstruction is shown in Fig. 7.

We can see from Fig. 7 that the use of the OMP algorithm combining with the GAPSO algorithm can enhance the convergence rate, i.e. it is faster than the original OMP algorithm to find the best atoms, and its RMSE of the signal reconstruction is also lower t for the same iteration number.

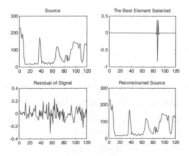

Fig. 4. Two-dimensional signals compression with $L =3$, CR $=3.3\%$

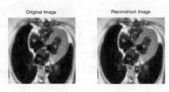

L=1997,CR=13.89%

Fig. 5. The original and the reconstructed image

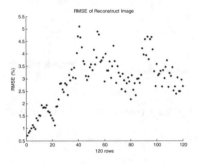

Fig. 6. RMSE of the reconstructed image

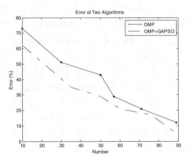

Fig. 7. Relationship between the number of iterations and the RMSE of the signal reconstruction

4   Conclusions

In this work, the sparse decomposition approach of biomedical signal compression was studied. An over-complete dictionary was constructed according to the characteristics of biomedical signals. Based on the dictionary, two approaches of sparse decomposition, the OMP algorithm and the OMP algorithm combing with GAPSO algorithm, were discussed. Simulation results showed that our approach was feasible for biomedical signal compression. With this genetic hybrid particle swarm algorithm, the convergence rate (CR) and the root-mean-square error (RMSE) were improved along with less distortion. For MCG signals in mid-length, simulation results showed that the standard error was 2.78%, when the compression ratio was 15%. However, the algorithm needs to be further improved for the ultra-large-scale (N= 10000 or more) biomedical signal compression.

Acknowledgments. This work obtained support from the National Natural Science Foundation of China (60771030), the National High-Technology Research and Development Program of China (2008AA02Z308), the Shanghai Science and Technology Development Foundation (08JC1421800), and the Open Project of State Key Laboratory of Function Materials for Information.

References

1. Malat, S., Zhang, Z.: Matching pursuits with time-frequency dictionaries. IEEE Trans. Signal Process. 41(12), 3397–3415 (1993)
2. Donoho, D.: Compressed sensing. IEEE Trans. Inform. Theory 52(4), 1289–1306 (2006)
3. Candès, E.J., Tao, T.: Reflections on compressed sensing. IEEE Information Theory Society Newsletter 58(4), 20–23 (2008)
4. Monaci, G., Jost, P., Vandergheynst, P., Mailhe, B., Lesage, S., Gribonval, R.: Learning Multi-Modal Dictionaries. IEEE Transactions on Image Processing 16(9), 2272–2283 (2007)
5. He, Z., Andrzej, C., Rafal, Z., Xie, S.: Improved FOCUSS Method with Conjugate Gradient Iterations. IEEE Transactions on Signal Processing 57(1), 399–404 (2009)
6. Qiu, C., Lu, W., Vaswani, N.: Real-time Dynamic MR Image Reconstruction using Kalman Filtered Compressed Sensing. In: IEEE Int. Conf. on Acoustics, Speech, and Signal Processing (ICASSP), Taipei, Taiwan (2009)
7. Wang, C., Liu, J., Sun, J.: Compression algorithm for electrocardiograms based on sparse decomposition. Front. Electr. Electron. Eng. China 4(1), 10–14 (2009)
8. Holland, T.H.: Adaptation in natural and artificial system. The University of Michigan Press, Michigan (1975)
9. Kennedy, J., Ebethart, R.C.: Particle swarm optimization. In: Proceedings of IEEE International conference on Neural Networks, Perth, Australia, pp. 1942–1948 (1995)
10. Biomedical image library: http://www.images.md/

A Comparative Study of Color Correction Algorithms for Tongue Image Inspection

Xingzheng Wang and David Zhang

Biometric Research Center, The Hong Kong Polytechnic University, Hong Kong
{csxzwang,csdzhang}@comp.polyu.edu.hk

Abstract. Color information is of great importance for the tongue inspection of computer-aided tongue diagnosis system. However, the RGB signals generated by different imaging device varied greatly due to dissimilar lighting conditions and usage of different kinds of digital cameras. This is a key problem for the tongue inspection and diagnosis. A common solution is to correct the tongue images to standard color space by the aid of colorchecker. In this paper, three general color correction techniques: polynomial regression, artificial neural network and support vector regression (SVR) are applied to the color correction of tongue image and compared for their performance of accuracy and time complexity. The experimental results of colorchecker correction show that when properly optimized, SVR performs the best among these three algorithms, with a training error of 0 and a test error of 0.68 to 3.03. The polynomial regression algorithm performs a little worse, but it is more robust to the fluctuations of the environmental illuminant and much faster than SVR to train the parameters. The ANN performs worst, and it is also time-consuming to train. Performance comparison to correct real tongue images shows that polynomial regression is better than SVR to achieve a close correction result to human perception. Finally, this paper is concluded that for tongue inspection in a computer-aided tongue diagnosis system, polynomial regression is suitable for online system correction to aid tongue diagnosis, while SVR technique offer a better alternative for the offline and automated tongue diagnosis.

Keywords: Color correction, polynomial regression, support vector regression, artificial neural network, tongue image inspection.

1 Introduction

Color information is of great importance in the computer-aided tongue diagnosis system of Traditional Chinese Medicine [1-2]. By using the digital imaging technology, we can briefly retrieve some physiological information of human body condition from the color features and color differences extracted from the digital tongue images. However, one problem with this process is that the color images produced by digital cameras are usually device-dependent. i.e., different digital cameras produce different RGB responses for the same scene. So there are literally thousands of different RGB spaces because of the dependent on the specific system, and this will be a great problem when we try to exchange the images among different imaging systems. Moreover, the color

D. Zhang and M. Sonka (Eds.): ICMB 2010, LNCS 6165, pp. 392–402, 2010.
© Springer-Verlag Berlin Heidelberg 2010

information also varied due to the inconstancy of lighting conditions (including lighting geometry and illuminant color characteristics). Therefore, further color correction on color images to determine the mapping algorithm between its unknown device-dependent RGB color spaces to a particular device-independent color space are necessarily needed in order to get consistent and standard color perception for the tongue inspection of computer-aided tongue diagnosis system.

In the color science area, many color correction algorithms have been proposed for various color correction tasks [2-6], and some of them have been applied to the color correction of tongue image. In [7], J.H. Jang adopted Trigonal Pyramid method for color reproduction. Yang Cai's system [8] adopted Munsell colorchecker and a linear color correction model to compensate the errors of the original colors of tongue image under various lighting conditions. Y. G. Wang, et al. [9] used polynomial regression based method and H. Z. Zhang, et al [10] adopted support vector regression (SVR) based method separately to conduct the color correction. Although some correction algorithms have been proposed, however, these algorithms were proposed and tested only in their specific imaging device, i.e. the previous researchers mainly focused on the own specific imaging camera and lighting condition, so they correction model may lack some generality. Moreover, comprehensive comparison and optimization which are the most important parts when developing a computer-aided tongue diagnosis system have not been developed. In view of this condition, further research is urgently needed to test and compare the performance of different color correction algorithms in a standard and generalized platform.

In this paper, a comparative study of the color correction algorithms for tongue image inspection in computer-aided tongue diagnosis system is conducted. We mainly compare three different kinds of techniques, i.e., polynomial regression, artificial neural network and support vector regression (SVR) to correct the image in device-dependent RGB color space to standard RGB (sRGB) color space. In order to conduct the generalized comparison among these three techniques, the Munsell Colorchecker, which is a checkerboard array of 24 scientifically prepared colored squares in a wide range of colors, will be adopted as supervised color samples to train the parameters in our experiment.

The organization of this paper is as follows. Section 2 briefly describes these three color correction algorithms. After that, in section 3, we elaborate the comparison experiments including the experimental setting and the performance evaluation methods. In section 4, we show the evaluation results and exhaustive analysis. Finally, section 5 offers the conclusion and discusses.

2 Color Correction Algorithms

2.1 Polynomial Regression

The first algorithm to be compared in this paper is the polynomial regression. It is widely used for color correction and characterization of different color devices. The detailed information of polynomial transform is as follows: Suppose the RGB value of the ith patch of the colorchecker can be represented as $X_i : (R_i, G_i, B_i)$ $(i = 1, 2, \ldots 24)$, and the corresponding standard device-independent value is $S_i : (SR_i, SG_i, SB_i)$ $(i = 1, 2, \ldots 24)$.

If we adopt the polynomial $\rho:[R,G,B,RGB,1]$ to derive the transformation model, it can be represented as the following formula:

$$\begin{cases} SR_i = a_{11}R_i + a_{12}G_i + a_{13}B_i + a_{14}R_iG_iB_i + a_{15} \\ SG_i = a_{21}R_i + a_{22}G_i + a_{23}B_i + a_{24}R_iG_iB_i + a_{25} \quad (i = 1,2,...\ 24) \\ SB_i = a_{31}R_i + a_{32}G_i + a_{33}B_i + a_{34}R_iG_iB_i + a_{35} \end{cases} \tag{1}$$

This equation also can be rewritten in the matrix format as:

$$S = A^T \bullet X \tag{2}$$

Where S is the matrix of the standard sRGB value of the 24 color patches

$$S = \begin{bmatrix} SR_1 & SR_2 & ... & SR_{24} \\ SG_1 & SG_2 & ... & SG_{24} \\ SB_1 & SB_2 & ... & SB_{24} \end{bmatrix} \tag{3}$$

A is the coefficient matrix in formula (),

$$A = \begin{bmatrix} a_{11} & a_{21} & a_{31} \\ a_{12} & a_{22} & a_{32} \\ \vdots & \vdots & \vdots \\ a_{15} & a_{25} & a_{35} \end{bmatrix} \tag{4}$$

And the X is polynomial matrix generated from the imaging data of the digital camera.

$$X = \begin{bmatrix} R_1 & R_2 & R_3 & ... & R_{24} \\ G_1 & G_2 & G_3 & ... & G_{24} \\ B_1 & B_2 & B_3 & ... & B_{24} \\ R_1G_1B_1 & R_2G_2B_2 & R_3G_3B_3 & ... & R_{24}G_{24}B_{24} \\ 1 & 1 & 1 & ... & 1 \end{bmatrix} \tag{5}$$

Actually we have tested different terms and orders of polynomials such as:

$$\rho:[R,G,B]$$
$$\rho:[R,G,B,RGB,1]$$
$$\rho:[R,G,B,RG,RB,GB]$$
$$\rho:[R,G,B,RG,RB,GB,RGB,1] \tag{6}$$
$$\rho:[R,G,B,RG,RB,GB,R^2,G^2,B^2]$$
$$\rho:[R,G,B,RG,RB,GB,R^2,G^2,B^2,RGB,1]$$
$$\vdots$$

According to the least square regression method, the solution to the coefficient matrix is as follows,

$$A = (X^T X)^{-1} X^T S \qquad (7)$$

A is the correction matrix, and we apply it to the real tongue image, suppose the image matrix for tongue image is I_{in}, the polynomial matrix for I_{in} is X_{in}, the output image matrix is:

$$X_{out} = A^T \bullet X_{in} \qquad (8)$$

2.2 SVR Based Algorithm

The second algorithm to be compared and optimized in this paper is Support Vector Regression (SVR), which uses a nonlinear kernel to transform the training data into a high dimensional feature space where linear regression can be performed. In this paper, we choose the Gaussian Radial Basis Function (RBF) kernel to optimize the parameters. The nonlinear SVR solution, using an ε-insensitive loss function

$$L_\varepsilon(y) = \max(0, |f(x) - y| - \varepsilon) \qquad (9)$$

and Gaussian RBF kernel with the form,

$$K(x - x_i) = \exp(-\frac{(x - x_i)2}{2\sigma^2}) \qquad (10)$$

is given by

$$\max_{\alpha,\alpha^*} W(\alpha, \alpha^*) = \max_{\alpha,\alpha^*} \sum_{i=1}^{l} \alpha_i^*(y_i - \varepsilon) - \alpha_i(y_l + \varepsilon)$$
$$-\frac{1}{2} \sum_{i=1}^{l} \sum_{j=1}^{l} (\alpha_i - \alpha_i^*)(\alpha_j - \alpha_j^*) K(x_i, x_j) \qquad (11)$$

subject to the conditions,

$$0 \le \alpha_i, \alpha_i^* \le C, i = 1, 2, \cdots, l \qquad (12)$$

$$\sum_{i=1}^{l} (\alpha_i - \alpha_i^*) = 0 \qquad (13)$$

Solving Equation with constraints Equation determines the Lagrange multipliers, α_i, α_i^*, and the regression function is given by,

$$f(x) = \sum (\overline{\alpha_j} - \overline{\alpha_j^*}) K(x_i, x_j) + \overline{b} \qquad (14)$$

After we generate the regression model, we can apply it to the color correction of colorchecker and tongue images.

2.3 Artificial Neural Network

The third algorithm mapping device-dependent RGB values and device-independent values is artificial neural networks (ANNs) in which a large number of inputs are mapped onto a small number of outputs, which makes it suitable for regression problems. There are many different types of ANN but in this paper we mainly focus on the optimization of BP neural networks. Fig. 1 shows the schematic diagram showing how a neural network can be used to find a mapping between the RGB values.

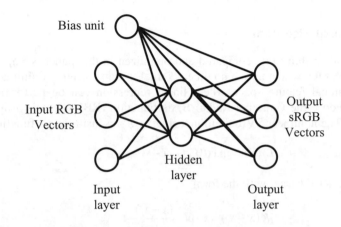

Fig. 1. Schematic diagram showing how a neural network can be used to find a mapping between the RGB values (input vectors) and the standard values (such as XYZ and sRGB)

3 Comparison Experiments

3.1 Correction Procedure

Fig. 2 shows the schematic diagram of the correction procedure, which is to utilize a reference target (Munsell Colorchecker in our experiment) that contains a certain number of color patches to train for the correction parameters, and then apply this correction algorithm to the uncorrected tongue images. To achieve this, we firstly capture the colorchecker images by using the camera and lighting condition which are the same with the acquisition of the tongue images, and extract the color checker RGB values in the source color space. We then use a spectrophotometer to measure the device-independent sRGB values in target color space. After both of the two values of the colorchecker are extracting, they will be inputted into different regression algorithms to train the correction parameters. Finally, we apply this color correction algorithm to the tongue images to get the corrected tongue images.

3.2 Experiment Setting

A three-chip CCD camera with 8-bit resolution for each channel and 1024*768 spatial resolutions was used as the imaging device in this study. Two standard D50 fluorescent

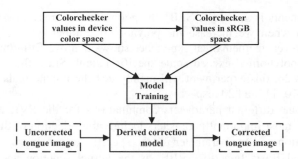

Fig. 2. Schematic diagram of the correction procedure. We utilize a reference target (Munsell Colorchecker in our experiment) that contains a certain number of color patches to train for the correction parameters, and then apply this correction algorithm to the uncorrected tongue.

tubes, whose color temperature are 5000K and color rendering index (CRI) is bigger than 90, were adopt as the illuminant of the experiment, and they were positioned approximately in a 0/45°illumination and viewing geometry. The whole imaging system was installed in a small man-made darkroom which only has one small open window. Figure 3 shows the appearance of tongue image acquisition device.

Fig. 3. Appearance of the tongue image acquisition device. A three-chip CCD video camera plus two standard D50 fluorescent tubes are placed inside of this equipment.

We used the Munsell colorchecker 24 as the supervised color samples to train parameters of correction algorithms. Moreover, since colorchecker images captured at different time have color distortions, it was reasonable to use the colorchecker images captured at different time as test samples. In this experiment, we captured two groups of colorchecker images by change the environmental illuminant. In group A, the device was putted under an office illuminant condition, while in group B the device was putted in a darkroom. In each group we captured five sample images in four different times: zero, three, five, and ten minutes after turning on the fluorescent light. This produced eight image groups which we labeled as A0-A3 and B0-B3, and each group has five images. We chose one as the training sample and the others were used for testing.

Experiments setting for three correction algorithms are listed as follows. Regarding the polynomial regression algorithm, obviously, the order of polynomial would affect the training and testing performance greatly. When the order is 1 and the number of polynomial elements is 3, i.e. [R, G, B], the polynomial regression would be a linear transform. And when we increase the polynomial number, the training error will decreased. However, mathematical experience shows that overfitting may occur when the number of polynomial exceed some specific extent. Since the total numbers of color patches is 24, this experiment will mainly test the polynomials whose element number is 3, 5, 10, 11, and 20 respectively.

There are some different parameters combinations for the SVR color correction algorithms. The most important parameters include the selection of the loss function, the kernel function and the parameter for the specific kernel. In this experiment, we chose the Radial Basis Function (RBF) as the kernel function and optimized the parameter combination for this kernel.

Fully connected multilayer perceptron networks were used in this study to derive mapping between the device-dependent RGB values and the standard sRGB values. The networks contained three input units to receive the camera responses, three output units to output the corrected values. Since the number of hidden layer will increase the time span for training largely, we will only use one hidden layer. The number of units N in the hidden layer was varied to be 3, 5, 10, 18, 27, or 40. The activation function of the units in the hidden and output layers is the sigmoid function. The sigmoid activation function can only output in the range [0 1], and therefore the output data were scale to this interval before they were used for training.

3.3 Evaluation of Correction Performance

Color differences between the corrected sRGB values and the corresponding standard ones for each color patch in the Munsell colorchecker were calculated according to formula (15) to evaluate the accuracy of these three algorithms. Since the color difference was defined in the CIE Lab color space, we first transferred the color values in sRGB color space to CIE Lab color space and then calculated the color differences.

$$\Delta E_{ab}^* = \sqrt{(L_1^* - L_2^*)^2 + (a_1^* - a_2^*)^2 + (b_1^* - b_2^*)^2} \tag{15}$$

The other criterion for the algorithm testing is the time complexity for training and testing color checker samples. Since the time span is crucial to the color correction of in tongue imaging system, and the complexity of the algorithm will affect the implementation of the system considerably, we need choose the algorithm which can perform the training and testing in a relatively short time. All computations were performed in the MATLAB programming environment.

Since human color perception is hard to describe and evaluate precisely, performance evaluation on correction of tongue images were conduct by experience TCM doctors. We asked some TCM specialist to evaluate the correction performance among the compared algorithms. Since this kind of evaluation was subjective and may have some difference between various TCM doctors, we would only take this evaluation as a reference.

4   Results and Discussion

4.1   Time Complexity Comparison

Table 1 shows the running time comparison of these three algorithms separately. From this table, we may find that the polynomial regression is much faster than its counterpart both in the training time and the test time, and both of them are less than 0.001s. While the SVR achieve the 2nd performance among the three tested algorithms, while the training time is around 30 times the test time is 14 times of the polynomial based method respectively. ANN need much more time than polynomial and SVR, which need over 3s to train the network. This is extremely time-consuming to train only one single image.

Table 1. Running time Comparison among three algorithms

	Train time(s)	Test Time(s)
Polynomial	0.0006	0.0001
SVR	0.0175	0.0014
ANN	3.1109	0.026

4.2   Accuracy Comparison on Colorchecker

Table 2 shows the performance test between the three regression algorithms when each of them is properly optimized. In this experiment, we chose the first group image A0 as the training sample. From the table, we can see that, the SVR based method achieves the best performance result including the training error of 0. The test errors on other images are around 0.68 to 11.71. The biggest test error is 11.71 when test the B0 image, this happens may be due to the big distortion of the illuminant. The polynomial achieve the train error at 4.26, and the test errors in all the other images are almost keep the same except the biggest one at image B0. Compare to the SVR method, this relative fluctuation is relatively smaller. This means the polynomial based method is more robust to the fluctuation of the environmental illuminant. The ANNs get the poorest performance among the three tested algorithms. Both of The train error and the test error are much bigger than the other two, so it was not suitable to the color correction of the tongue image.

Table 2. Performance comparison between three techniques

Algorithm		Polynomial	SVR	ANNs
Train Error		4.26	0	9.70
	A0	4.52	0.68	9.92
	A1	5.11	1.74	10.27
	A2	5.23	2.03	10.88
Test Error	A3	5.87	3.03	10.90
	B0	9.23	11.71	16.39
	B1	8.30	5.54	10.63
	B2	6.72	3.87	9.61
	B3	4.90	1.58	10.74

We also conduct the cross validation of this color correction experiment. Since we have eight groups of images, we may choose one of them as the training sample and test the other images. In this experiment, we calculate the mean color difference of all the other images when chose one group as the training sample. Table 3 and Figure 4 show the comparison of the SVR and polynomial algorithms when changing the training sample. From this figure, we can get that although the SVR algorithm performs better than polynomial regression does; the polynomial regression algorithm is more robust to the illuminant variation than SVR algorithm. This is also very important in a real computer-aided tongue diagnosis system.

Table 3. Cross Validation when Change the Training Sample

Training Sample	A0	A1	A2	A3	B0	B1	B2	B3
Polynomial	6.96	7.49	7.68	8.11	10.67	9.03	7.60	6.91
SVR	3.78	4.12	4.35	4.92	8.93	5.91	4.55	3.85

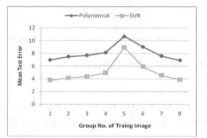

Fig. 4. Cross validation when change the training sample. Polynomial regression algorithm is more robust than SVR when training sample changes, i.e., the imaging environment changes.

4.3 Correction on Tongue Images

In order to compare the algorithms performance when they are applied to tongue images, we do further experiment on real tongue image correction. We firstly capture colorchecker images and tongue images under two imaging conditions respectively: the source one is normal imaging condition (common camera and illuminant), the target one is the standard imaging condition including the standard CIE lighting condition and industrial camera. In this experiment, we corrected the tongue images captured under the source imaging condition by the use of polynomial regression and SVR, and after correction, they are supposed to be close to the images captured under the target image condition. Fig. 5 (a) shows the tongue images captured under the target imaging condition (we call these target tongue images), and Fig. 5 (b)-(c) respectively show the tongue images corrected using polynomial regression and SVR based algorithms.

From these images in figure 5, we can get that the polynomial regression algorithms can achieve the closest perception to the target tongue image among all these three algorithms, and it will also be selected and optimized as our final optimized algorithms. Also we can see that there have remarkable color differences between the target image and the images corrected by SVR algorithm.

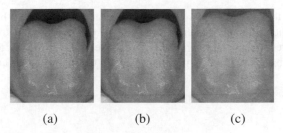

(a)   (b)   (c)

Fig. 5. Performance comparisons between different algorithms. (a) The target image. (b) Corrected image by polynomial regression. (c) Corrected image by SVR based algorithm.

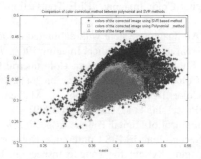

Fig. 6. Distribution of images corrected by SVR and polynomial regression in chromatic diagram

Figure 6 shows the CIEXYZ distribution in chromatic diagram which are transformed from images in figure 5. we can get that the image corrected by polynomial regression algorithm is much closer to the target image than the image corrected by SVR based algorithm.

From the experimental results, we can get that although the SVR methods achieve the best precision performance among all the three tested algorithms, but it take more time to train the parameters. Moreover, when it was applied to color correction on tongue image, it cannot achieve a good color perception with the target tongue image. Meanwhile, although the polynomial based method performs a litter worse than SVR does, it can get better performance to correct the tongue image and can be implemented much more quickly. Therefore, in a real application for tongue imaging system, polynomial is more suitable for a online color correction, and SVR based techniques is more applicable for an offline and automated tongue diagnosis.

5   Conclusions

The purpose of this study is to investigate which color correction algorithm among the mostly general ones, i.e. polynomial transform, artificial neural network and support vector regression, is the most suitable one for the color correction task of tongue inspection. The study shows that polynomial regression and support vector regression are capable of producing acceptable results on color correction of colorchecker when the parameters are properly optimized. However, given that that SVR can be time-consuming to train, and it cannot correct the tongue image to human perception very well. We therefore recommended the use of the polynomial transforms for the online color correction of the tongue imaging system to aid for the tongue diagnosis, while the SVR based method can be applied to the offline and automated tongue image analysis.

References

1. Kirschbaum, B.: Atlas of Chinese Tongue Diagnosis. Eastland, Seattle (2000)
2. Maciocia, G.: Tongue Diagnosis in Chinese Medicine. Eastland, Seattle (1995)
3. Sharma, G., et al.: Digital Color Imaging. IEEE Transactions on Image Processing 6(7) (July 1997)
4. Haeghen, Y.V., et al.: An imaging system with calibrated color image acquisition for use in dermatology. IEEE Transactions on Medical Imaging 19(7), 722–730
5. Hong, G.W., et al.: A Study of Digital Camera Colorimetric Characterization Based on Polynomial Modeling. Color Research and Application 26(1), 76–84
6. Vrhel, M.J., et al.: Color Device Calibration: A Mathematical Formulation. IEEE Transactions on Image Processing 8(12) (December 1999)
7. Jang, J.H., et al.: Development of Digital Tongue Inspection System with Image Analysis. In: Proceedings of the Second Joint EMBS Conference, Houston, TX (October 2002)
8. Yang, C.: A Novel Imaging System for Tongue Inspection. In: Proceeding of the 19th IEEE Conference on Instrumentation and Measurement Technology, IMTC/2002, May 2002, vol. 1, pp. 159–163 (2002)
9. Wang, Y.G., et al.: Research on Color Reproduction of Tongue Image Analysis Instrument. China Illuminating Engineering Journal 12(6) (2001)
10. Zhang, H.Z., et al.: SVR based color calibration for tongue image. In: Proceedings of 2005 International Conference on Machine Learning and Cybernetics, vol. 1-9, pp. 5065–5070 (2005)

An Mean Shift Based Gray Level Co-occurrence Matrix for Endoscope Image Diagnosis

Yilun Wu[1], Kai Sun[1], Xiaolin Lin[2], Shidan Cheng[2], and Su Zhang[1]

[1] Biomedical Engineering Dept., Shanghai Jiao Tong Univ., Shanghai, China
[2] Gastroenterology Dept., Ruijin Hospital, School of Medicine, Shanghai, China

Abstract. Endoscope is important for detecting gastric lesions. Computer aided analysis of endoscope images is helpful to improve the accuracy of endoscope tests. In this paper, Mean Shift-Gray Level Co-occurrence Matrix algorithm (MS-GLCM), an improved algorithm for computing Gray Level Co-occurrence Matrix (GLCM) based on Mean Shift, is presented to solve the problem that computing GLCM costs too much time. MS-GLCM is used in Color Wavelet Covariance(CWC) as a substitute for classical GLCM. The new CWC algorithm is applied to extract texture features, which are classified by AdaBoost, in endoscope images. Experiment shows that MS-GLCM saves the time cost and partly prevents from data redundancy, with a similar output like GLCM. And it decreases the final error rate in lesion detection of endoscope images.

Keywords: endoscope image, Gray level co-occurrence matrix, Mean Shift, Color wavelet covariance.

1 Introduction

Endoscope is a major device of clinical gastric examinations. The diagnosis of physicians mostly depends on reading endoscope images, which causes many subjective mistakes [1]. So computer aided detection is helpful. However, there are little clear boundary, complex color information, abundant texture information and much interference, like deformation, spectral reflection, etc, in endoscope images. It was found that texture information is an important component, as a local repeated pixel series in a certain pattern [2]. Texture feature is a key point of endoscope images.

Gray level co-occurrence matrix (GLCM) is one of the major methods in statistics texture feature extraction [3], which was used to identify objects in the photos captured by earth resources technology satellite (SRTS). Many works about its improvements of algorithm speed and optimization appeared. Ulaby selected four most key features (Energy, Entropy, Contrast, Inverse Difference Moment), which were proved irrelevant, from the fourteen statistical features [4]. It simplified the procedure of GLCM as well as held a relatively good classification performance. Baraldi showed that the contrast and entropy are the most important feature functions [5]. Kandaswamy proposed a method with similarity assessment and approximate textural features to improve the efficiency of texture analysis, based on the possession statistics model [6]. The distance parameter was determined as one through tests and

D. Zhang and M. Sonka (Eds.): ICMB 2010, LNCS 6165, pp. 403–412, 2010.

analysis [7]. Now, GLCM has a satisfactory recognition performance in the fields like lesion detection [8], treatment prediction [9] and computer aided diagnosis [10].

Mean Shift is a nonparametric estimation algorithm of density gradient [3]. Cheng defined kernel group of Mean Shift, leading to different contributions of each point in sample regions to Mean Shift vectors [11]. Comaniciu and Peter Meer introduced Mean Shift into image processing especially filtering and segmentation [12, 13]. It is a powerful tool to analyze all kinds of feature spaces.

In this paper, we filter the pixel pairs to reduce the time cost and data redundancy. The filtering method is based on Mean Shift algorithm. The improved GLCM is called MS-GLCM. Meanwhile, MS-GLCM is used in CWC algorithm, a texture extraction method based on GLCM [14]. And it speed up the calculating procedure in feature extraction of endoscope images.

2 Methodology

2.1 Procedure Analysis of GLCM

GLCM estimates the second order joint conditional probability density in two dimensional feature space like a gray level image [3]. If $Lx=\{1,2,\ldots,m\}$ and $Ly=\{1,2,\ldots,n\}$ are the X and Y axis in the spatial domains, the gray level image I is a set of $m \times n$ pixels with some gray value $V \in \{1,2,\ldots,L\}$. Each point in the image can be presented by a function $V(Lx,Ly)$. If point $v(Lx, Ly)$ in I has the other point $v'(Lx', Ly')$ in a distance of d on an direction angle of θ, there is a probability density function $P(i,j,d,\theta)$ which indicates the number of the point pairs in the whole image. So for the image, a L*L matrix can be computed,

$$W = [P(i,j,d,\theta)]_{L*L} .$$
(1)

It is called as Gray level co-occurrence matrix (GLCM) of the image with certain distance and direction parameters, d and θ, which are determined with the priori knowledge. The variances, i and j, are the gray levels of the point pair. Various GLCM can be obtained with different d or θ from one image. The function should commonly be normalized by follow:

$$p(i,j,d,\theta) = \frac{P(i,j,d,\theta)}{S}$$
(2)

S is the sum of pixels in the image.

GLCM is a matrix about direction, distance and difference of pixels in an image, which is regarded as texture information. But as a matrix, it is rare to be used as feature vector, a second order statistics is necessary. From fourteen feature functions, only four ones, contrast, inverse difference moment, correlation and energy, are selected for uncorrelation and the most importance [4].

GLCM is quite effective in texture feature recognition. However, in GLCM, each point is calculated on every direction, which leads to an amazing time cost. It is believed a pixel is the member of only one kind texture, so there is one direction that is helpful to texture classification. Eight-direction calculating wastes space and carries

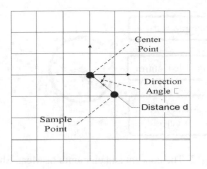

Fig. 1. The procedure of GLCM consisting of two points and two parameters

information interference. So research on its improvement is kept going. Up to now, there is no ideal method to prevent from its heavy cost.

2.2 Mode Search Based on Mean Shift Algorithm

Mean Shift algorithm is a nonparametric density gradient estimation technique [11]. Based on various kernels, it has been used in many applications including image filtering, image segmentation, clustering, object tracking, etc. It has been approved a robust technique to analyze feature spaces [13].Dense regions in the feature space are considered as local maxima of the probability density function (PDF). These local maxima points are called modes [13]. As a tool to analyze feature space, Mean Shift algorithm is used to pursue and determine the location of these modes of the density, from any points in the feature space. Modes will be the center to grouping together all the clusters associated, center of a region.

Given n feature vectors X_i, i=1,…,n, in the d dimensional space R^d. If each vector represents a point in the feature space, the vector contains a variance in each dimension. In the feature space, the nonparametric estimation of a multivariate probability density is defined as follow [15]:

$$\hat{P}_N(X) = \frac{1}{Nh^n} \sum_{j=1}^{N} k\left(\frac{X-X_j}{h}\right) \tag{3}$$

where N is the number of samples and k(.) is the kernel. What Mean Shift computes is the gradient of it. In applications, a normal kernel is always used and provides good performance. Then the function becomes

$$\hat{\nabla}_x P_N(X) = \frac{1}{N} \sum_{j=1}^{N} \frac{X_j - X}{(2\pi)^{\frac{n}{2}} h^{n+2}} e^{-(X-X_j)^T (\frac{X-X_j}{2h^2})} \tag{4}$$

where h is the kernel bandwidth, which affects the resolution of the mode detection. If kernel g(x) is defined from k(x) as follow:

$$k'(Y) = -g(Y), G(Y) = cg(\|Y\|^2) \tag{5}$$

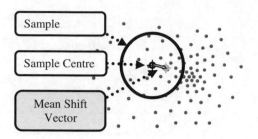

Fig. 2. The concept and elements of Mean Shift

The density gradient estimator is obtained

$$\bar{\nabla}_x P_N(X) = \frac{2c}{Nh^{n+2}} \left[\sum_{j=1}^{N} g \left(\left\| \frac{X-X_j}{h} \right\|^2 \right) \right] \left[\frac{\sum_{j=1}^{N} X_j g \left(\left\| \frac{X-X_j}{h} \right\|^2 \right)}{\sum_{j=1}^{N} g \left(\left\| \frac{X-X_j}{h} \right\|^2 \right)} - X \right] \tag{6}$$

The Mean Shift at X is defined as:

$$m(X) = \frac{\sum_{j=1}^{N} X_j g \left(\left\| \frac{X-X_j}{h} \right\|^2 \right)}{\sum_{j=1}^{N} g \left(\left\| \frac{X-X_j}{h} \right\|^2 \right)} - X \tag{7}$$

where different kernel g leads to different weighted mean. From Equ. 12, the Mean Shift is the difference between the sample mean at x and the data point x, as a vector always points toward the direction of maximum increase in the density. The procedure, data point repeatedly moves to the sample mean, is called the Mean Shift algorithm [11]. The algorithm cycles, to compute the Mean Shift vector and translate the kernel by $m(X)$, until it is converge to zero, i.e., $m(X) = 0$. A path leading to the mode, a stationary point of the estimated density where the estimate has zero gradient, will be defined through the Mean Shift algorithm.

2.3   Mean Shift-Gray Level Co-occurrence Matrix (MS-GLCM)

In classical GLCM, point pairs depend on the parameter, d and θ , which are determined artificially with prior knowledge. And each point will be computed one time on every direction. But it is obvious that there is only one texture-related direction for each point in an image. These lead to time waste and data redundancy. We present a method based on Mean Shift to improve GLCM by selecting the computed point pairs.

After comparing Fig. 2 and Fig. 3, the similarity was found that both Mean Shift and GLCM contain point pairs. Each sample point is with a mode point in Mean Shift algorithm, and GLCM is the frequency statistics of point pairs with different gray level in an image. Mean Shift vector point out the changing trend of texture in local

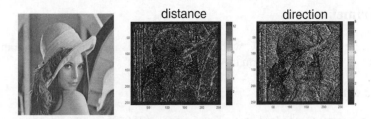

Fig. 3. Mean Shift distance and direction maps. Left: original image; Right-up: distance map; Right-bottom: direction map

regions, which is most probably an effective direction. So it can help GLCM screen one pair point for each point in an image. Through the search procedure of Mean Shift, the distance and direction are determined automatically. The new method computing GLCM is named as Mean Shift based Gray Level Co-occurrence Matrix. The only parameter is the size of Mean Shift sample region.

If there is an $m \times n$ image with gray level x (0-255), the process of MS-GLCM is as follows. First, the sample parameter of Mean Shift, h, should be determined according the size and general contrast of the image. To obtain a mode map, consisting of local maximum of probability density in feature space, Mean Shift algorithm is used to each point. If GLCM of 4 directions is wanted, a direction map of image can be computed from the relationship of each point and its mode with following function:

$$\theta(i,j) = \begin{cases} 0, 0 < \cos\dfrac{j-1}{d(i,j)} < 1, i > k \\[2mm] 90, -1 < \cos\dfrac{j-1}{d(i,j)} < 0, i > k \\[2mm] 180, -1 < \cos\dfrac{j-1}{d(i,j)} < 0, i < k \\[2mm] 270, 0 < \cos\dfrac{j-1}{d(i,j)} < 1, i < k \end{cases} \tag{8}$$

where i and j are the coordinates of each point, when k and l are the coordinates of the mode point of point x(i,j), as $mode(i,j) = x(k,l)$. d(i,j) is the distance between x(i,j) and mode(i,j), computed with follows:

$$d(i,j) = \sqrt{(i-k)^2 + (j-l)^2} \tag{9}$$

Based on the direction map θ, the GLCM $W = [p(i,j,\theta)]_{L*L}$ can be obtained. If the point $X(i,j) = l_1$, $l_1 = 0,1,...,255$, have a value $\theta(i,j) = \alpha$, $\alpha = 0,90,180,270$, in the direction map and its mode is $M(i,j) = X(m,n) = l_2$, $l_2 = 0,1,...,255$, the relationship of this point pair can be recorded as $p(l_1,l_2,\alpha) = p(l_1,l_2,\alpha)+1$ in GLCM of α direction. After recording the relationship of all the points in the image, the gray level co-occurrence matrixes of all directions, $W = [p(i,j,\theta)]_{L*L}$, are obtained.

3 Experiments

The tests were executed on a PC with C4 core and 2 G memory. The programs were coded and run on Matlab platform in Windows XP.

3.1 Standard Test Images

Thirty four common standard test color images (256*256, http://decsai.ugr.es/cvg/dbimagenes) were selected. This test dataset contains natural scenes, human beings, animals, subjects and textures. All test samples were saved in PPM. GLCM and MS-GLCM were used on these images. R, G and B components of images were computed separately. So there are 3 compartments, 3 256*256 matrixes, for each color image.

The program of GLCM and MS-GLCM were coded without any algorithm optimization. And the distance parameter of GLCM was set to 1. The h of sample region in MS-GLCM was 8. Both algorithms were calculated on 4 directions. Therefore, from each color image, 4*3 gray level co-occurrence matrixes were obtained. Then fourteen feature statistics [3] were computed further so that a 1*14 vector can be extracted from each co-occurrence matrixe. Finally, on each compartment of a color image, 4 vectors were got. The comparison of output vectors between GLCM and MS-GLCM are illustrated in Fig. 4.

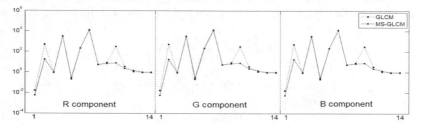

Fig. 4. The feature vector comparison between MS-GLCM (blue-solid) and GLCM (red-dotted) in 34 test image

To quantify the similarity of the vectors from these two algorithms, a Euclidean distance in feature space was calculated as follow:

$$\text{Dist}(i) = \frac{1}{M} \sum_{j=1}^{M} \left(\frac{V1(i,j) - V2(i,j)}{V1(i,j) + V2(i,j)} \right)^2, i = 1, 2, ..., N \tag{10}$$

Where $\{V1(i,j)\}_{j=1,2,...,M}^{T}$ and $\{V2(i,j)\}_{j=1,2,...,M}^{T}$ are two normalized vectors by GLCM and MS-GLCM. i is the number of images and j is the dimension of vectors. M is the number of space dimension and N is that of vectors. Function Dist quantifies the difference between two vectors. The mean of Dist is

$$\overline{\text{Dist}} = \frac{1}{N} \sum_{i=1}^{N} \text{Dist}(i) \tag{11}$$

So the general consistency can be defined:

$$\overline{\text{Hom}} = 1 - \sqrt{\overline{\text{Dist}}} \tag{12}$$

From the equations above, $\overline{\text{Dist}}$ of 34 images is 0.1524 and $\overline{\text{Hom}}$ is 0.6126. The time cost of GLCM and MS-GLCM are compared in Fig. 5 and Fig. 6.

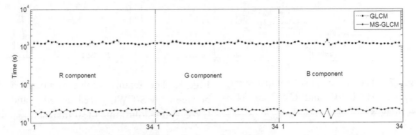

Fig. 5. The time cost comparison of gray level co-occurrence matrix computing between MS-GLCM (blue-solid) and GLCM (red-dotted) in 34 test images

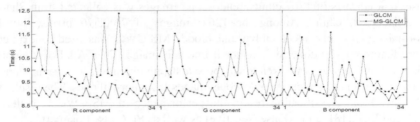

Fig. 6. The time cost comparison of statistics function calculating between MS-GLCM (blue-solid) and GLCM(red-dotted) in 34 test images

From figures above, MS-GLCM has an obvious advantage on computing rate, in the computing of gray level co-occurrence matrix , with a similar output with GLCM.

For further study, MS-GLCM took the place of GLCM in CWC process [14]. Both classical GLCM and MS-GLCM adopted CDF9/7 color wavelet transform and 5 feature statistics functions (the angular second moment, the correlation, the inverse difference moment, entropy and energy) [16]. The test set was still thirty four standard test images. The result is as follows:

The Dist and the Homology of output vectors, calculated by two CWCs, are as follows:

$$\overline{\text{Diff}} = \frac{1}{34} \sum_{i=1}^{35} \sum_{j=1}^{120} \left(2 \frac{V1(i,j) - V2(i,j)}{V1(i,j) + V2(i,j)} \right)^2 = 0.0514$$

$$\overline{\text{Hom}} = 1 - \sqrt{\overline{\text{Diff}}} = 0.7733 = 77.33\%$$

That means the output vector nearly don't change but has a obvious decrease on time cost.

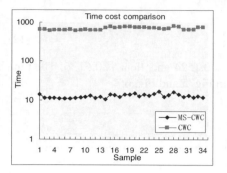

Fig. 7. The time cost comparison between MS-CWC and CWC in 34 test images

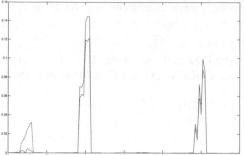

Fig. 8. The feature vector comparison between MS-CWC (red-dotted) and CWC (blue-solid) in 34 test images

3.2 Endoscope Images

Based on the above work, MS-CWC algorithm was applied to endoscope images. A dataset that consists of 1267 clinical endoscope images was collected from Shanghai Ruijin Hospital, China. Among the 1267 images of 768*576 pixels, 746 have abnormal regions, like ulcer, polyp and blood. MS-CWC was used to extract the texture features,a 120-dimension vector from each image. Then AdaBoost algorithm [17], as a training tool, was taken to compose a strong classifier, which can recognize the feature vectors of abnormal endoscope images from normal ones. The results were as Table1. From classification result, the performance of MS-CWC is better than CWC and LBP [18] in FN (False Negative) as well as FP (False Positive).

Table 1. Classification results of different feature extractions

Feature extraction	Classifier	FN (%)	FP (%)	Error Rate(%)
LBP	Adaboost	28.15	54.23	38.86
CWC	Adaboost	26.01	56.13	38.39
MS-CWC	Adaboost	25.47	52.31	36.49

4   Discussion and Conclution

In this paper, GLCM was proposed to lesion detection in endoscope images. After analyzing GLCM, its major problem is the time cost of computing gray level co-occurrence matrix. Therefore, Mean Shift algorithm was introduced and used to improve GLCM. This new method was named Mean Shift Gray Level Co-occurrence Matrix (MS-GLCM). The proposed algorithm (MS-GLCM) replaced GLCM in CWC, which extracted the texture features in endoscope images. A future perspective for the extension of this work has several fields. More special image sets could be used to further application tests. And some other improved GLCM algorithms could be compared with MS-GLCM. The theory support is necessary.

Acknowledgements. This work is funded by the National Basic Research Program of China (973 Program, No.2010CB732506) and Medical-Engineering Joint Research Funding of Shanghai Jiaotong University (No.YG2007MS02).

References

1. Rong, W., Wu, J.: The Research in Progress of Diagnosis of Early Gastric Cancer. Chinese Journal of Clinical Oncology and Rehabilitation 13, 469 (2006)
2. Jain, A.K., Karu, K.: Learning Texture Discrimination Masks. IEEE Transactions on Pattern Analysis and Machine Intelligence 18, 195–205 (1996)
3. Haralick, R.M., Shanmugam, K., Dinstein, I.: Textural Features for Image Classification. IEEE Transactions on Systems, Man and Cybernetics smc 3, 610–621 (1973)
4. Ulaby, F.T., Kouyate, F., Brisco, B., Williams, T.H.L.: Textural Information in Sar Images. IEEE Transactions on Geoscience and Remote Sensing GE-24, 235–245 (1986)
5. Baraldi, A., Parmiggiani, F.: An Investigation of the Textural Characteristics Associated with Gray Level Cooccurrence Matrix Statistical Parameters. IEEE Transactions on Geoscience and Remote Sensing 33, 293–304 (1995)
6. Kandaswamy, U., Adjeroh, D.A., Lee, M.C.: Efficient Texture Analysis of Sar Imagery. IEEE Transactions on Geoscience and Remote Sensing 43, 2075–2083 (2005)
7. Javier, A., Papa Neucimar, J., Silva Torres Alexandre, X.F.: Learning How to Extract Rotation-Invariant and Scale-Invariant Features from Texture Images. EURASIP journal on advances in signal processing 2008, 18 (2008)
8. Liu, B., Cheng, H.D., Huang, J., Tian, J., Liu, J., Tang, X.: Automated Segmentation of Ultrasonic Breast Lesions Using Statistical Texture Classification and Active Contour Based on Probability Distance. Ultrasound in Medicine & Biology 35, 1309–1324 (2009)
9. El Naqa, I., Grigsby, P.W., Apte, A., Kidd, E., Donnelly, E., Khullar, D., Chaudhari, S., Yang, D., Schmitt, M., Laforest, R., Thorstad, W.L., Deasy, J.O.: Exploring Feature-Based Approaches in Pet Images for Predicting Cancer Treatment Outcomes. Pattern Recognition 42, 1162–1171 (2009)
10. Chen, C.-Y., Chiou, H.-J., Chou, S.-Y., Chiou, S.-Y., Wang, H.-K., Chou, Y.-H., Chiang, H.K.: Computer-Aided Diagnosis of Soft-Tissue Tumors Using Sonographic Morphologic and Texture Features. Academic Radiology 16, 1531–1538 (2009)
11. Cheng, Y.: Mean Shift, Mode Seeking, and Clustering. IEEE Transactions on Pattern Analysis and Machine Intelligence 17, 790–799 (1995)
12. Comaniciu, D., Meer, P.: Robust Analysis of Feature Spaces: Color Image Segmentation. In: Proceedings of the IEEE Computer Society Conference on Computer Vision and Pattern Recognition, San Juan, PR, USA, pp. 750–755 (1997)
13. Comaniciu, D., Meer, P.: Mean Shift: A Robust Approach toward Feature Space Analysis. IEEE Transactions on Pattern Analysis and Machine Intelligence 24, 603–619 (2002)
14. Iakovidis, D.K., Maroulis, D.E., Karkanis, S.A., Flaounas, I.N.: Color Texture Recognition in Video Sequences Using Wavelet Covariance Features and Support Vector Machines. In: Proceedings of 29th Euromicro Conference 2003, pp. 199–204 (2003)

15. Cacoullos, T.: Estimation of a Multivariate Density. Annals of the Institute of Statistical Mathematics 18, 179–189 (1966)
16. Siew, L.H., Hodgson, R.M., Wood, E.J.: Texture Measures for Carpet Wear Assessment. IEEE Transactions on Pattern Analysis and Machine Intelligence 10, 92–105 (1988)
17. Freund, Y., Schapire, R.E.: Experiments with a New Boosting Algorithm. International Conference on Machine Learning, ICML 1996 (1996)
18. Ojala, T., Pietikainen, M., Maenpaa, T.: Multiresolution Gray-Scale and Rotation Invariant Texture Classification with Local Binary Patterns. IEEE Transactions on pattern analysis and machine intelligence 24, 971–987 (2002)

A Novel Criterion for Characterizing Diffusion Anisotropy in HARDI Data Based on the MDL Technique

H.Z. Zhang[*], T.M. McGinnity, S.A. Coleman, and M. Jing[*]

Intelligent Systems Research Centre, University of Ulster at Magee
Derry, BT48 7JL, Northern Ireland, UK
{h.zhang,tm.mcginnity,sa.coleman,m.jing}@ulster.ac.uk

Abstract. Based on the spherical harmonic decomposition of HARDI data, we propose a new criterion for characterizing the diffusion anisotropy in a voxel directly from the SH coefficients. Essentially, by considering the Rician noise in diffusion data, we modify the Rissanen's criterion for fitting the diffusion situation in a voxel. In addition, the minimum description length (MDL) criterion has been employed for interpreting information from both the SH coefficients and the data. The criterion obtained can make use of the diffusion information so as to efficiently separate the different diffusion distributions. Various synthetic datasets have been used for verifying our method. The experimental results show the performance of the proposed criterion is accurate.

Keywords: HARDI, Spherical harmonic, MDL, Diffusion anisotropy.

1 Introduction

In contrast to the 2nd order diffusion tensor imaging (DTI) model, the high angular resolution diffusion imaging (HARDI) technique has shown promise for identifying intravoxel multiple fiber bundles. The spherical harmonics (SH) technique provides a natural way to describe noisy HARDI data; a spherical function is constructed to characterize the diffusion anisotropy along any given angular direction using the available measurements. Diffusion anisotropy in HARDI data can be estimated in two ways: the apparent diffusion coefficients (ADC) profile is estimated directly from HARDI data [9,3,5]; reconstruction of the diffusion distribution is implemented by applying various transforms [15,11,6]. In the SH framework, these techniques can be described as a SH expansion of diffusion anisotropy in HARDI data, such as the ADC estimation of HARDI data [3] and the orientation distribution function (ODF) reconstruction in Q-ball imaging [11,6]. Thus, the diffusion anisotropy in a voxel is usually determined from the maximum order L of the SH expansion (isotropic: $L=0$; single fiber: $L=2$; two fiber $L=4$) [5,9].

---

[*] Supported by the Computational Neuroscience Research Team under the N. Ireland Department for Education and Learning "Strengthening the All-island Research Base" project.

D. Zhang and M. Sonka (Eds.): ICMB 2010, LNCS 6165, pp. 413–422, 2010.

Mathematically, the SH expansion can approximate any ADC value and describe any bounded ODF by giving a sufficiently large order L according to numerical approximation theory [1]. In practice, the number of diffusion measurements is limited. Moreover, there is a compromise between robustness and the reconstructed data for selection of the maximal order L. If L is large, spurious peaks in the reconstruction of HARDI data may be produced even if the possible higher angular resolution of the ADC or ODF can be achieved and vice versa [3]. Thus, the SH series must be truncated and then the choice of the harmonic model order L in the SH basis is very important for characterizing the diffusion anisotropy [3,9]. Some studies have addressed these problems. Alexander [1] constructed a statistic for the F-test of the hypothesis so that lower and high order models can be compared for the selection. Subsequently, various measures based on the SH coefficients have been proposed for providing the truncation evidences [9,5]. However, due to intrinsic deficiencies such as insufficient information extracted from diffusion data in a measure, these methods cannot characterize diffusion anisotropy properly or identify fiber population correctly in accordance with various diffusion distributions.

In this paper, we consider the orthonormality of the SH bases and the Rician noise nature of HARDI data [10]. A novel criterion is proposed based on the MDL criterion [13] so that prior information from both the data and the SH function can be incorporated into the selection scheme. Moreover, the use of the MDL criterion frees us from any subjective parameter setting. It is particularly important for diffusion data where the noise level is difficult to obtain or estimate a priori [10]. In Section 2, related work is discussed. The proposed method based on MDL is introduced in Section 3. The application of this proposed method in synthetic HARDI data is implemented in Section 4 and the conclusion and future work are presented in Section 5.

2   Related Work

HARDI imaging samples gradient directions uniformly on a sphere and then calculates the ADC profile along each gradient direction. Hence, at each voxel, a spherical function can be employed to describe the diffusion anisotropy within the voxel. This reconstruction formulation was first used in the context of diffusion imaging by Frank [9] to determine ADC as follows:

$$D(g) = -\frac{1}{b} \log \frac{s(g)}{s_0} = \sum_{l=0}^{L} \sum_{m=-l}^{l} a_{lm} Y_l^m(\theta, \varphi) \tag{1}$$

where $Y_l^m(\theta, \varphi)$ are the spherical harmonics of complex valued functions being defined on the unit sphere, θ and φ are the elevation and azimuthal angles respectively, $D(g)$ and $s(g)$ are the ADC value and the signal measurement with respect to the diffusion gradient $g = (\theta, \varphi)$, s_0 is the measurement in the absence of any gradient and b is the diffusion-weighting factor. Particularly, L denotes the model order and l denotes the item order in SH expansion in this paper.

Equation (1) is a truncated SH series and the reconstruction of ADC in HARDI data becomes a truncation problem of SH expansion. Various schemes have been proposed to deal with this reconstruction problem, which requires selection of the harmonic model order L and the determination of the coefficients $\{a_{lm}\}$. For a given L, Frank [9] calculates the coefficients directly using an inverse SH transform. Recently, Descoteaux [5] presented a robust estimate of the coefficients by employing the Laplace-Beltrami regularization. In this scheme, a modified SH basis Y with terms Y_j is used in accordance with the real diffusion signals as follows:

$$Y_j = \begin{cases} \sqrt{2} \cdot Re(Y_k^m), & -k \leq m < 0; \\ Y_k^0, & m = 0; \\ \sqrt{2} \cdot Img(Y_k^m), & 0 < m \leq k. \end{cases} \quad (2)$$

where $Re(Y_k^m)$ and $Img(Y_k^m)$ represent the real and imaginary parts of Y_k^m respectively, index $j = (k^2+k+2)/2+m$ for $k = 0, 2, \ldots, l$ and $m = -k, \ldots, 0, \ldots, k$. It is easy to conclude that this basis is symmetric, real and orthonormal. Thus, at each of the N gradient directions $g_i = (\theta_i, \varphi_i)$, the ADC profile $D(\theta_i, \varphi_i)$ can be represented by reformulating equation (1) in terms of equation (2):

$$D(\theta_i, \varphi_i) = \sum_{j=1}^{R} c_j Y_j(\theta_i, \varphi_i) \quad (3)$$

where $R = (L+1)(L+2)/2$ is the number of terms in the modified SH basis Y of order L. Letting $(D(\theta_i, \varphi_i))_{N \times 1}, c = (c_j)_{R \times 1}$ and $B = (Y_j(\theta_i, \varphi_i))_{N \times R}$, the problem becomes that of solving an over-determined linear system ($i = 1, \ldots, N$). By applying the Laplace-Beltrami regularization, the desired SH series coefficient vector can be obtained: $c = (B^T B + \lambda U)^{-1} B^T d$ where U is the $R \times R$ matrix with entries $l_j^2(l_j + 1)^2$ along the diagonal, λ is the weight on the regularization term and $d = (d_j)_{N \times 1}$ is the diffusion weighted signal vector.

As previously discussed, the selection of the harmonic model order L is still a challenging problem. Usually, it is a compromise between the model order and the model robustness in SH decomposition. When the order is high, the presence of negative peaks in the ODFs resulting from noise is more likely to occur than when the order is low. Various schemes have been proposed for determining the order L. In practice, the SH order L must be chosen according to the signal-to-noise ratio (SNR) in the diffusion-weighted data. In [5], the SH order was empirically set as $L = 4$, leading to 15 coefficients to be estimated per ODF. There are several measures defined on the set of coefficients for describing the anisotropy profile of HARDI data. Frank [9] defined a fractional multifiber index (FMI) for determining the model order that is the ratio of the sum of squared high-order coefficients over order-2 coefficients, with which the significance of a given order (such as a comparison between order-4 terms and order-2 terms) can gauge of whether or not a voxel is of isotropic or single fiber or multifiber [9].

$$FMI = \frac{\sum_{\{j:l \geq 4\}} |c_j|^2}{\sum_{\{j:l = 2\}} |c_j|^2} \quad (4)$$

where l is the order of items in the SH series. However, this is merely a measure of comparison and means of separating single and multifiber populations and the question of which threshold is chosen for the significance of a multiple-fiber population exists as described in [9]. Furthermore, it is quite inaccurate and often leads to incorrect results as highlighted in Section 4. An improved measure has been given for describing various fiber populations within a voxel [3,5]:

$$R_0 = \frac{|c_0|}{\sum_j |c_j|}, R_2 = \frac{\sum_{\{j:l=2\}} |c_j|}{\sum_j |c_j|}, R_{multi} = \frac{\sum_{\{j:l\geq4\}} |c_j|}{\sum_j |c_j|} \tag{5}$$

Usually, application of this method needs to consider the variance of the measurements. If R_0 is large, or the variance is small, the isotropic diffusion happens in a voxel. If R_2 is large, the one fiber diffusion is characterized in a voxel. For two or more fibers, R_{multi} is used for characterizing the diffusion anisotropy using a three-step algorithm [5]. We refer to this as Chen's method throughout this paper. However, due to obtaining information from coefficients, this method is invalid in high-order cases although it can work well in zero or order-2 cases for ADC decomposition as shown in Section 4.

Due to the uncertainties of diffusion anisotropy in fibrous tissues, quantifying the resultant SH expansion is still a prominent issue. A reasonable measure or criterion can present accurate information of fiber populations so that a logical microstructure can be inferred in fibrous tissues. These uncertainties should be incorporated into a solution scheme so as to evaluate confidence in the reconstruction of diffusion profiles.

3   Proposed MDL-Based Method

As discussed in Section 2, the coefficients of the SH series are obtained by solving a linear system (3) based on a given regularization scheme. Thus, the ADC can be estimated from (3) in accordance with the SH decomposition. The observed measurements $\{d_{ij}, i = 1 : N\}$ can be separated into a meaningful part and a bias part as follows:

$$d_i = \sum_{j=1}^{R} c_j y_{ij} + e_i \quad or \quad d = Yc \tag{6}$$

where e_i is the bias of the ith measurement between the observed and the estimated measurements, $\{y_{ij}, i = 1 : N, j = 1 : R\}$ is the matrix of discrete SH basis and $\{c_j, j = 1 : R\}$ are the estimated coefficients. Due to the signal from two channels having Gaussian noise characteristics, Gudbjartsson [10] demonstrates that the bias follows the Rice distribution, which can be expressed as follows:

$$P_M(D|A, \sigma) = \frac{D}{\sigma^2} e^{\frac{-(D^2+A^2)}{2\sigma^2}} I_0\left(\frac{AD}{\sigma^2}\right) \tag{7}$$

where D is the measured signal, A is the estimated signal, σ^2 is the variance of the noise and I_0 is the modified zeroth order Bessel function of the first kind. However, the Rician and Gaussian distributions become identical while the SNR $D >> 3\sigma$ [10].

For the model in equation (6), we wish to find out which subset of coefficients $\{\hat{c}_j, j = 1 : R\}$ is the most accurate for preserving the diffusion information. Then, the best order of SH series can be obtained for characterizing the diffusion anisotropy in a voxel. In fact, we can consider equation (6) as a denoising model. Dom [7] investigated this problem by applying normalized maximum-likelihood (NML) for interpreting the stochastic complexity of the model and obtaining the MDL estimation. In brief, based on the MDL criterion [14,13], a two-part encoding scheme is proposed for defining the stochastic complexity of the model in equation (6). Thus, the model selection can be achieved using the following criterion:

$$L^*(x^n) = \min_{\xi}\{-\ln \hat{P}(x^n|\xi) + L(\xi)\} \tag{8}$$

where x^n is the set of n measurements, ξ is the model structure, $L(\xi)$ is the code length of model structure and $\hat{P}(x^n|\xi)$ is the NML density function. The Gaussian noise in equation (6) is discussed by Rissanen [14] and an exact criterion is obtained by applying the encoding scheme (8) in an orthonormal basis. In the case of $R = N$ in equation (6), the best subset γ of the coefficient set c can be found using the following criterion (E_γ named information entropy of subset γ):

$$\min_{\gamma} E_\gamma = \min_{\gamma}\{(N - k) \ln \frac{c^T c - \hat{Y}_k}{N - k} + k \ln \frac{\hat{Y}_k}{k} - \ln \frac{k}{N - k}\} \tag{9}$$

where $\hat{Y}_k = \hat{d}^T \hat{d} = \hat{c}^T \hat{c}$, $\hat{d}$ and $\hat{c}$ are the components in d and c with indexes in γ.

However, in equation (6), the noise distribution actually is the Rician and usually the matrix R is not square. So, we need to modify this criterion to fit these circumstances. Due to the complexity of the Rice distribution, it is difficult to obtain an analytical solution for the criterion. A feasible strategy is to produce a suitable criterion by comparing the Gaussian with the Rician based on their statistical properties and then the criterion used in the Gaussian case can be modified to meet the requirement in the Rician case. To this end, we compare the negative logarithm of Gaussian likelihood with the negative logarithm of Rician likelihood so that equation (9) can be modified for the Rice distribution. Firstly, considering the variance of the Rice distribution $\sigma_D^2 = A^2 - (ED)^2 + 2\sigma^2$ where ED is the expectation of observed data D and σ^2 is the Gaussian variance [10], we use $2\sigma^2$ as the approximation of σ_D^2 when D is not much greater than σ (otherwise, the Rician starts to approximate the Gaussian if $D \geq 3\sigma$ [10,12]). Then, the subset of coefficient set $\hat{C}_k$ is employed to substitute the $\hat{Y}_k$ in equation (9) because we concentrate on the investigation of the SH order and the number of measurements N is not usually equal to the number of basis components R in practice. Thus, we obtain the criterion as follows:

$$\min_{\gamma} E_\gamma = \min_{\gamma}\{(N - k) \ln \frac{(c^T c - \hat{C}_k)}{2(N - k)} + k \ln \frac{\hat{C}_k}{k} - \ln \frac{k}{N - k}\} \tag{10}$$

We will apply this criterion to characterize the diffusion anisotropy and estimate the number of fibers in a voxel. The outline of the algorithm is as follows:

1. Choose an initial SH model order for ADC decomposition ($L = 6$ in our case) and compute the SH coefficients $c = \{c_i\}$ based on the Laplace-Beltrami regularization from equation (3).
2. Construct the subset $\gamma = \{c_k, k \leq i\}$ in accordance with each c_i.
3. Calculate the information entropy E_γ of each the subset of coefficient set c using equation (10).
4. Obtain the minimum E_γ and get the corresponding coefficient c_i. Obtain the order l in the ith item with the coefficient c_i because an order compartment is decomposed into several items in the SH series, such as order-2 compartment having 6 items.
5. The acquired l represents the diffusion characteristic. If $l = 0$, the diffusion at such voxels is classified as isotropic. If $l = 2$, the diffusion at such voxels is characterized as single-fiber diffusion. If $l = 4$, the corresponding voxels have two-fiber diffusion, etc.

4   Implementation and Results

Experiments have been performed to determine whether the criterion equation (10) can accurately characterize diffusion anisotropy using synthetic data. Datasets were generated by the Camino software [2]. There are 64 gradient direction acquisitions, the first three gradients are zero and the b-value is $3146 \ s/mm^2$. The Rician noise is incorporated by default in the synthetic data construction in [2]. In addition, noisy data is generated by adding Gaussian noise according to the SNR value. Thus, spin-echo attenuation and thence ADC measurements in each of the 61 sampled directions are then calculated. For reconstruction of the datasets, the initial SH order is assumed as $L = 6$ for avoiding the spurious peaks in the SH expansion as previously discussed.

As introduced in Section 2, there are 28 terms in the SH decomposition with order $L = 6$. Thus, 27 E_γ values will be produced for each voxel in the experiments because we need not consider the no truncation case. In the following tables, we merely present the minimum values of E_γ which correspond to the

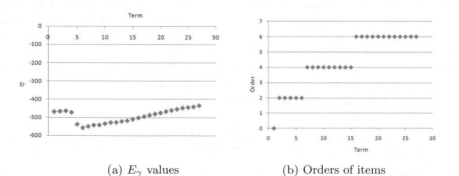

(a) E_γ values   (b) Orders of items

Fig. 1. An example for E_γ values and orders of items in the SH expansion

optimal order for the truncation of the SH decomposition as shown in our proposed algorithm in Section 3. In Fig. 1, for demonstration purpose, we present an example for showing these values for voxel 10 in Table 2. Fig .1(a) shows the E_γ values for the corresponding 27 items in the SH expansion. The minimum value is -556.4952 which corresponds to item 6. Fig. 1(b) presents the orders of the corresponding 27 items in the SH expansion, which shows item 6 corresponding to order 2 (refer to step 4 in the proposed algorithm in Section 3). Thus, the result of our proposed method is 2 for voxel 10 in Table 2.

Fig. 2. Isotropic diffusion with Gaussian noise (first one with no noise)

Fig. 2 visualizes the synthetic data with 11 voxels. The first voxel has no noise and the remaining voxels have randomly generated added Gaussian noise (SNR=8). Table 1 shows the results of applying our proposed method, which can estimate the orders correctly in all voxels. Due to R_0 values being sufficiently large, Chen's judgement in these data for the ADC case is correct for all voxels. We have not presented the results for Frank's method as it aims to separate the single and multi-fiber populations only as shown in equation (4).

Table 1. The results with isotropic data (ADC)

Voxel	Proposed MDL-based method		Chen's method					
	E_γ	Result(l)	R_0	R_2	R_4	R_6	Var	Result(l)
1	-1260	0	1	0	0	0	1.87E-19	0
2	-578.89	0	0.738	0.072	0.098	0.092	2.79E-08	0
3	-570.70	0	0.731	0.074	0.124	0.070	3.07E-08	0
4	-581.09	0	0.760	0.050	0.116	0.075	2.22E-08	0
5	-552.54	0	0.676	0.103	0.140	0.080	3.64E-08	0
6	-582.13	0	0.783	0.030	0.084	0.103	3.76E-08	0
7	-546.57	0	0.640	0.141	0.148	0.071	4.95E-08	0
8	-569.09	0	0.722	0.077	0.104	0.097	3.21E-08	0
9	-574.71	0	0.728	0.085	0.110	0.077	2.92E-08	0
10	-586.51	0	0.771	0.058	0.071	0.100	2.29E-08	0
11	-556.50	0	0.699	0.116	0.122	0.063	3.38E-08	0

In Fig. 3, the diffusion data in a voxel is rotated in various directions in order to obtain 11 voxels with different fiber orientations and Table 2 presents the results when three methods are applied to this dataset. From Table 2 we can see that the proposed method obtains the correct result for all voxels. By applying Frank's method, the FMI column shows whether or not a fiber is preferred [9]. We set the threshold $T = 1$. If FMI<1, $l = 2$ and if FMI>1, $l = 4$. Obviously, each value in the FMI column is very small ($\ll 1$) so that it prefers to order 2 and all

Fig. 3. Single-fiber data with various rotations

Table 2. The results with single-fiber data (different orientations-ADC)

Voxel	Proposed MDL-based method		Frank's method		Chen's method					
	E_γ	Result(l)	FMI	Result(l)	R_0	R_2	R_4	R_6	Var	Result(l)
1	-723.53	2	7.1E-06	2	0.543	0.454	0.0016	0.0011	2.03E-07	0
2	-615.89	2	4.97E-04	2	0.452	0.526	0.0134	0.0092	2.03E-07	2
3	-554.82	2	0.012	2	0.468	0.427	0.0625	0.0043	2.03E-07	0
4	-714.66	2	5.57E-06	2	0.467	0.531	0.0014	0.001	2.03E-07	2
5	-716.96	2	5.01E-06	2	0.453	0.545	0.0013	0.0009	2.03E-07	2
6	-549.93	2	0.011	2	0.439	0.466	0.0564	0.0384	2.03E-07	2
7	-617.15	2	4.72E-04	2	0.455	0.523	0.0130	0.0087	2.03E-07	2
8	-582.46	2	0.002	2	0.433	0.521	0.0270	0.0183	2.03E-07	2
9	-582.22	2	0.002	2	0.457	0.495	0.0286	0.0195	2.03E-07	2
10	-556.50	2	0.008	2	0.436	0.483	0.0484	0.033	2.03E-07	2
11	-664.56	2	5.40E-05	2	0.453	0.540	0.0042	0.0029	2.03E-07	2

resultant orders are 2. Thus, Frank's method can work well. For Chen's method, there are two incorrect estimations in voxels 1 and 3 because the relevant R_0 is greater than R_2 and the variances are equal.

A bi-Gaussian model is employed to obtain non-Gaussian ADC profiles corresponding to mixed tissue and fiber crossings. Fig. 4 depicts a two-fiber dataset which is generated using a mixed Gaussian model. We choose different values of mixing parameters where partial volume fractions change from (0.2,0.8) in voxel 1 to (0.8,0.2) in voxel 10. The results of applying the proposed, Frank's and Chen's methods are presented in Table 3. In this case, Chen's method is generalized by Descoteaux for more than two directions (multifiber case) [5]. The performance of our method works very well with no incorrect judgement. However, Frank's method prefers to choose order-2 for all voxels because FMI<1. Chen's method makes incorrect judgements for all voxels because R_0 is much greater than the others and also the variances are very small.

Fig. 4. Two-fiber data with different values of mixing parameters

Table 3. The results with two-fiber data (mix Gaussian tensor model-ADC)

Voxel	Proposed MDL-based method		Frank's method		Chen's method					
	E_γ	Result(l)	FMI	Result(l)	R_0	R_2	R_4	R_6	Var	Result(l)
1	-582.80	4	0.151	2	0.597	0.290	0.1029	0.0099	5.42E-08	0
2	-587.63	4	0.213	2	0.613	0.267	0.1096	0.0101	4.62E-08	0
3	-598.24	4	0.362	2	0.647	0.224	0.1192	0.0099	3.64E-08	0
4	-608.20	4	0.511	2	0.707	0.159	0.1267	0.0076	3.17E-08	0
5	-606.94	4	0.491	2	0.685	0.181	0.1251	0.0085	3.23E-08	0
6	-603.41	4	0.436	2	0.665	0.203	0.1230	0.0096	3.38E-08	0
7	-598.53	4	0.362	2	0.645	0.224	0.1199	0.0105	3.65E-08	0
8	-593.17	4	0.284	2	0.628	0.246	0.1159	0.0111	4.06E-08	0
9	-587.88	4	0.213	2	0.611	0.267	0.1107	0.0113	4.63E-08	0
10	-583.06	4	0.151	2	0.595	0.290	0.1039	0.0110	5.44E-08	0

5   Discussions and Conclusion

The SH technique is a model-free reconstruction tool and can be used to reconstruct both the ADC profile and the diffusion ODF from HARDI data. Various solution schemes have been successfully proposed for the application of SH techniques [1,9,11,5]. However, how to characterize the diffusion anisotropy accurately from the SH expansion (ADC profile or diffusion ODF) is still unclarified in this community. Actually, it is a problem how the SH series is truncated appropriately by finding an optimal compromise between approximation accuracy and robustness to noise. There are two issues regarding this problem: how to interpret the information of diffusion anisotropy in the data using the SH (the reconstruction of HARDI data); how to analyze the reconstruction result for obtaining the fiber information in the data (the characterization of diffusion anisotropy in HARDI data). Our proposed method is relative to the latter. The fiber populations need to be clearly known after the reconstruction of the diffusion ODF. Thus, We have proposed a novel criterion for characterizing the diffusion anisotropy in a voxel of HARDI data based on the previous work [5,9,3]. The method incorporates the Rician distribution of noisy data into the solution scheme by applying the MDL technique, which is based on an information-theoretic analysis of the concepts of complexity and randomness [13]. The proposed criterion can combine the SH coefficients with the data structure so that it can describe the diffusion anisotropy reliably in a voxel. Furthermore, it can directly calculate the fiber number while the existing methods need to depend on the combination of multi-channel measurements and often judge unsuccessful. Based on synthetic data, the experimental results demonstrate that our method performs accurately and reasonably.

Future work will be carried out on three main aspects. Firstly, the investigation of fiber orientation should be meaningful. The powerful properties of the SH representation can lead to some valuable postprocessing application about the diffusion anisotropic orientation. Secondly, we need to further apply our method to real datasets with benchmark results. Finally, some new methods have recently been proposed for characterizing the diffusion anisotropy such as cumulative residual entropy (CRE) in [4] and the Bayesian information criterion (BIC) method [8]. We will compare our proposed method to these criteria and improve our method further.

References

1. Alexander, D., Barker, G., Arridge, S.: Detection and modelling of non-Gaussian apparent diffusion coefficient profiles in human brain data. Magn. Reson. Med. 48, 331–340 (2002)
2. Cook, P.A., Bai, Y., Nedjati-Gilani, S., Seunarine, K.K., Hall, M.G., Parker, G.J., Alexander, D.C.: Camino: Open-Source Diffusion-MRI Reconstruction and Processing. In: International Society for Magnetic Resonance in Medicine, Seattle, WA, USA, May 2006, p. 2759 (2006)
3. Chen, Y., Guo, W., Zeng, Q., Yan, X., Huang, F., Zhang, H., Vemuri, B., Liu, Y.: Estimation, smoothing, and characterization of apparent diffusion coefficient profiles from high angular resolution DWI. In: Washington, D. (ed.) Proc. IEEE CVPR, Washington, D.C, USA, pp. 588–593 (2004)

4. Chen, Y., Guo, W., Zeng, Q., Yan, X., Rao, M., Liu, Y.: Estimation, smoothing, and characterization of apparent diffusion coefficient profiles from high angular resolution DWI. In: Proceedings of the 19th ICIPMI, USA, pp. 246–257 (2005)
5. Descoteaux, M., Angelino, E., Fitzgibbons, S., Deriche, R.: Apparent diffusion coefficients from high angular resolution diffusion imaging: Estimation and applications. Magn. Reson. Med. 56, 395–410 (2006)
6. Descoteaux, M., Angelino, E., Fitzgibbons, S., Deriche, R.: Regularized, fast, and robust analytical Q-ball imaging. Magn. Reson. Med. 58, 497–510 (2007)
7. Dom, B.: MDL estimation for small sample sizes and its application to linear regression. IBM Corp., Res. Rep. RJ 10030, June 13 (1996)
8. Fonteijn, H., Verstraten, F., Norris, D.: Probabilistic inference on q-ball imaging data. IEEE Trans. Medical Imaging 26(11), 1515–1524 (2007)
9. Frank, L.: Characterization of anisotropy in high angular resolution diffusion-weighted MRI. Magn. Reson. Med. 47(6), 1083–1099 (2002)
10. Gudbjartsson, H., Patz, S.: The Rician distribution of noisy MRI data. Magn. Reson. Med. 34(6), 910–914 (1995)
11. Hess, C., Mukherjee, P., Han, E., Xu, D., Vigneron, D.: Q-ball reconstruction of multimodal fiber orientations using the spherical harmonic basis. Magn. Reson. Med. 56, 104–117 (2006)
12. Karlsen, O., Verhagen, R., Bovee, W.: Parameter estimation from Rician distributed data sets using a maximum likelihood estimator: application to T1 and perfusion measurements. Magn. Reson. Med. 41, 614–623 (1999)
13. Rissanen, J.: Modeling by shortest data description. Automatics 14, 465–471 (1978)
14. Rissanen, J.: MDL denoising. IEEE Trans. Information Theory 46(7), 2537–2543 (2000)
15. Tuch, D.: Q-ball imaging. Magn. Reson. Med. 52, 1358–1372 (2004)

Author Index

Printing: Mercedes-Druck, Berlin
Binding: Stein+Lehmann, Berlin